Methods in Molecular Biology

For further volumes:
http://www.springer.com/series/7651

For over 35 years, biological scientists have come to rely on the research protocols and methodologies in the critically acclaimed *Methods in Molecular Biology* series. The series was the first to introduce the step-by-step protocols approach that has become the standard in all biomedical protocol publishing. Each protocol is provided in readily-reproducible step-by-step fashion, opening with an introductory overview, a list of the materials and reagents needed to complete the experiment, and followed by a detailed procedure that is supported with a helpful notes section offering tips and tricks of the trade as well as troubleshooting advice. These hallmark features were introduced by series editor Dr. John Walker and constitute the key ingredient in each and every volume of the *Methods in Molecular Biology* series. Tested and trusted, comprehensive and reliable, all protocols from the series are indexed in PubMed.

Transforming Molecular Biology with Emerging Technologies

Edited by

Robert E. Dempski

Department of Chemistry and Biochemistry, Worcester Polytechnic Institute, Worcester, MA, USA

Humana Press

Editor
Robert E. Dempski
Department of Chemistry and
Biochemistry
Worcester Polytechnic Institute
Worcester, MA, USA

ISSN 1064-3745 ISSN 1940-6029 (electronic)
Methods in Molecular Biology
ISBN 978-1-0716-5319-7 ISBN 978-1-0716-5320-3 (eBook)
https://doi.org/10.1007/978-1-0716-5320-3

This Humana imprint is published by the registered company Springer Science+Business Media, LLC, part of Springer Nature.
The registered company address is: 1 New York Plaza, New York, NY 10004, U.S.A.

Preface

Methods in molecular biology are being reshaped by an unprecedented convergence of emerging computational approaches including, but not limited to, artificial intelligence, experimental innovation, and automation. Machine learning and large language models, including generative tools, have rapidly become integral to hypothesis generation, experimental planning, and data interpretation. These emerging technologies are redefining not only how we conduct experiments but how we train the next generation of scientists.

This volume brings together methods that reflect this transformation. Contributors describe state-of-the-art laboratory practices ranging from Python-based interactive notebooks, a virtual reality-based approach to biochemistry, as well as open-source in silico drug screening. At the same time, new approaches including measuring aptamer-protein interactions help to inform our experimental approaches.

By presenting computational and experimental methods side by side, this volume aims to equip researchers with a practical and forward-looking toolkit. The chapters emphasize rigor, reproducibility, scalability, and ethical integration of new technologies. These are core principles in an era where AI algorithms, robotic platforms, and analytical instruments operate as partners in scientific discovery. The hope of this compendium is that this collection inspires investigators to adopt, adapt, and expand these approaches in ways that accelerate workforce development, foundational research, and translational impact.

Ultimately, the rapid evolution of molecular biology ensures that today's emerging technologies will become tomorrow's standard practices. By documenting this moment of transformation, we aim to provide a resource that not only captures current capabilities but also encourages creative and responsible innovation across fields.

Worcester, MA, USA *Robert E. Dempski*

Contents

Contributors

LUCIANO A. ABRIATA • *Laboratory for Biomolecular Modeling and Protein Structure Core Facility, School of Life Sciences, École Polytechnique Fédérale de Lausanne (EPFL) and Swiss Institute of Bioinformatics (SIB), Lausanne, Switzerland*

ANANTA ARDYANSYAH • *Chemistry Department, Faculty of Mathematics and Science, Universitas Negeri Malang, Malang, Indonesia*

OSCAR CLAUDIO ARES • *Department of Chemistry, University of Puerto Rico-Río Piedras, San Juan, Puerto Rico*

DARIYA BAIZHIGITOVA • *Department of Chemical Engineering and Materials Science, Stevens Institute of Technology, Hoboken, NJ, USA*

OWEN BORDEN • *Department of Chemistry, University of Puerto Rico-Río Piedras, San Juan, Puerto Rico*

ELIZABETH BRUNK • *Department of Chemistry, University of North Carolina at Chapel Hill, Chapel Hill, NC, USA; Department of Pharmacology, University of North Carolina at Chapel Hill, Chapel Hill, NC, USA; Integrative Program for Biological and Genome Sciences (IBGS), University of North Carolina at Chapel Hill, Chapel Hill, NC, USA; Computational Medicine Program, University of North Carolina at Chapel Hill, Chapel Hill, NC, USA*

TAI-YEN CHEN • *Department of Chemistry, University of Houston, Houston, TX, USA*

FABIO J. CORTÉS RODRÍGUEZ • *Laboratory for Biomolecular Modeling and Protein Structure Core Facility, School of Life Sciences, École Polytechnique Fédérale de Lausanne (EPFL) and Swiss Institute of Bioinformatics (SIB), Lausanne, Switzerland*

MATTEO DAL PERARO • *Laboratory for Biomolecular Modeling and Protein Structure Core Facility, School of Life Sciences, École Polytechnique Fédérale de Lausanne (EPFL) and Swiss Institute of Bioinformatics (SIB), Lausanne, Switzerland*

MAYARA DE CARVALHO SANTOS • *Institute of Chemistry, State University of Campinas (UNICAMP), São Paulo, Brazil*

PAULO C. M. DE LACERDA MIRANDA • *Institute of Chemistry, State University of Campinas (UNICAMP), São Paulo, Brazil*

WILTON JOSÉ DIOLINDO DO NASCIMENTO JÚNIOR • *Institute of Chemistry, State University of Campinas (UNICAMP), São Paulo, Brazil*

LALLA FALL • *Department of Chemistry and Biochemistry, California State University East Bay, Hayward, CA, USA*

LUIZ B. FERNANDEZ • *Department of Chemistry, University of Houston, Houston, TX, USA*

AUDREY G. FIKES • *Chemistry of Life Program, NC State University, Raleigh, NC, USA; Department of Chemistry, NC State University, Raleigh, NC, USA*

GILDO GIROTTO JUNIOR • *Institute of Chemistry, State University of Campinas (UNICAMP), São Paulo, Brazil*

RAHUL KAPUR • *Department of Biological and Chemical Sciences, New York Institute of Technology, Old Westbury, NY, USA*

JACQUELINE KEIGHRON • *Department of Biological and Chemical Sciences, New York Institute of Technology, Old Westbury, NY, USA*

LUCIEN F. KRAPP • *Laboratory for Biomolecular Modeling and Protein Structure Core Facility, School of Life Sciences, École Polytechnique Fédérale de Lausanne (EPFL) and Swiss Institute of Bioinformatics (SIB), Lausanne, Switzerland*
PIN-KUANG LAI • *Department of Chemical Engineering and Materials Science, Stevens Institute of Technology, Hoboken, NJ, USA*
LUIS A. LANDRAU CORREA • *Department of Biology, University of Puerto Rico-Río Piedras, San Juan, Puerto Rico*
BENJAMIN J. LEAR • *Department of Chemistry, Penn State University, University Park, PA, USA*
ZEKUN LI • *Department of Chemical Engineering, Worcester Polytechnic Institute, Worcester, MA, USA*
ADAM EE XIAN LOH • *Department of Chemistry, Faculty of Science, Universiti Malaya, Kuala Lumpur, Malaysia*
RENATO LOMBARDO • *Dipartimento di Scienze e Tecnologie Biologiche, Chimiche e Farmaceutiche, Università degli Studi di Palermo, viale delle Scienze, Palermo, Italy*
MAUREEN MCKEAGUE • *Department of Chemistry, Faculty of Science, McGill University, Montreal, QC, Canada; Department of Pharmacology, Faculty of Medicine and Health Sciences, McGill University, Montreal, QC, Canada*
FERNANDO TEIXEIRA PINTO MEIRELES • *Laboratory for Biomolecular Modeling and Protein Structure Core Facility, School of Life Sciences, École Polytechnique Fédérale de Lausanne (EPFL) and Swiss Institute of Bioinformatics (SIB), Lausanne, Switzerland*
MURILO NÍCOLAS MOBELLI • *Institute of Chemistry, State University of Campinas (UNICAMP), São Paulo, Brazil*
YANNICK BELO NUAPIA • *Pharmacy Department, School of Healthcare Sciences, University of Limpopo, Polokwane, South Africa*
KEVIN PARK • *Department of Digital Art and Design, New York Institute of Technology, Old Westbury, NY, USA*
WOJCIECH POKRZYWA • *Laboratory of Protein Metabolism, International Institute of Molecular and Cell Biology in Warsaw, Warsaw, Poland*
SRI RAHAYU • *Chemistry Department, Faculty of Mathematics and Science, Universitas Negeri Malang, Malang, Indonesia*
AA'ISHAH RIAZ • *Department of Chemistry and Biochemistry, California State University East Bay, Hayward, CA, USA*
MAIRA RIVERA • *Department of Chemistry, Faculty of Science, McGill University, Montreal, QC, Canada*
KRITI SHUKLA • *Department of Chemistry, University of North Carolina at Chapel Hill, Chapel Hill, NC, USA*
MONIKA SOMMERHALTER • *Department of Chemistry and Biochemistry, California State University East Bay, Hayward, CA, USA*
MELISSA C. SROUGI • *Biotechnology Program, NC State University, Raleigh, NC, USA; Department of Molecular Biomedical Sciences, NC State University, Raleigh, NC, USA*
KRISHNA MOHAN SURAPANENI • *Department of Biochemistry, Panimalar Medical College Hospital & Research Institute, Chennai, Tamil Nadu, India; Department of Medical Education, Panimalar Medical College Hospital & Research Institute, Chennai, Tamil Nadu, India*
NATALIA A. SZULC • *Laboratory of Protein Metabolism, International Institute of Molecular and Cell Biology in Warsaw, Warsaw, Poland*

KHENG SOO TAY • *Department of Chemistry, Faculty of Science, Universiti Malaya, Kuala Lumpur, Malaysia*
GÜNTHER THIELE • *Albert-Ludwigs-Universität Freiburg, Institut für Anorganische und Analytische Chemie, Freiburg im Breisgau, Germany*
REIA THOMAS • *Department of Biological and Chemical Sciences, New York Institute of Technology, Old Westbury, NY, USA*
JOANA THOMSON • *Department of Biological and Chemical Sciences, New York Institute of Technology, Old Westbury, NY, USA*
ARTHUR D. TINOCO • *Department of Chemistry, University of Puerto Rico-Río Piedras, San Juan, Puerto Rico*
HLANGANANI TUTU • *Molecular Sciences Institute, School of Chemistry, University of the Witwatersrand, Johannesburg, South Africa*
MORGAN A. VINCENT • *Department of Chemistry, Penn State University, University Park, PA, USA*
CLARA A. VON RANDOW • *Albert-Ludwigs-Universität Freiburg, Institut für Anorganische und Analytische Chemie, Freiburg im Breisgau, Germany*
GUANGJIE YAN • *Department of Chemistry, University of Houston, Houston, TX, USA*
ERIC M. YOUNG • *Department of Chemical Engineering, Worcester Polytechnic Institute, Worcester, MA, USA; Department of Biology and Biotechnology, Worcester Polytechnic Institute, Worcester, MA, USA*

Chapter 1

The MolecularWeb Universe: Web-Based, Immersive, Multiuser Molecular Graphics and Modeling

Fabio J. Cortés Rodríguez, Lucien F. Krapp, Fernando Teixeira Pinto Meireles, Matteo Dal Peraro, and Luciano A. Abriata

Abstract

For decades, software for molecular visualization has been a cornerstone of research and education in the chemical and structural sciences. It is unfortunate, however, that consumer devices enable only flat 2D inputs and outputs, which brings two problems. First, traditional 2D screen-mouse interfaces struggle to convey the inherently three-dimensional nature of molecules and their interactions. Second, probably worse, the thoughtful manipulation of molecular structures in three dimensions is very hard to achieve with 2D peripherals. Immersive technologies under the umbrella of eXtended Reality (XR, including augmented and virtual reality, AR and VR) promise to revolutionize how we learn, teach, and conduct research in the chemical and structural sciences, by assisting both how we see and how we manipulate molecules in 3D. This chapter chronicles the development of the "MolecularWeb" ecosystem, a suite of web-based tools designed to make immersive molecular visualization and interactive modeling and simulations widely accessible to a broad audience. Centralized at https://molecularweb.org, these tools have found wide applicability in education and science communication, promising to directly assist research as well. We describe first MolecularWeb, which delivers activities and tools for education in pass-through and mirror-like AR using regular devices like phones, tablets, and computers. We then cover MolecularWebXR, a platform for multiuser immersion using WebXR-enabled headsets—yet remaining accessible with simpler devices—that brings content about chemistry, materials sciences, and structural biology in formats engaging for educational and science outreach activities, also serving for immersive scientific discussions. A short visit to PDB2AR will show how users can create tailored content for MolecularWebXR and also stand-alone AR and VR experiences. We finally delve into our HandMol prototypes, which allow for immersive visualization of molecular systems and their modeling with bare hands in immersive 3D by multiple concurrent users assisted by on-the-fly molecular mechanics calculations, seamless file exchange, a language model module that interprets users' inputs without them having to learn any menus or scripting, alternative immersive access with consumer devices, and more. With the various tools presented, we offer a glimpse into the present and near future of accessible, interactive, and intuitive molecular science on the web.

Key words Molecular graphics, Molecular modeling, Augmented reality, Virtual reality, WebXR, Web programming, STEM education, Science communication, Interactive simulation, Multiuser VR

Robert E. Dempski (ed.), *Transforming Molecular Biology with Emerging Technologies*, Methods in Molecular Biology, vol. 3043, https://doi.org/10.1007/978-1-0716-5320-3_1,

1 Introduction: The Challenge of Visualizing and Interacting with Virtual Representations of Molecular Systems

Understanding the three-dimensional structure, dynamics, and interactions of molecules is fundamental to a vast array of scientific disciplines, including chemistry, drug discovery, materials science, and molecular biology. For generations, scientists and science students have relied on physical models and two-dimensional (2D) computer graphics to visualize molecular systems and other entities of inherently 3D nature, such as molecular orbitals, surfaces, volumetric maps, etc. [1–4] While computer graphics have undeniably been invaluable tools, they present inherent limitations. On the one hand, there is the difficulty of representing and perceiving true depth and complex spatial relationships on a 2D display, [3, 4] which traditional molecular graphics programs have tackled this by allowing users to use cuing, clipping, cutting planes, etc. In fact, for users who gain dexterous use of such tools after months or years of work, the problem of realistic 3D perception on flat screens is probably overcome. However, this is not the case for novice users and for occasional users such as students or the public.

On the other hand, there is the difficulty in manipulating 3D objects with a 2D input peripheral such as a mouse, sometimes also with a keyboard, all of which is often cumbersome and counterintuitive even for experienced users. Achieving more natural manipulation of 3D objects in real 3D space can be particularly useful for intricate tasks such as molecular docking, detailed conformational analysis, or deciphering stereochemistry and chirality [5–7]. In fact, we have argued in a recent perspective that the toughest limitation of traditional molecular graphics software is not in 3D visualization but in 3D manipulation [8].

Immersive technologies, termed also XR after eXtended Reality and encompassing virtual reality (VR), which plunges the user into a completely digital environment, and augmented reality (AR), which overlays digital information onto the real world, offer a paradigm shift in human-computer interaction. Particularly for molecular graphics and modeling, XR-based applications promise a more natural and intuitive way to interact with 3D molecules and data associated with them [6, 7, 9–21]. The potential benefits for education are immense, offering students a more tangible, hands-on experience with abstract concepts that are traditionally difficult to grasp; for research, in turn, these technologies could accelerate discovery by allowing more fluid exploration and discussion of molecular interactions, properties, and dynamic behaviors [6, 11, 22–30].

The literature on AR/VR systems for molecular graphics and XR in science is very extensive, with various recent reviews summarizing different parts of it [14, 23]. Historically, however, the widespread adoption of immersive molecular modeling has been hampered by several factors. Firstly, high-end VR/AR systems were, until recently, prohibitively expensive for many end users, and some still are even in wealthy countries. Second, now largely alleviated, headsets used to entail bulky hardware tethered to external computers and, in some cases, also to towers equipped with instrumentation for body tracking. Third, the setup and use of these systems frequently involved, and in many cases still today involve, software configurations that represent a very steep barrier for adoption. Fourth, many early or simpler AR solutions were restricted to static visualizations, lacking the dynamic interactivity crucial for deeper understanding and research. Finally, a fragmented landscape of diverse hardware and software ecosystems made it difficult to develop universally accessible and standardized tools; in fact even today, most devices have different operating systems that require specialized compilations for software to run on them—think Apple vs. Android systems for smartphones and tablets, Windows vs. Mac vs. Linux for computers, or the bunch of different operating systems for the different AR/VR headsets. To the rescue and as we advocate for, [31–34] web programming and in particular the WebXR standard offer out-of-the-box instant access to immersive molecular graphics and modeling without installs and with supreme device compatibility [31, 35]. All our tools from the MolecularWeb universe benefit from this technology, as we will cover in more detail throughout this chapter.

In what follows, we present the MolecularWeb ecosystem in a chronological order that shows how we progressively enhanced, with a fully web-based spirit, visual immersion, seamless interactivity, and human-human plus human-computer collaboration applied to the molecular sciences. Our resources and platforms are all delivered through the web browser, thus being available on every computer, phone, tablet, and the latest generations of XR headsets from major vendors. Centralized at https://www.molecularweb.org, they include MolecularWeb, the MolecularWebXR platform, the PDB2AR tool, the HandMol client-only prototype, and the upcoming MolecularWebXR-integrated HandMol (Table 1).

2 Getting Virtual Molecules into the Real World: MoleculARWeb for Education in Chemistry and Structural Biology Using Commodity AR

2.1 Motivation

Our initial foray into democratizing immersive molecular experiences focused on what we term "commodity augmented reality," that is, AR applications that run on readily available, everyday consumer devices such as webcam-equipped computers, tablets, and smartphones, without requiring any specialized headsets. This endeavor, inspired by the work and demonstrations made a decade earlier by Arthur Olson, [5] culminated in the creation of

Table 1
Web apps and web platforms of the MolecularWeb universe

Website/tool (subheading in this chapter)	Purpose/use cases	Direct link(s)	References
MolecularWeb (Subheading 2)	Passthrough and mirror AR for webcam-enabled devices (laptops, smartphones, tablets) using printed fiducial markers (or no markers in some activities) Education in general and organic chemistry; introductory materials sciences and structural biology	https://molecularweb.epfl.ch https://molecularweb.org/	[27, 36, 37]
PDB2AR (Subheading 3)	Content generator for molecular visualization in AR with or without fiducial markers, in stand-alone VR on headsets, or inside MolecularWebXR rooms	https://molecularweb.epfl.ch/pages/pdb2ar.html	[44]
MolecularWebXR (Subheading 4)	Multiuser discussions in rooms with preset content or with content customized via PDB2AR. Virtual models are static only Preset content targets introductory chemistry and structural biology	https://molecularwebxr.org	[60]
Client-only HandMol prototype (Subheading 5 intro)	Immersive visualization of PDB files by 1 or 2 concurrent users hand-driven molecule manipulation. On-request minimization with molecular mechanics Immersive substitute of molecular graphics programs; suitable for small operations, education, etc.	https://go.epfl.ch/handmol	[47]
MolecularWebXR-integrated HandMol prototype (Subheading 5)	Like MolecularWebXR but with molecules loaded from PDB files and parametrized for interactive simulations Immersive substitute of molecular graphics software, enhanced with molecular mechanics, tools for collaboration, AI assistant for seamless control For self-education, single-person work, teacher-student(s) sessions, interactive discussions, and concurrent work	Coming up at https://molecularwebxr.org/ (most likely https://molecularwebxr.org/create)	(coming up)

MolecularWeb (https://molecularweb.epfl.ch), a free, open-source website hosting a diverse suite of AR web applications. These applications are primarily designed for chemistry and structural

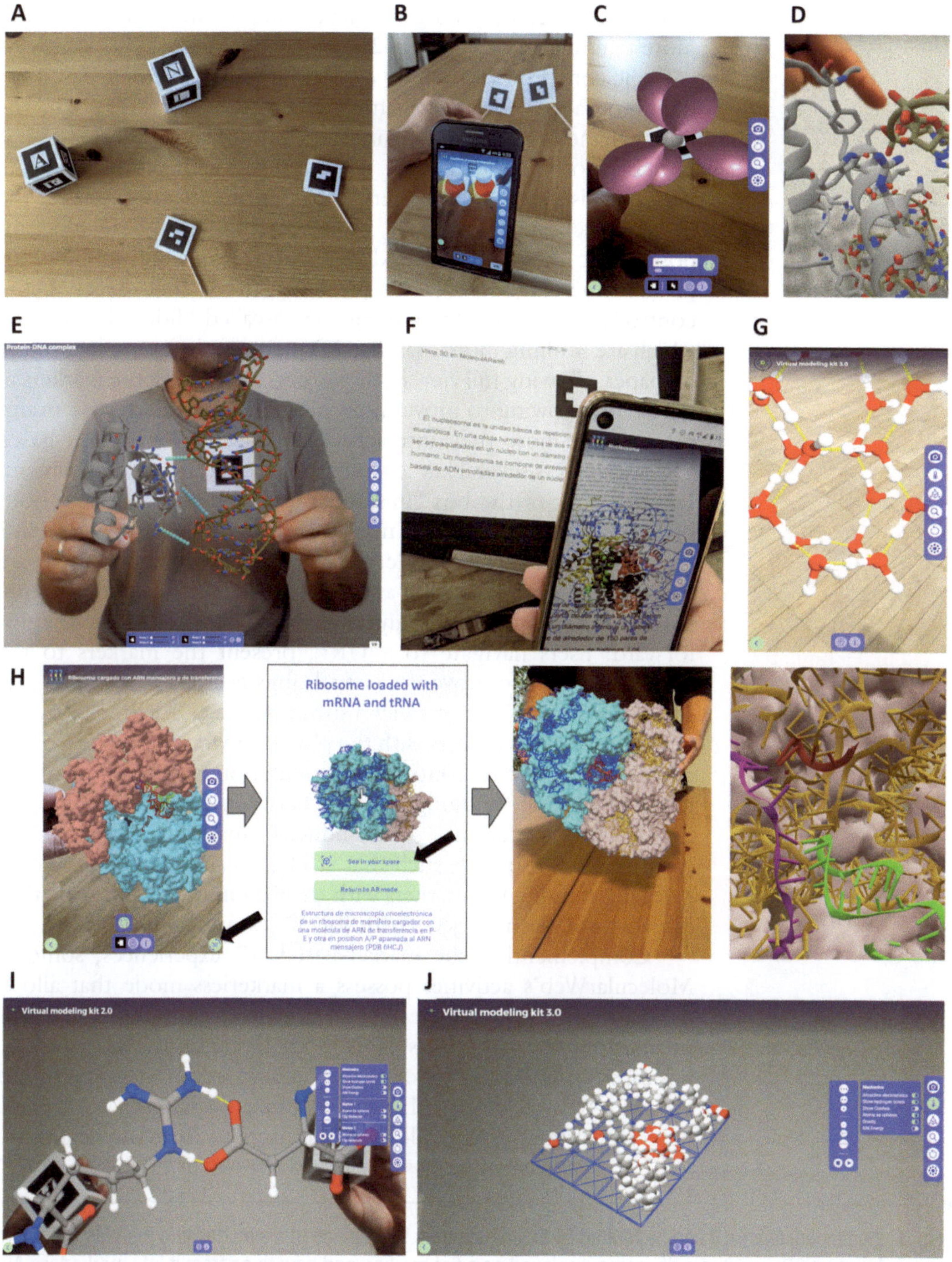

Fig. 1 Representative examples of MolecularWeb's activities and playgrounds. (**a**) The 4 main kinds of fiducial markers used in MolecularWeb, created with a regular printer and folded into 3D cubes (markers for VMK 2) or glued to sticks for convenient handling (flat markers for all other activities). (**b**) The flat markers working in pass-through AR on a phone, here to test how two water molecules interact by hydrogen bonding and exchange protons. (**c**) An sp^3d hybrid orbital on a flat marker, from the activity on molecular shapes. (**d**) AR "in the user's space" without any markers, and the user pointing at a phosphate group in the DNA backbone of a protein-DNA complex (more in panel **h**). (**e**) Docking a protein molecule into a cognate DNA, guided by visual cues. (**f**) One of the regular flat fiducial markers, embedded in a text to show the 3D molecule on the phone

biology education, aiming to make complex molecular concepts more tangible and engaging [27, 36, 37].

The core principle behind MolecularWeb is a marker-based AR user interface. Users print simple 2D fiducial markers (e.g., the "Hiro" and "Kanji" patterns, or two foldable cube designs, Fig. 1a) from provided PDFs (3D printing with black and white plastic ink is also possible); then, by presenting these markers to the device's webcam, virtual 3D molecular models appear superimposed onto them and adapted to their positions and orientations in real time (Fig. 1b–f). This results in the virtual objects being controllable naturally by moving the so-called "fiducial" markers, which are nothing more than flat 2D pictures, for example, printed on paper, allowing full view of the objects by rotating the markers in space and allowing to move molecules in 3D, useful to, for example, explore how they interact (Fig. 1b–e). This marker-based approach offers several key advantages for educational deployment. It works "out-of-the-box" in standard web browsers across various operating systems, requiring no software installation beyond what's typically available on modern devices. The primary hardware requirement is a device with a webcam and an internet connection, making it highly affordable and available. The workflow is straightforward: users navigate to a URL, present the markers to the camera in order to move the virtual objects around, using regular controls with a mouse or touch buttons to execute actions. Manipulating the virtual objects with the physical marker provides a direct and tangible way to control the orientation and position of the virtual molecule, bridging the gap between physical models and digital representations. The fiducial markers can also be incorporated into texts such that students can visualize the molecules by scanning the markers with their phones, in which case they can then rotate the objects with touch gestures (Fig. 1f).

Complementing the marker-based AR experiences, some of MolecularWeb's activities possess a markerless mode that allows users to view the molecules "in their space" via look-through AR in smartphones and tablets. This mode leverages the spatial tracking capabilities of smartphones and tablets, fed into the app by the WebXR API, to display the virtual objects anchored to a plane detected by the system, allowing the users to freely move around

Fig. 1 (continued) (where it can be rotated by touch gestures). (**g**) A piece of ice loaded into the VMK 3.0 (more in panel **j**). (**h**) A 3D model of a ribosome displayed on a flat marker, and how to go from it into markerless AR: first tap on "3D" at the bottom right, then on "See in your space"; then scan a flat surface onto which the molecule will get anchored. (**i**) The VMK 2.0 being used in mirror AR mode to explore interactions between a positively charged arginine sidechain on the left and a negatively charged aspartate sidechain on the right; the app displays hydrogen bonds in real time (yellow dotted lines) because this is selected on the molecular mechanics pad. (**j**) The VMK 3.0 being used to show how polar and non-polar molecules phase-separate if polar forces are on, as controlled in the molecular mechanics pad

the object to inspect it (Fig. 1d and video linked to it, explanations in Fig. 1h).

2.2 MolecularWeb's Modules, Activities, and Playgrounds

In terms of content, MolecularWeb is structured into thematic modules, each containing several activities. A module on *Atomic and Molecular Orbitals and Molecular Shapes* allows users to visualize *s*, *p*, *d*, *f*, and various hybrid atomic orbitals; it also includes activities for exploring idealized VSEPR (valence shell electron pair repulsion) geometries and the molecular orbitals of simple molecules (e.g., H_2O, CO_2, NH_3) (exemplified in Fig. 1c). Next, a module on *Hydrogen Bonding, Acids and Bases* allows students to interactively explore hydrogen bond formation, proton transfer equilibria (such as water autoionization or the reaction of an acid/base with water), and the dynamic nature of these interactions (Fig. 1b).

Two modules on structural biology follow. *Atomic Structure of Biological Macromolecules* features visualizations of protein secondary structures (α-helices, β-sheets), protein-DNA complexes (including an activity where a protein and its cognate DNA are fitted by manual operations, Fig. 1e), and related content. Then, with activities in the module on *Large Biomolecular Assemblies,* users can explore complex structures like viruses, ribosomes, and nucleosomes, many of these activities supporting the markerless AR mode on compatible smartphones (example in Fig. 1h).

At the top of the MolecularWeb website, a special module of *Playgrounds* presents AR tools that behave as whole small programs rather than specific activities. The Virtual Modeling Kits (VMKs [36]) allows users to load any molecule via PDB coordinates that can be pasted into a box, fetched from databases, or drawn using an embedded JSME [38] applet based on hack-a-mol [39]. In the VMK 2.0, two molecules can be loaded, each displayed on a manipulable cube marker; each cube can then be moved independently to drive the molecules around as required to compare them, make them interact, etc. In the VMK 3.0, any number of molecules can be loaded to populate a box where they are contained. Both VMKs feature rudimentary molecular mechanics powered by the Cannon.js physics engine hacked to simulate thermal fluctuations and to emulate electrostatic forces that can be switched on and off at will. Additional controls allow users to switch between ball-and-stick and volume representations of the molecules, toggle display of hydrogen bonds, swap molecule chirality, emulate gravity (in the VMK 3.0, example in Fig. 1g and j), and control the view (switching camera, cutting the background, etc.). While our dedicated papers explore several applications of both VMKs, we outline in the next two paragraphs some example use cases.

With the VMK 2.0, having the molecules attached to the markers allows exploration of their internal degrees of freedom and dynamics as the rotational and translational diffusion are removed (the molecule is re-centered at the cube in each frame), and driving interactions between pairs of molecules, where the app can show hydrogen bonds and atom-atom clashes. With these capabilities, teachers can show (and students can explore by themselves) how atoms rotate around single bonds, how non-planar rings exchange conformations, how aromatic and conjugated moieties remain planar, how molecules interact, etc. (example of the latter in Fig. 1i). Another application of driving molecules with two cubes is in comparing stereoisomers, for example, by loading the same molecule on two cube markers but inverting only one.

In turn, the VMK 3.0 behaves as a simulation box that allows teachers to show or students to explore how multiple water molecules stick together through polar forces to form a drop; one can even start with an example 3D model of ice to understand its peculiar water arrangement (Fig. 1g) and how this structure is lost upon melting, in turn increasing the density. The melted piece of ice then behaves as a drop where the water molecules remain cohesive, yet some water molecules escape from time to time, illustrating vapor pressure that increases with temperature until eventually the system boils. Another interesting application uses the feature to toggle polar forces at will, to illustrate, for example, how miscibility and immiscibility work (Fig. 1j).

The *Playground* module also features a marker-less commodity AR version of all the static content available on the website [37]. Built by using the Handsfree library, in this web app, users can set 3D virtual molecules, orbitals, etc., and place them on the webcam's feed, to then move them with their hands.

Finally, the *Playground* module at https://molecularweb.org includes a web app called PDB2AR, useful to create custom 3D material that can be visualized with the cube marker, with AR in the user's space, or in AR/VR headsets in a stand-alone fashion or inside our website for multiuser molecular graphics, MolecularWebXR. PDB2AR is treated in Subheading 3, and MolecularWebXR is subsequently treated in Subheading 4.

2.3 User Reception, Usage Statistics, and Feedback

MolecularWeb relies solely on standard client-side web technologies: HTML5 for structure, CSS for styling, and JavaScript for interactivity. 3D graphics rendering is handled by Three.js, while marker tracking leverages AR.js, which itself is a wrapper for ARToolKit. This all makes the website entirely functional in regular laptops and computers equipped with a webcam, as well as in smartphones and tablets. The website is thus extremely easy to utilize, as attested by the high success documented by continuous access from all around the world (Fig. 2a).

A very interesting application of MolecularWeb reported from the University of San Luis in Argentina [40] shows how it can be integrated with other tools for AR visualization, in this case closing a cycle that connects chemistry with physics and biology by starting

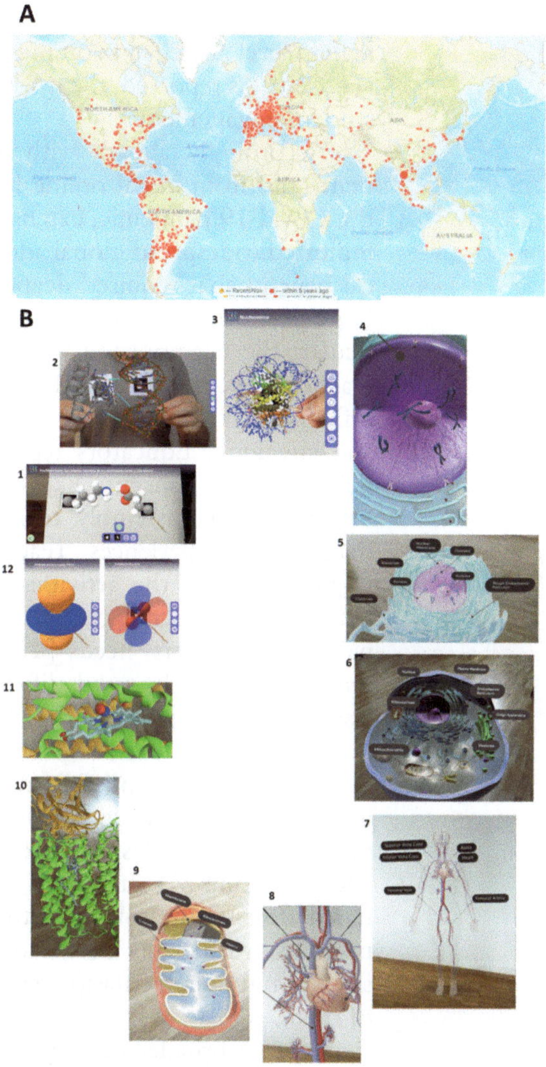

Fig. 2 MolecularWeb usage statistics and example integration with other AR tools. (**a**) Accesses since launch, as tracked at https:/clustrmaps.com/site/1ar2f. (**b**) Integration of MolecularWeb bringing AR views of chemistry and structural biology with Google-based AR views of systems, organs, tissues, cells, and organelles. This is part of an activity connecting physics, chemistry, and biology, all in AR, starting from electrostatic interactions (1) relevant in biology, for example, for protein-DNA complexation (2) and compaction into nucleosomes (3) and chromosomes (4) inside the cell's nucleus (5). A full view of the cell (6) then leads to the systems level (7), and then the activity zooms back in going through the heart as an organ (8), a mitochondrion (9) where cytochrome c oxidase (10) carries out that last step of respiration at its heme group (11) where oxygen interacts with a specific orbital with a specific orbital at the heme's iron site (12)

with two small molecules, then zooming out to the organismal system level and then zooming back in into molecules and orbitals (Fig. 2b). As reported, the novelty brought by AR produces strong student engagement.

Accordingly, already with our first tests on high-school students and from initial online feedback, all right in the middle of the Covid-19 pandemic, extensive user feedback and web analytics indicated widespread global adoption and some positive pedagogical benefits that required deeper study yet clearly demonstrated a huge positive effect on student engagement [25, 27, 41]. In particular, the accumulated evidence from over 4 years of online teacher surveys confirms the capacity of MolecularWeb to make difficult chemistry and structural biology topics more approachable and engaging. Educators repeatedly highlight how the platform helps simplify concepts like molecular bonding, geometry, and macromolecular assembly, turning abstract topics into interactive, manipulable experiences. Teachers appreciate not only the pedagogical value of the activities but also the platform's flexibility, being able to integrate custom content tailored to their curriculum, pace, and students' needs [42, 43]. Such adaptability allows the same activity to be used differently in a high school chemistry lesson, an undergraduate biochemistry class, or informal science outreach. In addition, teacher testimonials consistently point to increased student enthusiasm and participation when activities include hands-on AR interactions—students who might be passive during traditional lessons become actively involved in exploring and discussing 3D structures. However, feedback has also illuminated areas for improvement that we tried to attend as it becomes possible—with many of the requested features actually available already, but documentation gaps mean some users remain unaware of them.

From a broader perspective, MolecularWeb exemplifies both the promise and the ongoing challenges of integrating AR into STEM education. Immersive tools reliably capture attention and spark curiosity, yet engagement alone does not guarantee deeper or longer-lasting learning. The novelty of AR may be a key driver of student interest today, but sustained educational value depends on integrating these tools into pedagogically sound activities that actively support comprehension, retention, and skill-building. Furthermore, although anecdotal reports suggest AR can accelerate spatial understanding of molecular structures—freeing instructional time for higher-level discussion—systematic, controlled studies remain scarce. These considerations point to an important next step: pairing widespread deployment of platforms like MolecularWeb with rigorous, longitudinal research that evaluates learning gains, not just engagement. This would involve interdisciplinary collaboration between developers, educators, and education researchers to design studies that measure outcomes such as conceptual understanding, retention over time, and transfer of spatial skills. Ideally, such research should compare AR-based instruction to both traditional 2D approaches and physical model use, and should investigate how factors like prior knowledge, spatial ability, and content complexity influence effectiveness. In the meantime, the teacher and student feedback collected already offers us a roadmap for iterative improvement: enhance discoverability of existing features, expand the content library, refine interaction methods for mobile users, and explore integration with complementary tools (e.g., simulation engines, AI-based assistants). Addressing these points, while preserving MolecularWeb's core strengths of accessibility and ease of use, could further solidify its role as a bridge between fascination and meaningful learning in molecular sciences education.

3 The PDB2AR Tool for Custom Content Creation for All Devices

3.1 Motivation

An important aspect of democratizing these immersive technologies is to empower educators, researchers, and science communicators to create their own customized content without requiring extensive programming expertise or specialized software development skills. We experienced this urge first-hand when a few months after MolecularWeb's release several teachers had asked about the possibility of adding specific content. To address this need, we developed an online utility integrated within MolecularWeb's *Playgrounds* module, named PDB2AR (or "Build your own webAR/VR views," Fig. 3) [44]. This tool, accessible directly at https://molecularweb.epfl.ch/pages/pdb2ar.html, provides a streamlined workflow for generating shareable AR and VR experiences that work on the cube marker (in all devices), in the user's space via WebXR if running in a smartphone, or in full VR if running on the web browser of an AR/VR headset. Besides, the virtual objects created with this tool can be placed in sessions of our MolecularWebXR platform, described in Subheading 4.

3.2 Using PDB2AR

As input to PDB2AR, users can upload a PDB file containing atomic coordinates from their local system, fetch a structure directly from the Protein Data Bank (PDB) using its unique PDB ID, or retrieve a model from the EBI's AlphaFold Protein Structure Database using its UniProt entry. Alternatively, users can load virtual objects generated with VMD (exported as WaveFront OBJ + MTL files); this allows them to visualize not just molecules but also isosurfaces, grids, volumetric maps, etc., and thus essentially any other 3D element visible in their starting VMD scene (such as electron density maps, molecular orbitals, and custom geometric objects).

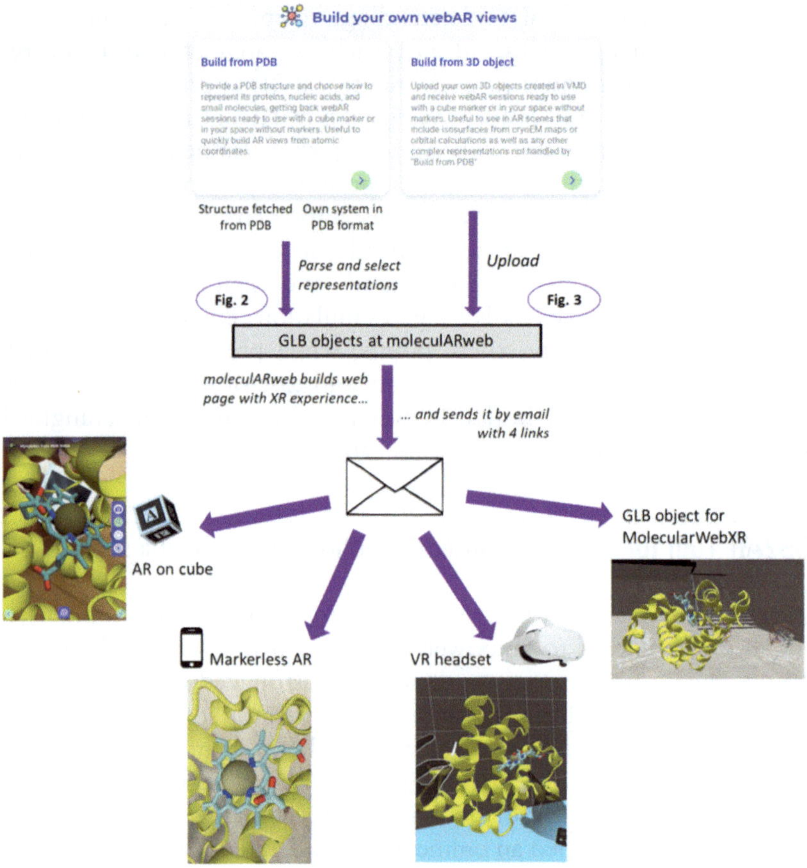

Fig. 3 Custom content generation within the MolecularWeb universe with PDB2AR. Starting from a PDB file that will be parsed by the app and converted into the user-selected representations, or from wavefront objects exported from VMD, PDB2AR creates AR/VR experiences that run using the cube fiducial marker, markerless AR "in the user's space," or in headsets, also providing validated links to add the elements into MolecularWebXR scenes

When a PDB file is provided as input, the web interface parses its molecules and then users can select how different molecular components (e.g., protein chains, nucleic acid strands, small molecules, ions, other molecules) should be displayed. Options include various standard representations like cartoons, sticks, spheres, surfaces, etc., along with a choice of color schemes (e.g., by chain, by element, by secondary structure); all following conventions of the VMD software [45]. In fact, PDB2AR will then generate a VMD script that will generate OBJ + MTL files, and the user can optionally edit this script in order to finely control the generated object.

Once the input and display options are set, the user provides a title for their scene and an email address. The PDB2AR tool then automatically processes the input on our server, generating the 3D virtual model in GLB format and with a special compression that makes the files light. The user then receives an email with various unique links to web pages that display the generated object in various forms, that is either anchored to the standard MolecularWeb cube marker, or in the user's space in pure augmented reality, or as a virtual object; besides, a unique link to the GLB file allows seamless addition of the element into a MolecularWebXR scene.

As we see with a number of submissions per month, the PDB2AR tool significantly lowers the barrier to entry for creating custom immersive molecular content. With it, educators can tailor visualizations precisely for their specific teaching objectives, and researchers can quickly generate VR-ready models for collaborative discussions, presentations, or personal exploration of complex structural data. To us, PDB2AR has been instrumental in expanding the content available at MolecularWeb and MolecularWebXR, as required for numerous educational and science communication activities.

4 Deeper Immersion Without Compromising Reach, Via WebXR: MolecularWebXR

4.1 Motivation

The maturation of the WebXR API presented a compelling opportunity to extend our web-based tools for molecular graphics into more deeply immersive AR and VR environments. The goal was to retain the core advantages of web delivery (no installation, broad accessibility) while offering a richer, more engaging and especially more immersive experience, that is also natively multiuser as all users in a session can see each other's avatars and talk naturally in the room as they manipulate virtual objects, possibly physically together or separated by thousands of kilometers. This is materialized in MolecularWebXR (https://molecularwebxr.org, Fig. 4a), a platform designed for multiuser discussions, educational sessions, and collaborative exploration within virtual "rooms" or "sessions" where users connect together (Fig. 4b and c) to concurrently view and manipulate models of relevance to chemistry, materials science and structural biology, that is not just molecular structures but also representations of molecular surfaces, orbitals, electron maps, etc. Each user can experience the session immersed in VR or AR with headsets (Fig. 4d), in pass-through AR or cardboard goggle-based VR with smartphones (Fig. 4e shows the latter) or in non-immersive way with every device.

4.2 Accessing and Using MolecularWebXR

To access the platform, free and without registration, users must direct the web browser of their device of choice (AR/VR headset, smartphone, tablet, computer) to https://molecularwebxr.org. Note, however, that the MolecularWebXR website, available as of early 2025, will become the "Science XR" branch of the website starting in 2026.

To use MolecularWebXR, an admin host must first create a session starting from a room with preset content or with an

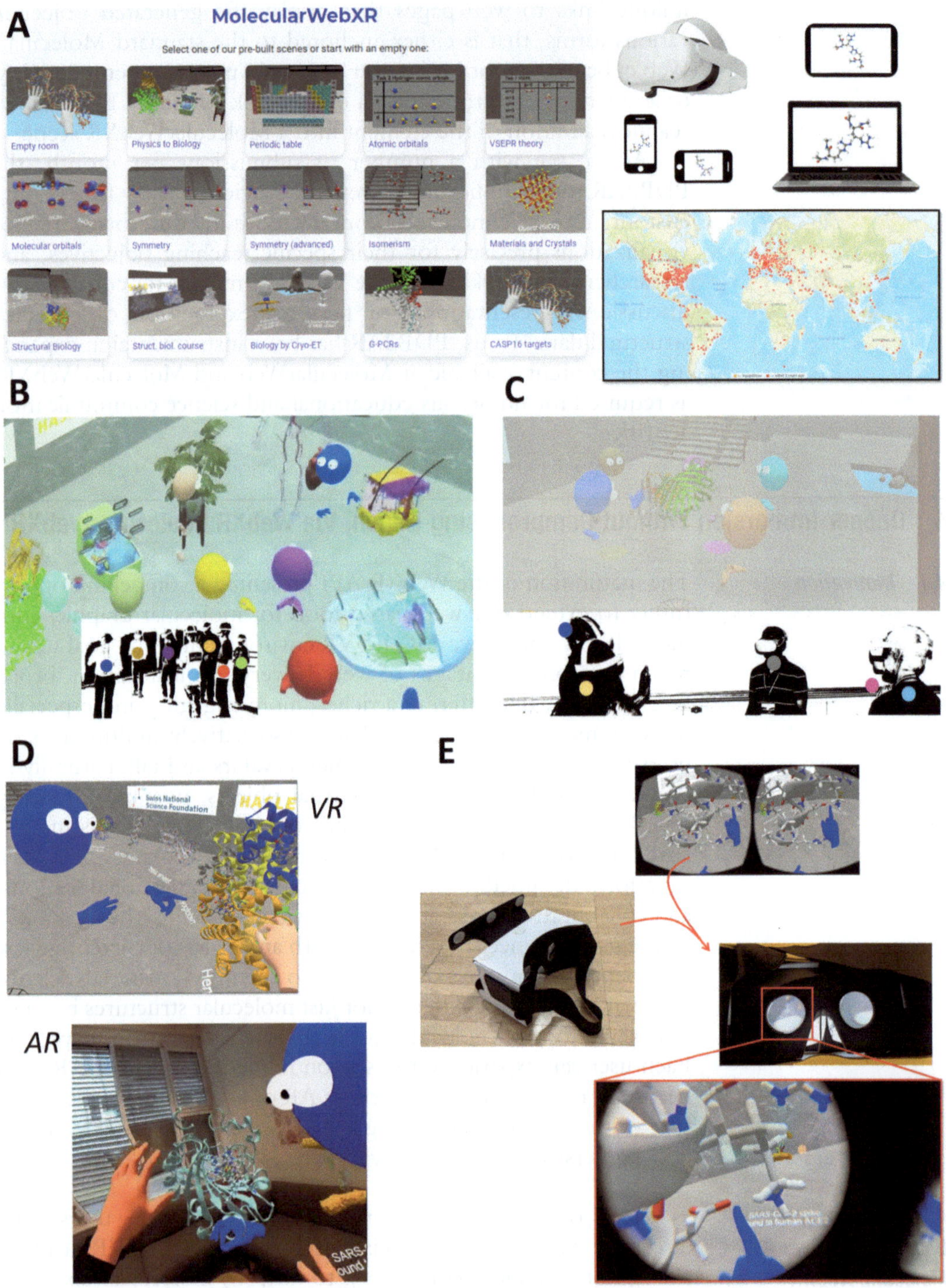

Fig. 4 A summary of MolecularWebXR. (**a**) The rooms with predefined content available as of mid-2025, filled with objects created by collaborators and using PDB2AR plus others obtained for free or purchased at Sketchfab.com. The map in the inset shows worldwide access since release. (**b**) A public communication session where a person (blue avatar) walks 6 guests around a virtual room connecting physics with chemistry and biology. Notice the presenter's hands pointing at sections of a model of a dermal tissue. Photograph taken at EPFL's Portes Ouvertes event in 2023, Lausanne. A short video demonstrating how this works is available at https:/www.youtube.com/watch?v=NnKq4L3NVIY. (**c**) Five scientists discussing a protein structure; notice

empty room to be populated with objects built with PDB2AR. The created session will have unique identifiers (four-letter codes under "Invite people") that the admin must send to those that should connect to it via the "Join room" button at the landing page. The admin user has access to several features to control the session, such as removing or adding objects, changing their sizes, resetting them, controlling which users can grab objects and what objects can be grabbed, enabling or disabling audio, etc. For a more comfortable use of these features, it is best to create the session on a computer.

As users join the session, they can enter AR or VR or simply stay outside of XR, depending on the device's capabilities (examples of the different modes in Fig. 4). Of course, the session is maximally experienced through AR/VR headsets, which offer the most immersive experience. Accessing with headsets, users are represented as avatars composed of their heads and hands or controllers, continuously updated as they walk and move their limbs around the virtual space. Users wearing headsets can interact with molecular models using built-in hand tracking or handheld controllers. By pinching with the index and thumb fingertips of one hand on a model (or pressing with the index finger on the controller), the user will grab and move the object. By pinching or pressing an object on two points, the user will change the object's size. Users can also talk with each other directly through the headset, a feature available in all devices that the host must disable if the users are also physically together.

Thanks to its web-based nature, MolecularWebXR also allows users to join sessions from smartphones and tablets, where they can navigate the 3D space using on-screen joysticks or touch gestures, and observe the avatars and actions of users who access with headsets. Moreover, they can also go into immersive VR by using cardboard goggles for a stereoscopic VR view (Fig. 4e) or in pass-through AR with 6 degrees of freedom. And of course, the website also works on regular computers, where participants can use keyboard and mouse moves for navigation and viewing.

4.3 Content Available at MolecularWebXR

MolecularWebXR includes several pre-defined virtual rooms containing curated content relevant to specific educational topics on chemistry, structural biology, and materials science. Since this content has evolved substantially since our paper presenting the website, we dedicate here quite some detail to all the content available as of mid-2025 (Fig. 4a).

Fig. 4 (continued) the presenter in blue pointing at the interface between two structures; photograph taken at the closing conference of the NCCR TransCure program in 2022, Bern. (**d**) Top: MolecularWebXR in VR mode, view of the user with the salmon avatar (notice the hand at the bottom right). Bottom: MolecularWebXR in AR mode, view of the user with the orange avatar. (**e**) MolecularWebXR experienced in immersive VR on phones, by tapping VR and plugging the device into cardboard goggles equipped with lenses

The first room, an "Empty Room," provides a virtual space without any objects, to be populated with material prepared with PDB2AR by clicking on the Elements menu and creating new ones by pasting the URLs to GLB objects received from PDB2AR by email. Next to this empty room, "Physics to Biology" is at the core of a science communication story connecting biology with chemistry and physics, described below in a specific section. Then, 8 rooms with material for introductory chemistry follow: a periodic table where the elements can be taken out one by one, useful for interactive quizzes and other activities (an XR version of the activities developed by Limpanuparb et al. [46]); a room with the atomic orbitals and another with the HOMO and LUMO molecular orbitals of molecules selected for pedagogic interest; a room on VSEPR theory; two rooms on molecules symmetry, one passive and one where students must grab objects representing symmetry axes and planes and use them to identify the elements in various molecules; one room on isomerism; and one with example structures of materials and crystals. Finally, five rooms are dedicated to different aspects of structural biology: a room that introduces protein and DNA structures; one used at a structural biology course and covered in more detail below; one on cryo-electron tomography covering very modern and interesting 3D views of viruses, cell interiors, and other structures; one on GPCRs focusing on its activation mechanics; and one with interesting targets from CASP16.

4.4 MolecularWebXR Case 1: Communicating Science "from Physics to Biology"

The room called "Physics to Biology" is at the core of a science communication story connecting biology with chemistry and physics that we carry out with a presenter and up to 4–6 visitors, all inside VR, with the headsets set to match the virtual and real environments so that all the people inside the session can move around freely. Overall, the presentation covers the same content as that presented in Fig. 2b but in a much more immersive and interactive fashion, as seen in the photograph in Fig. 4b. The room contains a model of a human body's interior, where the story starts, followed by a model of a heart, a piece of dermal tissue, and an animal cell through which the story continues. Next to the human cell and almost attached to its membrane, there is a tiny model of a SARS-CoV-2 virus whose interior can be inspected upon zooming. Back to the cell and focusing on the nucleus and its chromosomes, the story continues down the line of a chromosome that unfolds slowly to reveal how DNA is compacted, including, along the way, a crystal structure of a histone nucleus with two turns of DNA wrapped around. Toward the most uncompacted region of the DNA, an X-ray structure of a transcription factor bound to DNA allows the story to present the shape of the DNA molecule with its minor groove and major groove, the bases stack and interacting with each other via hydrogen bonds, the transcription factor's alpha helix docked into the DNA's major groove with some unspecific salt bridges to the backbone phosphates and some specific interactions to the bases.

After talking in such detail about the X-ray structure of the transcription factor bound to the DNA molecule, the story goes on with a piece of an X-ray structure where the electron density map is overlaid, serving to talk briefly about how scientists get to know the structures of molecules. The story then briefly refers to the structure of human carbonic anhydrase bound to a sulfonamide-based inhibitor as an example of how small-molecule therapeutics work.

Then, back to the start, a model of a plant next to that of the human body sets the stage to talk about plant cells and their differences from animal cells. When talking about the thylakoid, the story can move on to the three oxygen molecules leaving the plant cell, each with a different molecular orbital being shown: sigma and pi bonding orbitals, and the antibonding orbital with which oxygen binds the orbital protruding out of the heme group in cytochrome c oxidase. The structure of the oxidase is shown, and the storyteller can grab the oxygen molecule and show how it "diffuses" into the oxidase and toward the heme group, where it will finally accept electrons at the end of the respiratory chain to become water.

4.5 MolecularWebXR Case 2: Illustrating a Course on Structural Biology

The room "Structural Biology Course" was designed to support a course focused on protein structure, protein dynamics, and the various experimental techniques used to explore them. Like in the previous case, in this room, a presenter talks with up to 4–6 visitors, all inside VR with the headsets set to match the virtual and real environments, a story devised to match the content.

In this case, the story opens with the same animal cell model as the above-described room, and it follows with the same models of the unfolding chromosome together with the nucleosome and protein-DNA complex (Fig. 5a). But then, once the structure of the protein-DNA complex is explained, the story moves on to the detailed structural and dynamical features of proteins, in the framework of the techniques utilized to determine them.

First of all, the interactions that held together alpha-helices and beta-strands are covered in detail with models that include the hydrogen atoms and thus allow to very explicitly show the hydrogen bonds at their cores (Fig. 5b). This can be accompanied by inspection of the other models to go to quaternary structure; more content about secondary, tertiary, and quaternary structure is available in the room titled "Structural Biology."

Next in the story, four models of malate synthase G cover the three main techniques for structure determination at atomic resolution, that is, NMR, X-ray diffraction, and cryo-EM. Malate synthase G is the only protein whose structure was solved with all three techniques, representing the high-size limit for NMR

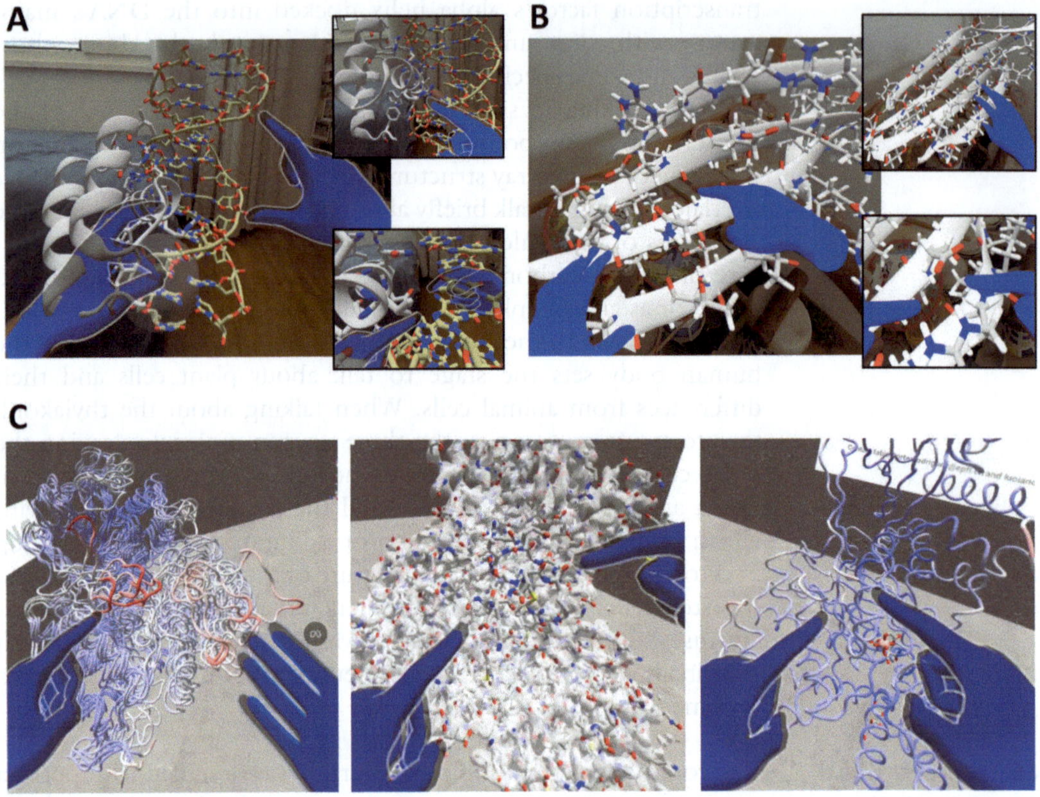

Fig. 5 MolecularWebXR's room presenting a course on structural biology. (**a**) A user showing with his hand the major groove of a dsDNA molecule, where a helix of the transcription factor docks. The minor groove is also easily spotted in this 3D model, as well as the non-specific salt bridges between DNA's backbone phosphate groups and positively charged sidechains of protein residues (top inset), and the hydrogen bonds between protein sidechains and DNA bases (bottom inset). (**b**) A user pointing at one of the hydrogen bonds that stabilizes pairing of beta strands in a beta sheet. In this model, it is also easy to spot charge-charge interactions across strands (bottom inset) and how one side of the sheet facing the interior of the full protein is hydrophobic (top inset) while the solvent-facing side is very polar (main panel). (**c**) Structure of Malate Synthase G as determined by (from left to right) NMR, cryo-EM, and X-ray diffraction; in the three cases, pointing at highly flexible loops that serve to explain to students how each technique works, what they can reliably solve, and more. In particular, the loop on the left lacks density in X-ray diffraction maps; hence, it is not modeled; it lacks restraints in NMR, hence appearing very flexible in the ensemble, and has very low density in the cryo-EM map

spectroscopy and the small-size limit for cryo-EM. Two representations are shown for the NMR structure: one displays in green a few examples of the long-range NOEs used to determine the structure, and the other model shows members of the ensemble in cartoon representations colored by the RMSD to the mean structure. In the ensemble, a highly flexible loop is shown that provides a chance to explain how this might be due to real dynamics and/or to a lack of assignments and restraints to define this region's details upon structure calculation (Fig. 5c). Moving on to the cryo-EM and

X-ray structures, the story explains how that same region is not modelled at all in the X-ray diffraction structure, and it is modeled despite missing density in the cryo-EM structure. The X-ray structure is colored by Cα B-factors, which provides a chance to identify other somewhat flexible regions and compare them with the NMR ensemble.

To conclude, the session also covers the same sulfonamide-bound human carbonic anhydrase covered by the room connecting physics to biology and also the structure of the ABL kinase with imatinib bound to it. At the level of the structural biology course, the story can go deeper to show, for example, how deeply imatinib goes into the kinase, pointing at some key contacts, and detailing the structure around human carbonic anhydrase's zinc ion.

4.6 User Reception, Usage Statistics, and Feedback

While the molecular models within these rooms are static in terms of their internal structural dynamics (unlike the dynamic simulations in the VMK presented in Subheading 2 or HandMol presented in the next section), the platform excels at providing shared, immersive viewing and discussion spaces. We have applied this tool at science communication activities, as shown with use case 1 in Fig. 1b, at bachelor and doctoral level courses, and at scientific conferences, as in Fig. 1c. MolecularWebXR is featured by Meta's web browser, was nominated for a PolyXR award in 2024, and is today widely accessible, as shown by statistics (inset in Fig. 4a).

5 Toward Interactive, Immersive, Multiuser Molecular Graphics and Modeling: HandMol

5.1 Motivation

With large numbers of users around the world, MolecularWeb and MolecularWebXR provide accessible web-based tools for immersive molecular visualization and shared immersive discussions. However, these tools primarily handle molecules as static 3D objects, or in the best case, such as in MolecularWeb's VMKs, with some flexibility but limited capabilities for immersive graphics and fine 3D manipulation. The ambitious next step was to move beyond toward truly intuitive, dynamic, physics-informed, and multiuser manipulation. We wanted to empower users to not just see a molecule in immersive multiuser 3D as possible with MolecularWebXR, but also to grab it, bend it, explore its energy landscape, and even simulate simple reactions with intuitive, bare-handed interaction. This vision required a fundamental architectural shift: instead of loading pre-rendered 3D models (GLB files), our application needed to treat molecules as collections of individual atoms and bonds, each capable of being rendered, processed, and displaced separately, as in MolecularWeb's VMKs but with MolecularWebXR's XR capabilities. This goal led us to web apps for genuine single- or multi-user molecular graphics and modeling with an option for immersive graphics, giving rise to the HandMol series of apps that consist of a client-only prototype [47] and a MolecularWebXR-based HandMol app, the former already available as a stand-alone web page without support and the latter coming up during 2026 in a stable release merged with the whole MolecularWebXR platform.

5.2 The Client-Only HandMol Prototype as a Proof of Concept

Our first implementation of the overarching idea was the HandMol client-only prototype, a system designed to validate several components required to couple WebXR with AI and molecular simulation engines. This prototype, available at https://go.epfl.ch/handmol, employs a two-part architecture where two web applications work in concert, connected wirelessly via the WebRTC protocol without any intermediate backend servers (i.e., being fully client-side except for the initial WebRTC connection).

The two web apps that make up the client-only HandMol open at different links within the main page: HandMol-VR and HandMol-Computer. HandMol-VR is meant to run in the web browser of a headset to provide immersive AR or VR experiences where users can grab and displace atoms at will, as well as get information about atom names, distances, etc. Like in MolecularWebXR, AR and VR can also be experienced, of course, without hand tracking capabilities. In turn, HandMol-Computer is meant to run on the web browser of a computer that will be used to move files easily into and out of the session, but it can also run in a second headset to support two-user immersive sessions.

The client-only HandMol prototype successfully integrated a suite of modern technologies to create a uniquely interactive experience. Three.js and the WebXR API provide immersive graphics and natural interaction capabilities. Meanwhile, the Cannon.js physics engine running inside the web browser provides real-time feedback on molecular flexibility and atomic clashes similar to those in MolecularWeb's VMKs, giving a tangible feel to manipulations. For more accurate simulations, the system can offload a given set of atomic coordinates to external APIs that perform energy minimizations, after which Cannon takes control back. Two such APIs are available, one running the ANI-2x neural network potential for fast, high-quality calculations on organic molecules, [48–50] and another running the AMBER14 force field (via an OpenMM backend) [51] that works on standard biomolecules. Energy minimizations can be triggered at will or automatically upon releasing an atom after moving it, providing immediate, physically realistic feedback.

The prototype also implements a rudimentary way to handle human-computer interaction through an LLM, in this case GPT-3.5-Turbo, for which the user must input an API key. Running in a Chrome browser on a computer, the HandMol-Computer app can use this browser's Speech Recognition API to capture voice commands in natural language (e.g., “hide the hydrogen atoms” or “show carbon atoms in magenta”). The LLM parses this command and translates it into a function call within the application, freeing the user from learning any menus or scripting languages. While the scope of this LLM test is narrow, it served as the basis for more comprehensive natural control of the MolecularWebXR-integrated HandMol program.

This prototype robustly demonstrated that highly interactive, AI-assisted, and computationally-grounded molecular modeling was feasible entirely within a web-based framework, with user tests (detailed in) showing significant gains in intuitiveness and task completion speed for complex 3D manipulations. More details available in a dedicated preprint [47] and in a blog post that discusses the implementation of the key technologies, including pieces of explained source code [52].

5.3 The MolecularWebXR-Integrated HandMol: From Regular but Easier to Use Molecular Graphics to Immersive Interactive Simulations and Multiuser Discussions

Building on the success of the HandMol prototype, the next evolution is the MolecularWebXR-based HandMol. It is intended as a regular molecular graphics program, i.e., where a molecule is seen against a black background and controlled with mouse gestures, but with the capability to spin up all those technologies tested in the HandMol prototype, and more. First of all, this (upcoming) web app can go immersive in AR or VR, again with head-mounted displays or on phones or tablets or as a first-person role game on computers (Fig. 6a). Second, multiple users can connect to see and act on the same molecular system, each from its own viewpoint, regardless of what device they use to access (Fig. 6a). Third, users can interact with atoms or beads of simulation by pinching atoms, while they can see atom names and atom-atom distances by pointing at them with index fingers (Fig. 6c). Fourth, actions and views can be controlled through a menu displayed inside AR/VR from the user’s left wrist (Fig. 6b), with regular menus as in traditional molecular graphics software (available to the Host only, as in Fig. 6d and e), and by typing or speaking up requests in natural language (Fig. 6d; note that the speech recognition feature is only available for the Host and when using Chrome web browsers). Figure 6e covers the controls and options available in the main box menus.

Currently at an advanced stage of development, HandMol is set to be integrated into the main MolecularWebXR platform as a production-ready public service in late 2025/early 2026. This new HandMol is not a two-part application but rather a single-web page application that works like MolecularWebXR, that is, where a Host creates a session—in this case, starting from atomic coordinates in PDB format—and shares the session’s ID for its own or other users’ access through immersive (or any regular) devices. This integration represents the culmination of our efforts, bringing dynamic simulation into the same space as our multi-user

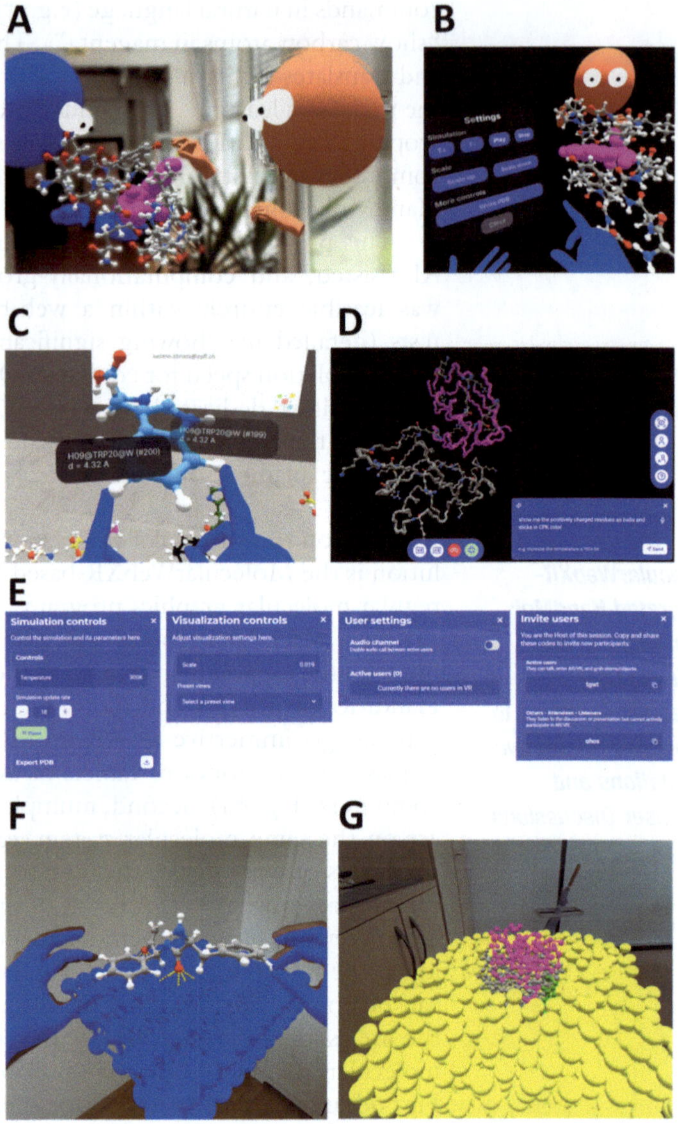

Fig. 6 Some features of the upcoming MolecularWebXR-integrated HandMol web app. (**a**) Two users acting on a molecular system in XR, as seen by a third one who accessed through AR on a smartphone (the background was blurred upon figure preparation, for clarity). (**b**) Same session as in A but seen from the viewpoint of the user in blue avatar. This user has unfolded a menu accessible on the left wrist, with quick commands that can be executed right from within the immersive experience. (**c**) On pointing at an atom, its name is displayed with format Atom@Residue@Chain (atom number). When two atoms are pointed at, the distance between them is shown under their names. (**d**) The web app being used as a regular molecular graphics program by a single user on a laptop, controlling the molecule's orientation with the mouse ("orbit controls" on). This panel also exemplifies spoken commands being recognized (AI assistant on the bottom right). (**e**) Overview of the box menus available to the Host: *Simulation* controls allow to start and stop simulations and control the temperature, as well as downloading in PDB format the coordinates as currently laid out. *Visualization* controls allow quick scaling of the molecular system and provide access to quick

presentation and education rooms. Like in the previous apps, this again supports AR or VR, in headsets or regular devices, and for single or multiple users.

The architecture of the new HandMol is more robust and scalable than that of the HandMol prototype, using web sockets to centralize virtual sessions and synchronize them with molecular simulations running on a server. When a Host user creates a session in the new HandMol, the server spins up a dedicated container for that specific session, which runs the chosen simulation engine to support interactive MD. Available MD engines as of the date of release will be the fast rigid body-based system based on the Cannon.js library (derived from MolecularWeb's VMKs [27, 36] and exemplified in Fig. 6f), the ANI-2x AI system [48] or OpenMM-based AMBER14 (with automatically generated parameters or with user-provided parameters) for atomistic simulations with higher physical accuracy, and Calvados [53] for coarse-grained simulations of proteins (inserted in membranes if chosen, example in Fig. 6g). One special container, called simply HandMol, provides visualization-only capabilities (with all other features enabled).

In the new HandMol, the XR sessions synchronize all users with each other and with simulation engines via a persistent socket connection that requires high bandwidth and computing power. Therefore, the backend manages resource allocation, preventing overload by tracking server load and limiting the number of concurrent computer-intensive simulations, and we have opted to limit each session to 15 min—those with special needs for longer sessions can contact us directly to discuss options.

On the client side, the application retains and refines all the core functionalities of the prototype: intuitive bare-hand manipulation,

Fig. 6 (continued) preset views. Under *User settings,* the Host can see who is connected to the session after they have been invited with the codes displayed in *Invite users*. The AI module works separately and is shown in action in panel **d**. (**f**) The rigid body-based MD engine can handle all elements from the periodic table plus some special-purpose atoms (HX, CX, NX, OX as color alternatives to the regular H, C, N, O atoms) and coarse-grained beads (under development); besides, specific interactions can be hard-coded into the PDB. In this example, a bilayer of Cu atoms is set to interact with O atoms, which helps the user to dock a molecule of biphenylalanine in preparation for subsequent studies via, for example, DFT calculations as in [59]. (**g**) Display of systems simulated with the CALVADOS engine utilizes specialized beads for the 20 amino acids plus two types of beads for the membranes. Shown in this panel is the X-ray structure of *T. thermophilus* cytochrome *c* oxidase (3 polymer chains) inserted in a membrane mimic. All the features shown in this figure are at the moment evolving and might look different when the actual first version of the platform is out. For more examples, including some of HandMol running in visualization-only mode or interactive MD with Amber and ANI engines, *see* Fig. 7

on-the-fly molecular mechanics feedback (where energy minimization was replaced by a continuously running simulation that can work as a continuous minimizer by setting its temperature to zero), and a LLM-powered voice/type assistant much improved from the original prototype, in this case powered by OpenAI's GPT-4o-mini. Furthermore, users can use in-VR menus to dynamically adjust parameters like simulation temperature or zoom level. The system also visualizes forces and interactions through "rubber bands" that appear when a user pulls on an atom. Additionally, pointing at an atom with the index fingertip displays the atom's name, and when two atoms are pointed the distance between them is displayed as well (Fig. 6c).

5.4 Example Use Cases of the HandMol Web Apps

By integrating dynamic, physics-based modeling directly into the multi-user MolecularWebXR platform, HandMol achieves our goal of creating a single, web-based ecosystem that spans the entire spectrum from simple AR visualization to shared, immersive discussions, and finally to collaborative, real-time, AI-assisted interactive molecular simulation—all in an accessible format that is just an URL away.

We have so far used our programs to get a "feel" on the strength of certain interactions, to manually dock small molecules into pockets, to explore possible conformations and effects of chemical modifications, to set up systems for more thorough downstream analyses using regular molecular simulations, to teach about molecular structure, and to discuss molecules. In this subheading, I present specific use cases, some illustrated in Fig. 7, moving beyond applications to education (showing just two examples: Fig. 7a and b) by exploiting actual applications to our research work (Fig. 7c–f plus utilization inside the cited works).

Through mechanics assisted by the rigid-body dynamics engine, we have used the client-side HandMol prototype and the upcoming MolecularWebXR-integrated HandMol to dock small molecules into proteins (example in Fig. 7c). We found this of utmost utility when clear sources of information and physicochemical patterns define pretty obvious poses, but regular docking programs do not produce compatible complexes. For example, we used the rigid-body engine to manually dock substrates into the two active sites of a newly discovered enzyme based on templates of similar small molecules bound to similar domains, to then relax the complexes by atomistic MD and derive residues studied by mutagenesis [54]. Similarly, we used interactive simulations to thread single-strand DNA molecules into a protein nanopore (Fig. 7d) and to place small molecules inside protein nanopores, [55] in both cases for downstream studies through extensive, regular MD simulations. Notably, both tasks are very difficult to achieve with regular 2D software and without the assistance of a forcefield, especially when threading the bulky DNA strand into the nanopore; here is

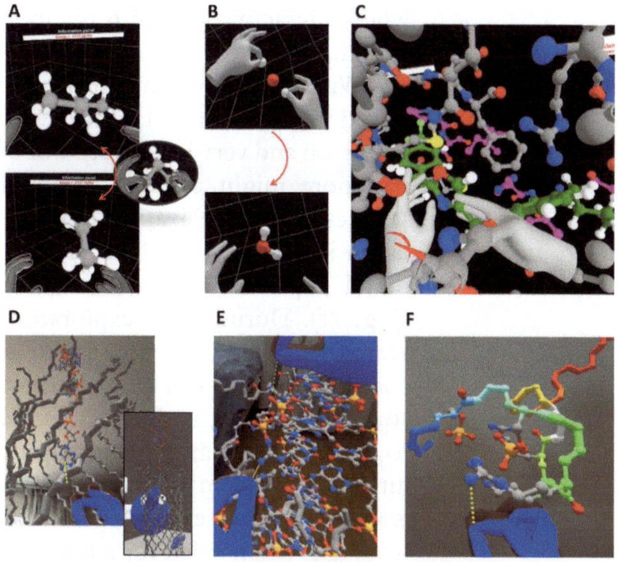

Fig. 7 Use cases of our HandMol programs. (**a**) Driving the dihedral angle around the two central atoms of a butane molecule as ANI displays energies (red in the back). This is taken from a preset example of the HandMol prototype with which teachers can explain and students can explore by themselves various aspects of molecular dynamics and conformations, for example, by comparing butane with 2-butene, isolated saturated rings with fused rings and aromatic rings, etc. (**b**) Building small molecules with ANI-2x and checking their geometries, to teach/learn about bonding, geometries, and more—as well as to build molecules that can then be exported for work outside of the app. In this specific case, the user approached two H atoms to an O atom in a linear arrangement; upon release ANI minimized the system, building a water molecule with the right shape. (**c**) A small molecule (green carbons) right after it was docked into a pocket in a protein using the HandMol prototype. This was part of a quick exploratory work done for a grant proposal submission. (**d**) Threading a strand of DNA into a protein nanopore in an interactively steered MD simulation assisted by the Amber14 forcefield inside the modern HandMol, as seen from the headset (main panel) and from the side by a user at a computer (inset). The result of this operation was a system that was later on used as the starting point for MD simulations testing how much the DNA molecules blocks ion transport across the nanopore. (**e**) Probing docking strength at a protein-DNA complex, here pulling a positively charged arginine to break its interaction with a DNA phosphate group. To achieve this view, the system was asked to show only the DNA and arginines as balls and sticks, leaving the rest as backbone representations only. This was part of a study testing how strong the different kinds of interactions feel. (**f**) Probing the binding of phosphorylated serines to arginine sidechains in an intrinsically disordered peptide that presumably folds upon phosphorylation. This used custom parametrization for the phosphorylated serines and was part of a discussion on how to interpret observations from various experiments

when interactive MD with the modern HandMol displays its main power.

With interactive simulations driven by an Amber force field, we found the upcoming HandMol useful to explore protein mechanics hands-on and very intuitively. For example, we tested how a protein nanopore might collapse under strong electric fields and other sources of destabilization, [56] we probed how a transcription factor interacts with DNA (Fig. 7e), and we modeled how a disordered peptide might close up on itself upon phosphorylation (Fig. 7f). During such explorations, it appears in our hands that seeing the yellow strings stretch and compress like rubber bands conveys a visual feel of the forces applied, consistent with two studies showing that interactive MD in VR allows human subjects to sense molecular properties purely through visual cues as they interact with them in VR [57, 58]—making haptic feedback, which is currently very expensive and hard to integrate into devices of mass use, less needed for a full immersive experience.

6 The MolecularWeb Universe: Current Standing, Impact, and Future Outlook

This book chapter summarizes how by developing the MolecularWeb ecosystem we materialize an effort to make immersive molecular science more accessible, intuitive, interactive, and collaborative. By consistently leveraging the power and ubiquity of standard web technologies (HTML, JavaScript, CSS, WebGL, WebXR, WebRTC) and by progressively integrating cutting-edge elements such as those based on AI, we have journeyed from relatively simple marker-based AR for largely static visualizations to highly dynamic, multiuser immersive environments that incorporate real-time physics and connections to sophisticated computational chemistry backends.

What next, then? While the MolecularWeb ecosystem has made significant strides, several exciting frontiers remain. The "dream tool" for molecular sciences is a seamlessly integrated, fully immersive, intuitively controlled, computationally powerful, and universally accessible collaborative environment, with features inspired in our recent perspective on "molecular modeling in the holodeck" and in our recently developed AI agent part of the PDB manipulation suite.

A key area for advancement is in pushing the boundaries of natural human-computer interaction. This includes extending controller-free hand tracking to commodity devices by leveraging web-based machine learning libraries like MediaPipe, a concept already prototyped in MolecularWeb's "Markerless AR" playground. Further enriching this interaction, the role of AI assistants can be expanded beyond simple command interpretation toward a more proactive partner, potentially even capable of generating and executing code within its code—a powerful but challenging goal that comes with significant security considerations—or perhaps "materializing itself" inside the XR sessions as one more user who then interacts with one naturally, as a virtual character in the Holodeck. The ultimate sensory goal, the integration of haptic feedback to let users "feel" molecular forces, remains on the horizon, awaiting the maturation of consumer-level hardware compatible with web standards, although, as discussed earlier, humans seem to be good at "extrapolating" haptic feedback from visual cues [57, 58].

Beyond the user interface, the scientific heart of the MolecularWeb ecosystem might also need to evolve from a tool primarily for exploring known structures and modifying some of its elements to one for genuine discovery. This certainly requires adding more features, as already possible in other immersive environments most notably UnityMol, [17] displaying more information about the molecular system like regular molecular graphics programs do, adding tools to mutate residues, change rotamers, add or remove atoms, etc., and possibly broadening the scope of integrated physics and chemistry by adding a wider array of simulation engines such as the MARTINI force field for coarse-grained simulations of biological systems, reactive potentials for simulating bond formation and breaking, specialized engines for materials science, and even quick quantum mechanical methods for high-accuracy studies of small molecules. Such advanced simulations also demand advances in the visualization methods, moving beyond simple ball-and-stick representations to include cartoon representations of proteins, dynamic molecular surfaces, and crucially, the ability to render and interact with volumetric data like cryo-electron microscopy maps, which would in turn open a window into extending our tools directly into modern structural determination workflows.

Perhaps the most critical future directions involve closing the loop between our virtual environment, experimental data, and proven efficacy. A powerful evolution is to enable the real-time calculation of experimental observables directly from the manipulated models. Imagine interactively fitting a molecule into a density map and seeing a cross-correlation score update in real-time, or adjusting a protein's conformation and instantly seeing how its simulated SAXS profile compares to experimental data. This would transform HandMol into an invaluable tool for hypothesis testing and model refinement.

Finally, as we build these increasingly powerful tools, we have a responsibility to rigorously evaluate their impact. While anecdotal evidence and user feedback strongly indicate enhanced engagement, formal, large-scale pedagogical studies are essential to quantify improvements in learning outcomes and retention. Likewise, systematic evaluation is needed to determine if these immersive and intuitive interfaces truly lead to faster, more effective, and more creative research.

The "MolecularWeb" universe, with HandMol at its current frontier, demonstrates that many key components of a next-generation platform for molecular science are now feasible. By continuing to innovate at the intersection of web technologies, immersive interfaces, AI, and human-computer interaction, and by remaining keenly responsive to the needs of the scientific and educational communities, the web can indeed serve as the foundation for a future where exploring, understanding, and teaching the complexities of the molecular world is more intuitive, engaging, and collaborative than ever before.

Acknowledgments

Most of the MolecularWeb universe was developed by Fabio José Cortés Rodríguez under supervision of Luciano Andrés Abriata, with substantial contributions from Lucien Fabrice Krapp, Fernando Augusto Teixeira Pinto Meireles and other members of EPFL's Laboratory for Biomolecular Modeling led by Prof. Matteo Dal Peraro, in Lausanne, Switzerland. Several other collaborators at EPFL and other institutions are acknowledged for their help in providing content and ideas for our platforms, as well as for testing and applying our tools to various problems and in various situations.

Several individuals who helped with initial prototypes, server setup, and the contribution of open libraries on top of which we built these systems, are also heartily acknowledged. Likewise, we thank various teachers, students, and researchers who provided valuable feedback after testing the tools.

We acknowledge the Swiss National Science Foundation, Hasler Stiftung (Bern), and Werner Siemens Stiftung for financial support through multiple grants specific for different projects within the MolecularWeb universe.

Competing Interest Statement The authors have no conflicts of interest to declare that are relevant to the content of this chapter.

References

1. Goodsell DS (2003) Looking at molecules—an essay on art and science. Chembiochem 4: 1293–1297
2. Verma VV, Vimal S, Mishra MK et al (2025) A comprehensive review on structural insights through molecular visualization: tools, applications, and limitations. J Mol Model 31:1–15
3. Wu H-K, Shah P (2004) Exploring visuospatial thinking in chemistry learning. Sci Educ 88: 465–492
4. Kozlíková B, Krone M, Falk M et al (2017) Visualization of biomolecular structures: state of the art revisited. Comput Graph Forum 36: 178–204

5. Gillet A, Sanner M, Stoffler D et al (2005) Tangible interfaces for structural molecular biology. Structure 13:483–491
6. O'Connor M, Deeks HM, Dawn E et al (2018) Sampling molecular conformations and dynamics in a multiuser virtual reality framework. Sci Adv 4:eaat2731
7. El Beheiry M, Doutreligne S, Caporal C et al (2019) Virtual reality: beyond visualization. J Mol Biol 431:1315–1321
8. Abriata LA (2026) Elements for molecular graphics and modeling "in the Holodeck". Methods Mol Biol. https://pubmed.ncbi.nlm.nih.gov/41556555/
9. O'Donoghue SI, Goodsell DS, Frangakis AS et al (2010) Visualization of macromolecular structures. Nat Methods 7:S42–S55
10. O'Donoghue SI, Gavin A-C, Gehlenborg N et al (2010) Visualizing biological data—now and in the future. Nat Methods 7:S2–S4
11. Bennie SJ, Ranaghan KE, Deeks H et al (2019) Teaching enzyme catalysis using interactive molecular dynamics in virtual reality. J Chem Educ 96:2488–2496
12. Jamieson-Binnie AD, O'Connor MB, Barnoud J et al (2020) Narupa iMD: a VR-enabled multiplayer framework for streaming interactive molecular simulations. In: ACM SIGGRAPH 2020 Immersive Pavilion, pp 1–2
13. Kuťák D, Vázquez P, Isenberg T et al (2023) State of the art of molecular visualization in immersive virtual environments. Comput Graph Forum 42
14. Fombona-Pascual A, Fombona J, Vicente R (2022) Augmented reality, a review of a way to represent and manipulate 3D chemical structures. J Chem Inf Model 62:1863–1872
15. Pietikäinen O, Hämäläinen P, Lehtinen J et al (2021) VRChem: a virtual reality molecular builder. Appl Sci 11:10767
16. Rau T, Sedlmair M, Köhn A (2024) chARpack: the chemistry augmented reality package. J Chem Inf Model 64:4700–4708
17. Doutreligne S, Cragnolini T, Pasquali S et al (2014) UnityMol: Interactive scientific visualization for integrative biology. In: 2014 IEEE 4th Symp. Large Data Anal. Vis. LDAV, pp 109–110
18. Stroud HJ, Wonnacott MD, Barnoud J et al (2025) NanoVer server: a python package for serving real-time multi-user interactive molecular dynamics in virtual reality. J Open Source Softw 10:8118
19. Cassidy KC, Šefčík J, Raghav Y et al (2020) ProteinVR: web-based molecular visualization in virtual reality. PLoS Comput Biol 16: e1007747
20. Xu K, Liu N, Xu J et al (2021) VRmol: an integrative web-based virtual reality system to explore macromolecular structure. Bioinformatics 37:1029–1031
21. Argüello JM, Dempski RE (2020) Fast, simple, student generated augmented reality approach for protein visualization in the classroom and home study. J Chemi Educ
22. O'Connor MB, Bennie SJ, Deeks HM et al (2019) Interactive molecular dynamics in virtual reality from quantum chemistry to drug binding: an open-source multi-person framework. J Chem Phys 150:220901
23. Baaden M, Glowacki DR (2025) Virtual reality in drug design: benefits, applications and industrial perspectives. Curr Opin Struct Biol 92:103044
24. Deeks HM, Zinovjev K, Barnoud J et al (2023) Free energy along drug-protein binding pathways interactively sampled in virtual reality. Sci Rep 13:16665
25. QuimicaViva vol20 num3
26. Ateş H (2025) Integrating augmented reality into intelligent tutoring systems to enhance science education outcomes. Educ Inf Technol 30:4435–4470
27. Rodríguez FC, Frattini G, Krapp LF et al (2021) MoleculARweb: a web site for chemistry and structural biology education through interactive augmented reality out of the box in commodity devices. J Chem Educ 98:2243–2255
28. Levy J, Chagunda IC, Iosub V et al (2024) MoleculAR: an augmented reality application for understanding 3D geometry. J Chem Educ 101:2533–2539
29. Chiu W-K (2021) Pedagogy of emerging technologies in chemical education during the era of digitalization and artificial intelligence: a systematic review. Educ Sci 11:709
30. Bullock M, Huwer J (2024) Using AR to enhance the learning of chirality. Educ Sci 14: 1214
31. Rodríguez FC, Dal Peraro M, Abriata LA (2021) Democratizing interactive, immersive experiences for science education with WebXR. Nat Comput Sci 1:631–632
32. Abriata LA (2020) Building blocks for commodity augmented reality-based molecular visualization and modeling in web browsers. PeerJ Comput Sci 6:e260
33. Abriata LA (2017) Web apps come of age for molecular sciences. Informatics 4:28
34. Abriata LA, Rodrigues JPGLM, Salathé M et al (2018) Augmenting research, education, and outreach with client-side web programming. Trends Biotechnol 36:473–476

35. MacIntyre B, Smith TF (2018) Thoughts on the Future of WebXR and the Immersive Web. In: 2018 IEEE Int. Symp. Mix. Augment. Real. Adjun. ISMAR-Adjun, pp 338–342
36. Rodríguez FC, Krapp LF, Dal Peraro M et al (2022) Visualization, Interactive handling and simulation of molecules in commodity augmented reality in web browsers using moleculARweb's virtual modeling kits. Chimia 76: 145–145
37. Rodríguez FC, Dualde F, Massiot P et al (2023) Updates on moleculARweb, the Swiss portal for chemistry and structural biology education using augmented and now also virtual reality: chemical education. Chimia 77: 264–265
38. Bienfait B, Ertl P (2013) JSME: a free molecule editor in JavaScript. J Cheminformatics 5:24
39. Hack-a-Mol
40. Cortez Rodriguez F, Aguirre Pranzoni CB, Principe A et al (2021) De átomos y sus orbitales a moléculas y estructuras biológicas, en realidad aumentada adaptada a teléfonos, tablets y computadoras sin instalar programas ni hardware especial
41. Abriata LA (2022) How technologies assisted science learning at home during the COVID-19 pandemic. DNA Cell Biol 41:19–24
42. Abriata LA (2023) Three years of positive teacher feedback on MoleculARweb, our website for chemistry education using.... Age Aware
43. Abriata LA (2023) Fostering discussion on the impact of virtual and augmented reality on STEM education. Age Aware
44. Rodríguez FC, Dal Peraro M, Abriata LA (2022) Online tools to easily build virtual molecular models for display in augmented and virtual reality on the web. J Mol Graph Model 114:108164
45. Humphrey W, Dalke A, Schulten K (1996) VMD: visual molecular dynamics. J Mol Graph 14:33–38, 27–28
46. Limpanuparb T, Chiranon W, Intaraprasit M (2024) Building words from chemical elements: a fun and inclusive approach to introduce the periodic table. Chem Teach Int 6: 311–321
47. Rodriguez FC, Krapp L, Dal Peraro M et al (2023) HandMol: coupling WebXR, AI and HCI technologies for immersive, natural, collaborative and inclusive molecular modeling
48. Devereux C, Smith JS, Huddleston KK et al (2020) Extending the applicability of the ANI deep learning molecular potential to sulfur and halogens. J Chem Theory Comput 16:4192–4202
49. Gao X, Ramezanghorbani F, Isayev O et al (2020) TorchANI: a free and open source PyTorch-based deep learning implementation of the ANI neural network potentials. J Chem Inf Model 60:3408–3415
50. Smith JS, Isayev O, Roitberg AE (2017) ANI-1: an extensible neural network potential with DFT accuracy at force field computational cost. Chem Sci 8:3192–3203
51. Eastman P, Swails J, Chodera JD et al (2017) OpenMM 7: rapid development of high performance algorithms for molecular dynamics. PLoS Comput Biol 13:e1005659
52. Abriata LA (2023) Coupling four AI models to deliver the ultimate experience in immersive visualization and modeling. Medium
53. Cao F, von Bülow S, Tesei G et al (2024) A coarse-grained model for disordered and multi-domain proteins. Protein Sci Publ Protein Soc 33:e5172
54. Rashan EH, Bartlett AK, Khana DB et al (2024) ACAD10 and ACAD11 enable mammalian 4-hydroxy acid lipid catabolism. 2024.01.09.574893
55. Agerova A, Juarez JFB, Abriata LA et al (2025) Sub-nanomolar detection and discrimination of microcystin congeners using aerolysin nanopores. 2025.07.15.664886
56. Mayer SF, Mitsioni MF, Robin P et al (2025) Lumen charge governs gated ion transport in β-barrel nanopores. 2024.09.26.615172
57. Roebuck Williams R, Barnoud J, Toledo L et al (2024) Measuring the limit of perception of bond stiffness of Interactive molecules in VR via a gamified psychophysics experiment. Ext Real 15027:190–198
58. Williams RR, Varcoe X, Glowacki BR et al (2020) Subtle sensing: detecting differences in the flexibility of virtually simulated molecular objects. In: Ext. Abstr. 2020 CHI Conf. Hum. Factors Comput. Syst., pp 1–8
59. Lingenfelder M, Tomba G, Costantini G et al (2007) Tracking the chiral recognition of adsorbed dipeptides at the single-molecule level. Angew Chem Int Ed 46:4492–4495
60. Rodríguez FJC, Frattini G, Meireles FTP et al (2023) MolecularWebXR: multiuser discussions about chemistry and biology in immersive and inclusive VR. ArXiv Prepr ArXiv231100385

Chapter 2

Python-Based Interactive Notebooks for Teaching AI and Data Science to Undergraduate Students in the Life Sciences

Kriti Shukla and Elizabeth Brunk

Abstract

This chapter presents a reproducible, cloud-based framework for integrating Python-driven data analysis into a large-enrollment undergraduate biochemistry course. Rapid growth in technologies has transformed biochemistry into a data-intensive discipline, yet most life-science students receive little formal training in computational skills. To address this gap, we developed a set of interactive Jupyter notebooks that embed introductory coding, protein structure visualization, metabolic modeling, and simple multi-omics analysis directly within the core biochemistry curriculum. The workflow is delivered entirely through the CyVerse Discovery Environment, which provides browser-based JupyterLab sessions with preconfigured Python environments, version-controlled notebooks, and standardized datasets. This infrastructure eliminates the need for local installation, supports hundreds of simultaneous users, and ensures consistent execution across diverse student hardware. The notebooks follow a scaffolded learn-apply-reflect design that links biochemical concepts with incremental coding tasks, enabling students with no prior experience to manipulate real datasets, generate visualizations, and interpret quantitative patterns. Assessment using validated pre/post surveys across more than 1000 matched responses demonstrated substantial gains in students' confidence with coding and their ability to interpret omics data. The protocol outlined here offers a scalable, equitable model for incorporating computational literacy into biochemistry education and can be readily adapted for related life-science courses.

Key words Data science, Artificial intelligence, Machine learning, Omics, Biochemistry, Informatics, Jupyter, Python, Education

1 Introduction

Modern biochemistry is being reshaped by the exponential growth of big biological data [1–4]. Advances in sequencing, genomics, proteomics, and metabolomics technologies have transformed the discipline into one that is driven as much by computation as by experiment. The ability to analyze, visualize, and interpret "omics" data now underlies progress across basic science and biomedical research [5–7]. Yet, opportunities for students and early career

Robert E. Dempski (ed.), *Transforming Molecular Biology with Emerging Technologies*, Methods in Molecular Biology, vol. 3043, https://doi.org/10.1007/978-1-0716-5320-3_2,

researchers to gain the necessary computational skills remain limited [8–11]. Despite increasing demand for data literacy, most undergraduate programs continue to treat coding and quantitative analysis as separate from the life sciences. This perspective delays exposure of these skills until graduate or professional training. The disconnect between the need for these skills and the opportunities for training limits not only the size of the future workforce but also the types of backgrounds, perspectives, and career trajectories.

The National Institutes of Health and other agencies have emphasized that sustaining innovation in the biological and biomedical data ecosystem will require early training opportunities in data science [12]. However, universities have struggled to translate this vision into scalable curricula. Bioinformatics or computational biology majors exist at only a small fraction of research-intensive institutions (currently ~18 of 146 R1 universities) [13], leaving the majority of undergraduates without a formal route into biological data analysis. Undergraduate research experiences can partially fill this gap, but capacity limits and the perception that coding is "too advanced" deter many students from participating. Expanding access, therefore, requires accessible points that lower the barrier to learning computational skills within existing courses.

This chapter presents a reproducible, cloud-hosted protocol for integrating Python-based interactive Jupyter notebooks into a large enrollment undergraduate biochemistry course [14], without requiring prior coding experience or local software installation. The protocol enables instructors to deliver four scaffolded *Data Exploration* (DE) notebooks aligned to standard biochemistry topics: introductory data handling, protein structure visualization, metabolism and flux mapping, and simple multi-omics comparisons. The approach prioritizes accessibility, reproducibility, and scalability by leveraging the CyVerse Discovery Environment [15] to provide browser-based JupyterLab sessions with preconfigured environments and version-controlled notebooks distributed through GitHub. The workflow is designed for courses enrolling several hundred students, including those with limited instructional computing support.

1.1 Conceptual Overview

1.1.1 Integrating Data Science Within Core Biochemistry Education

Biochemistry occupies a unique position at the intersection of chemistry, biology, and quantitative reasoning; this intersection makes it a natural gateway for introducing applied computational skills. However, programming has historically been absent from standard biochemistry textbooks and curricula. Traditional instruction emphasizes molecular mechanisms (e.g., enzyme catalysis, protein structure, energy metabolism, DNA replication, and RNA processing) but rarely provides opportunities to apply these concepts to real biological data. As a result, many graduates leave with strong conceptual understanding but little confidence in translating

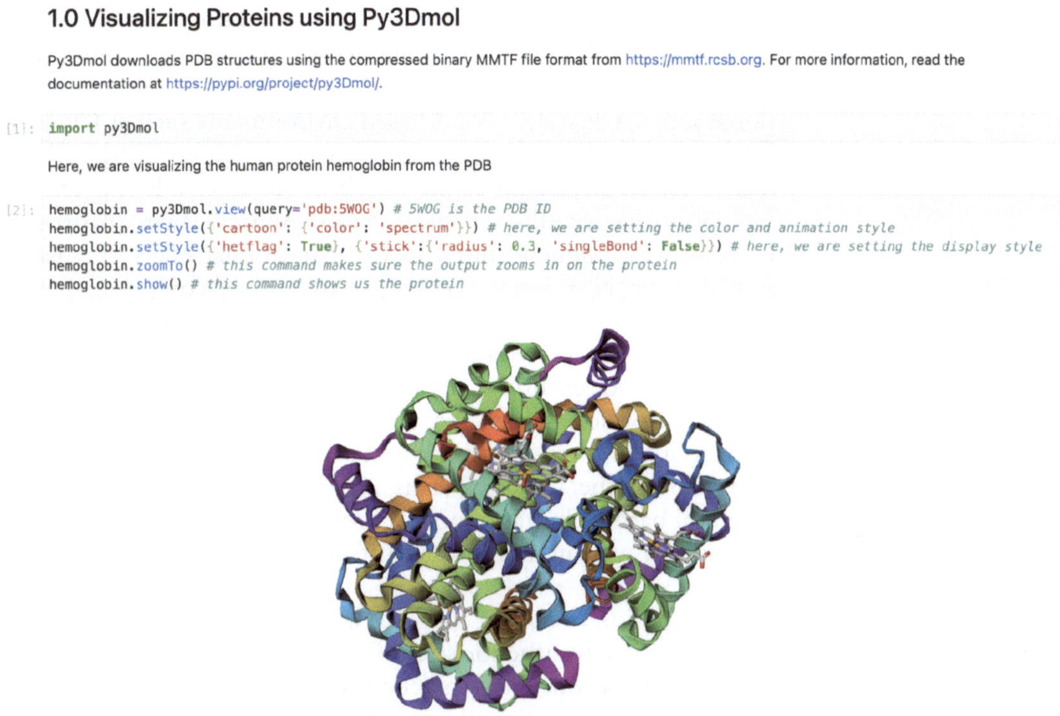

Fig. 1 An example overview of a Data Exploration, showing markdown sections with instructions for students, example code sections, and a protein structure output

theory into practice. The rapid expansion of omics datasets in research and medicine makes this gap increasingly untenable.

In recent years, chemical-education research has begun exploring ways to embed coding into undergraduate courses [16]. Faculty have implemented Jupyter notebooks to teach quantitative concepts in general and physical chemistry [17], reporting improved engagement and conceptual comprehension. However, similar efforts in biochemistry remain rare, particularly at the large-enrollment scale typical of gateway courses serving 400–600 students per semester. Existing computational modules [18] are often restricted to small upper-level classes, assume prior coding experience, and require local software installations; factors that limit scalability and inclusivity.

Our approach [14] addresses these challenges by embedding Python-based DE notebooks directly into a required large-enrollment biochemistry course (Fig. 1). Four DEs were designed to align with major biochemical topics (protein structure, metabolism, and gene expression) so that coding exercises reinforce material already being taught. This integration allows students from diverse STEM majors, including premedical, biological, and chemical engineering tracks, to gain meaningful exposure to computational tools and data analysis before graduation.

1.1.2 Leveraging Open and Cloud-Based Infrastructure

Introducing coding to hundreds of non-computer science majors poses practical hurdles, including access to adequate computing resources, managing software installations, and providing instructional support at scale. Cloud-based environments such as CyVerse [15], Google Colab [19], and Amazon SageMaker [20] address many of these challenges by offering browser-accessible JupyterLab interfaces that require no local setup. Among these platforms, CyVerse was uniquely suited to our educational goals because it provides free, NSF-funded access to standardized computing environments. This allows every student to run identical notebooks with preloaded data and verified dependencies, eliminating technical disparities and ensuring reproducibility across a large cohort. CyVerse also allows instructors to maintain version control through GitHub integration, enabling easy updates to datasets and instructions during the semester. This infrastructure not only reduces technical barriers but also models the same cloud-based workflows increasingly used in modern scientific research.

1.1.3 Pedagogical Framework: From Observation to Inquiry

The integration of coding followed a scaffolded learn-apply-reflect cycle that mirrors authentic scientific practice (Fig. 2). Each Jupyter notebook pairs conceptual biochemistry content with incremental coding challenges designed to reinforce both theoretical and practical understanding. Students first observe code execution during in-class demonstrations, then apply similar logic independently

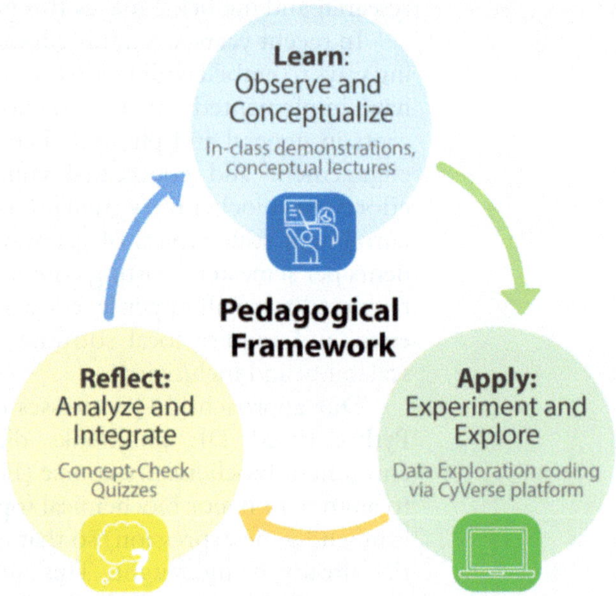

Fig. 2 Scaffolded learn-apply-reflect cycle in the context of the digital biochemistry course

within their own notebooks, and finally reflect on observed patterns through short concept-check quizzes.

This progression encourages the transfer of theoretical knowledge into applied reasoning. For example, when learning about protein structure and function, students use the Python package py3Dmol [21] and the Protein Data Bank [22] to visualize macromolecular structures in three dimensions. By manipulating residues and domains and linking these visualizations to amino acid sequence data parsed in pandas [23] DataFrames, students move from memorizing structural concepts to exploring them dynamically. This combination of conceptual framing, guided coding, and open-ended reflection transforms coding from an abstract task into a tool for biochemical inquiry.

1.1.4 Broader Educational Significance

Beyond this single course, the framework presented here provides a scalable model for integrating data science into life-science education without requiring curriculum overhauls or specialized programming expertise. The Jupyter-based modules can be easily adapted to other disciplines (such as introductory biology, genetics, or physics) and shared openly through GitHub for collaborative refinement. This openness aligns with the FAIR principles [24] of data and educational resource sharing, promoting reuse and interoperability across institutions.

The intention of this approach is not to train professional programmers but to cultivate computationally fluent biochemists (i.e., students who can read, execute, and interpret code sufficiently to engage in data-driven science). Early exposure to these skills demystifies modern research tools, helping students recognize that coding is not a specialized pursuit but a foundational scientific competency, much like graphing or pipetting.

1.2 Objectives of This Chapter

This chapter documents the development, implementation, and evaluation of Python-based interactive notebooks within a large-enrollment undergraduate biochemistry course. Our objectives are threefold:

To integrate coding and omics data analysis into the required biochemistry curriculum.

To demonstrate the feasibility of using cloud infrastructure to support scalable, equitable instruction.

To measure changes in students' attitudes toward computing, omics data, and research engagement using a validated survey instrument.

By sharing both the instructional materials and the assessment outcomes, we aim to provide a reproducible model for instructors seeking to bridge biochemistry and data science on scale. Empowering undergraduates to navigate and analyze large biological datasets fosters the analytical literacy essential for the next generation of scientists, clinicians, and educators operating in a data-saturated world.

2 Materials

All notebooks and data sheets used in this course are available at: https://github.com/Brunk-Lab/Digital-Upskilling-BioScience

Implementing interactive coding exercises in a large-enrollment biochemistry course requires careful coordination of computational infrastructure and classroom design. The essential components include the cloud accounts used to host the exercises, the hardware through which students access the platform, the software environment supporting the notebooks, and the instructional logistics that integrate these resources into the curriculum.

2.1 Cloud-Based Computing Accounts and Resources

All computational activities were conducted through the CyVerse Discovery Environment [15], a cloud platform that provides browser-based access to preconfigured JupyterLab workspaces. At the start of each semester, instructors created a teaching "workshop" on the CyVerse website. Once approved, students received an enrollment link and registered using their university email addresses. Instructors then verified enrollment and granted access through an institutional teaching partnership with CyVerse.

Participation in the workshop automatically provisioned each student with professional-tier CyVerse resources, including dedicated compute nodes and temporary cloud storage. This configuration provided consistent, reliable performance for several hundred simultaneous users, which is an essential feature for large-enrollment courses. The centralized setup also allowed instructors to monitor class participation, manage resource allocations, and maintain uniform computing environments across all users (*see* **Notes 1** and **4**).

2.2 Hardware

Students accessed CyVerse from their own laptops or tablets using standard web browsers such as Chrome, Firefox, or Safari. When users logged in, CyVerse automatically launched a virtual machine (VM) within the Discovery Environment, typically allocating two CPU cores and eight gigabytes of RAM per student. This configuration was sufficient for all course activities, including three-dimensional protein visualization and small-scale omics analyses.

Because the course relied entirely on browser-based computing, no specialized graphics processors or local installations were required. This hardware-independent design ensured equitable participation for students regardless of their personal computing resources, removing the financial and technical barriers that often limit engagement in computational biochemistry.

2.3 Software

All instructional modules were delivered as Jupyter notebooks hosted in the CyVerse Discovery Environment. CyVerse was chosen because it provides a stable, preconfigured Python environment that eliminates the need for local software installation. Each student, therefore, worked within an identical computing setup, ensuring reproducibility and minimizing troubleshooting across a cohort of several hundred learners.

Instructors could preload datasets, fix package versions, and test workflows before release, creating a controlled "sandbox" environment in which students could freely experiment without risk of damaging files or dependencies. The environment included the following Python 3 libraries, which together supported both introductory coding practice and biochemical data analysis (Table 1).

Students interacted with these libraries directly through guided code cells embedded within each notebook. Markdown explanations alternated with executable sections to connect biochemical reasoning with computational logic. To maintain consistent performance across semesters, instructors generated and froze a requirements.txt file using the same CyVerse base image each term, preventing software drift caused by version changes.

All course materials (including notebooks, data files, and updates) were maintained in a private GitHub repository managed by the instructional team (https://github.com/Brunk-Lab/Digital-Upskilling-BioScience). At the start of the semester, students cloned the repository into their CyVerse workspace (Fig. 3), allowing immediate access to all materials. Synchronization between GitHub and CyVerse ensured that any mid-semester patches or dataset revisions automatically propagated to student accounts, keeping every participant on the same version of the materials (*see* **Notes 1** and **4**).

Students completed four themed Data Exploration (DE) modules over the semester. Each DE was a self-contained Jupyter notebook designed to align with a major unit in the course: (1) introductory Python and data handling, (2) protein structure analysis, (3) metabolomics and flux mapping, and (4) transcriptomics and proteomics integration. Each module included a

Table 1
Python 3 libraries included in the environment

Utility	Packages
Core Scientific Stack	numpy, pandas, matplotlib, seaborn, scipy
Structural Biology Tools	py3Dmol, nglview, biopython
System Utilities	os, pathlib, requests, json, io

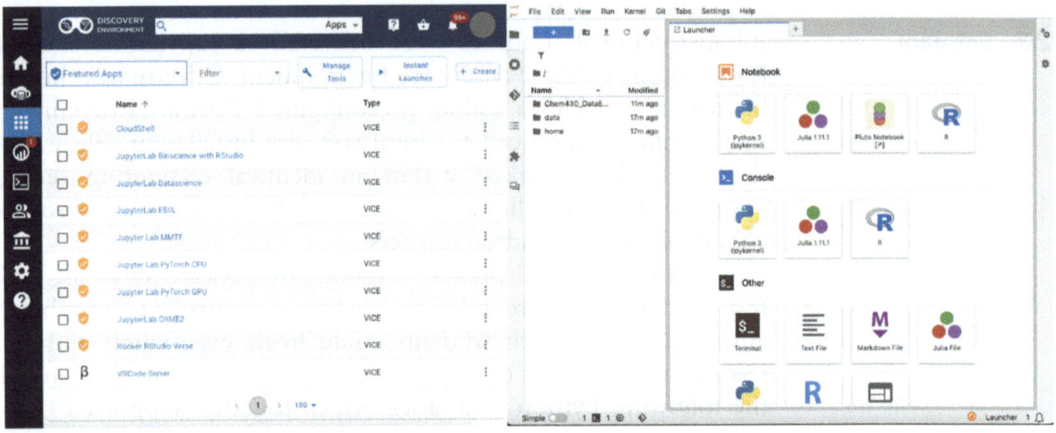

Fig. 3 *Left:* An example of the CyVerse Discovery Environment. *Right*: The CyVerse JupyterLab DataScience module

workflow notebook guiding students through the analysis, a short setup notebook that verified dependencies, and all input data required for the activity. This modular organization allowed every DE to run independently, to be updated without overwriting prior work, and to be version-tracked through GitHub for transparency and long-term reuse. Examples of these DEs are described in *see* **Note 2**.

2.4 Classroom Logistics

The protocol was implemented in CHEM 430: Introduction to Biological Chemistry, a required upper-level course enrolling 400–600 students per semester across two sections. The course combines lecture-based instruction with active learning, making it well-suited for the integration of short, interactive coding exercises within traditional biochemical instruction.

To link the DEs with lecture content, the instructional team adopted a hybrid active-learning model that alternated between conceptual lectures, live coding demonstrations, and independent data analysis. Each DE was introduced in class through a brief presentation that explained its scientific context and demonstrated key commands in real time. Students then completed the notebook independently in CyVerse, applying coding principles to biochemical datasets that paralleled lecture topics.

A graduate teaching assistant (TA) provided structured support throughout the semester. The TA led several in-class demonstrations (e.g., releasing each new Data Exploration module, clarifying workflow steps, and addressing common troubleshooting issues) and also held twice-weekly virtual office hours. Each in-class session followed three integrated phases: (1) background discussion, introducing the biological and computational objectives of the upcoming exercise; (2) live demonstration, during which the instructor or TA executed and explained code in real time; (3) reflection and questions, allowing students to discuss observations and clarify misunderstandings before working independently.

This structured approach helped students build confidence in coding by first observing correct workflows, then practicing with immediate feedback, and finally applying the same logic in self-guided sessions. Across both sections, this consistent framework fostered engagement, minimized frustration, and maintained a clear connection between computational analysis and biochemical reasoning.

3 Methods

3.1 Course Context and Setup

The intervention was implemented in CHEM 430: Introduction to Biological Chemistry, a required upper-division course that introduces the molecular basis of biological processes to students across the life sciences. The course is offered every semester and typically enrolls between 400 and 600 students across two large lecture sections.

3.1.1 Overview of Course Implementation

CHEM 430 uses a flipped-classroom model, in which students engage with core content asynchronously before attending in-person lectures, which are devoted to problem-solving, discussion, and application. Before each class, students watch short pre-recorded lecture videos and complete an online concept-check quiz on Canvas. These quizzes assess understanding of key ideas (such as enzyme mechanisms, bioenergetics, or macromolecular structure) so that lecture time can focus on higher-order reasoning rather than rote delivery (*see* **Notes 1** and **2**).

This structure creates the flexibility needed to integrate computational exercises at scale. Because students are familiar with asynchronous learning, class meetings can shift fluidly between conceptual discussion and hands-on demonstrations. The DE notebooks are introduced during in-person lectures and are completed by students independently. This pedagogical structure encourages active participation, peer interaction, and iterative practice, which are key elements of building confidence in both biochemical reasoning and computational skills.

The course is organized into four major modules, each representing a distinct area of biochemistry and building cumulatively from molecular fundamentals to systems-level integration (Table 2).

The four DEs align with these modules and are introduced at points where students can connect new biochemical knowledge to data-driven exploration. For example, DE1 appears during the introductory unit to teach basic data manipulation and visualization; DE2 follows lectures on protein structure and function; DE3 complements the metabolism block through pathway modeling;

Table 2
A description of the four Data Explorations (DEs)

DE	Subject	Description
1	Foundations of biochemistry and data handling	Chemical properties of water, amino acids, peptides, and macromolecular structure; introductory data literacy and Python fundamentals are introduced here
2	Structure and function of proteins	Protein folding, catalysis, enzyme kinetics, and allosteric regulation, paired with visualization and analysis of macromolecular structures
3	Metabolism and bioenergetics	Glycolysis, the citric acid cycle, oxidative phosphorylation, and flux analysis; this module introduces quantitative reasoning about pathway regulation and energetic coupling
4	Genetic information and omics integration	Nucleic acids, DNA replication, transcription, translation, and systems-level processes that link molecular mechanisms to cellular outcome

and DE4 coincides with the nucleic acid and gene expression unit, where students examine transcriptomic and proteomic data. The sequencing ensures that each DE extends rather than interrupts course content, reinforcing existing concepts through applied computation (*see* **Note 2** for more details).

3.1.2 Grading

Assessment in CHEM 430 is designed to promote consistent engagement and formative learning rather than high-stakes testing alone. Students complete concept-check quizzes before every lecture (32 assessments across the semester), which collectively account for 15% of the final grade. These low-stakes assessments ensure that students stay current with pre-class materials and are prepared to participate in in-class discussions and demonstrations.

The four DEs account for 10% of the overall grade. Each DE culminates in a short, graded quiz that evaluates both conceptual understanding and basic coding competency. These quizzes function analogously to laboratory reports in a traditional lab course but require no physical experimentation, focusing instead on interpretation of data and clarity of reasoning.

Three midterm exams (each closed-note, in-person) collectively contribute 35% of the final grade, and the final exam contributes 32%. The lowest of the three midterm scores dropped. The final exam includes a capstone project, in which small teams apply their understanding of biochemical principles and data analysis to an open-ended problem. The project requires submission of a Jupyter notebook with a short, recorded presentation, emphasizing collaborative reasoning, data visualization, and communication.

Grading policies, office-hour structures, and accommodations follow standard university policies. Peer mentors and a graduate teaching assistant provide regular support through structured office hours and review sessions, ensuring accessibility and continuity throughout the term.

Example Concept-Check Quizzes

Every DE concluded with a 10-question, open-book, asynchronous concept-check quiz designed to assess applied understanding rather than rote memorization. These short quizzes prompted students to modify a small segment of code or interpret an output figure in the context of the biological question explored in the DE notebook. Because they were completed independently and with full access to course materials, the quizzes emphasized reasoning and problem-solving over recall (Table 3). Responses provided immediate feedback to students and instructors alike, functioning as formative checkpoints within the broader learning cycle (*see* **Note 3**).

Table 3
Example concept-check quiz questions

DE	Example question	Answer options (correct answer bolded)
1	In the first DE, you learned about series and DataFrames and how they store data. What is the most representative statement that describes these entities?	(a) They both store data in 1D formats and can store thousands of data points (b) Dataframes are generally smaller in size compared to series (c) A series is a one-dimensional array; a dataframe can contain multiple dimensions of data (d) None of these are correct
2	In Section 2.0: how would you change the following command line code to visualize Heme groups instead of water molecules using larger spheres? hemoglobin.setStyle({'resn': 'HOH'}, {'sphere': {'radius':0.5}})	(a) **Change {'resn': 'HOH'}, to {'resn': 'HEM'}, and change {'radius':0.5} to {'radius':1.5}** (b) Change {'resn': 'HOH'}, to {'resn': 'HEME'}, and change {'radius':0.5} to {'radius':1.5} (c) Change {'resn': 'HOH'}, to {'resn': 'HEME'}, and change {'radius':0.5} to {'radius':1.5} (d) None of these is correct
3	How many metabolites have been measured in the metabolomics dataset from Section 4.0? How many total measurements have been made (all metabolites * time points)?	(a) 78,1204 (b) **86, 1204** (c) 86, 1104 (d) 78, 1104
4	Based on the heatmaps you generated in Section 4.0, select the gene whose RNA is most highly expressed in I3 at 1 h	(a) b0812 (b) b1562 (c) b112 (d) **b3510**

Python & Metabolic Engineering Tutorial: A Complete Introduction for Beginners

Microbial organisms are engineered to make commodity chemicals like drugs and biofuels. Genes from other organisms like yeast and plants are integrated into the E coli genome to create novel pathways for chemical production. In this tutorial, we will analyze actual E coli strain engineering data and understand how engineering impacts RNA, protein and metabolic networks.

1.0 Load in data files

A helpful summary of all of the data that has been loaded into the jupyter notebook:

```
print("number of proteins in study:", len(I1_prot_data_cobra.columns))
print("number of metabolites in study:", len(I1_data.columns[1:]))
print("number of strains in study:", len(list_strains))
print("numer of time points in study:", len(all_time_points))
print("number of unique heterologous reactions:", len(mevalonate_rxns_list.keys()))
print("number of unique heterologous metabolites:", len(mevalonate_mets_list))
```

```
number of proteins in study: 55
number of metabolites in study: 91
number of strains in study: 1
numer of time points in study: 14
number of unique heterologous reactions: 12
number of unique heterologous metabolites: 35
```

Note: a "heterologous reaction" refers to a reaction that is not endogenous to E coli. In other words, these are the genes that have been genetically added to the E coli genome

Exercise

Try on your own

1. Identify the full names of all of the heterologous metabolites.

2. Identify the full names of all of the heterologous proteins.

Hint: the BiGG IDs (the model IDs) have been given to you in the form of a list, and you are provided a dataframe to translate their model IDs to full names.

Use syntax that you learned from DE3 that will enable you to get a slice of this dataframe with the heterologous metabolites. You can apply the same idea for the protein names.

Fig. 4 A "try-on-your-own" section from DE4. Here, students are guided through example code, then encouraged to modify this code and build on what they have learned from previous DEs to answer exercise questions

3.1.3 Notebook Design Principles

Introducing Jupyter Notebooks as a Teaching Tool

The Jupyter notebook is at the heart of this instructional approach because it unites explanation, computation, and visualization in a single, interactive document. Within one window, students can read contextual information, execute code, and immediately see the outcome. This transparency makes Jupyter an ideal bridge between conceptual and computational thinking: learners can watch the logic of an analysis unfold cell by cell, rather than working in a black-box interface. In biochemistry education, where quantitative reasoning, visualization, and interpretation converge, this open, iterative format mimics the way scientific reasoning naturally proceeds.

Pedagogically, Jupyter functions like a digital lab manual. Text and code are interwoven; each analysis is documented in real time alongside its interpretation. For students, this fosters an intuitive connection between what they are learning and what they are doing. For instructors, the notebooks provide a reproducible and modifiable framework that can scale easily to hundreds of learners without requiring local installations or specialized software.

To make Jupyter notebooks most effective as a learning environment, they were written as self-guided tutorials that combine narrative explanation, structured examples, and short independent tasks. Each notebook reads like a conversation with the instructor: a brief conceptual framing introduces a topic, example code illustrates how to perform a task, and "try-on-your-own" exercises (Fig. 4) that invite students to modify the code that they have just seen. This structure of guided demonstration followed by independent practice turns each notebook into a scaffolded learning experience that builds confidence and mastery step by step. The specific design elements of this approach are described below (*see* **Notes 2** and **3**).

Stepwise Progression from Basic Coding to Omics Data Visualization

Because most undergraduates have limited experience with programming, the notebooks were organized as a deliberate progression of skills that gradually increase in complexity while maintaining biological relevance. Students first learn to view data not as static tables but as dynamic, manipulable objects that can reveal patterns when explored interactively.

In the introductory DE, students encounter small, spreadsheet-like datasets in comma-separated (CSV) format. They learn how to read these files into a DataFrame, inspect columns and rows, and perform simple operations such as sorting, filtering, and summarizing values. These early exercises replicate what many students already know from Excel but transfer those skills into a reproducible, Python-based, computational framework. Once students are comfortable navigating and plotting basic data, later notebooks introduce domain-specific applications that place the same logic into a biochemical context.

After mastering table manipulation in DE1, students move to DE2, where they apply similar logic to visualize three-dimensional protein structures. The idea of selecting and filtering columns becomes selecting and displaying residues or chains; plotting values becomes rendering molecular views. In DE3, students extend these skills to metabolic pathway data, analyzing flux distributions and visualizing network-level patterns. The final notebook, DE4, integrates multiple "omics" layers, combining transcriptomic and proteomic measurements to explore relationships between molecular processes. At each stage, students reuse earlier commands while adding new analytical tools. This recursive design strengthens retention and helps them perceive coding as a cumulative language rather than a series of disconnected tasks.

The stepwise sequence, therefore, mirrors how a scientist's competence evolves; from handling small, controlled datasets to reasoning about complex, multidimensional systems. By the end of the course, students can import, transform, visualize, and interpret real-world biological data, using skills built systematically from their very first lines of code.

Designing for Students with No Prior Coding Experience

The notebooks were written for absolute beginners, emphasizing approachability and momentum over technical detail. Each opens with a brief warm-up section that verifies the environment and produces an immediate, visible result (such as printing a variable or plotting a simple figure). When students can experience early success, it can reduce anxiety and establish good computational habits.

Instruction alternates between example code blocks and short "try-on-your-own" exercises. After a concept is introduced, students are asked to adapt the example (i.e., changing a variable, filtering a different column, or altering a plot type) to confirm their understanding. Because the modifications are incremental, students experience the satisfaction of solving a problem without being overwhelmed by new syntax. This repetition with slight variation builds confidence and intuition (*see* **Notes 2** and **3**).

Every notebook includes explanatory markdown written in plain language that describes what the upcoming code does and why it is being executed. Students can read the rationale before running a cell, making the workflow transparent and intentional. Checkpoints inserted throughout the notebook allow learners to confirm that their output matches expectations, turning potential frustration into quick feedback loops. Collectively, these features transform the notebooks into low-pressure, high-engagement spaces, where experimentation is encouraged and errors become part of the learning process.

Integrating Narrative and Computation for Conceptual Flow

A defining strength of each DE is its capacity to tell a story. Every notebook was written as a narrative that guides students from theory to observation to interpretation. Markdown sections frame each problem, often including a short explanation of the biological context or an equation that underlies the computation. Code cells follow immediately, executing the analysis that the text describes and produces a visual output (e.g., such as a scatterplot, a metabolic map, or a molecular rendering) that anchors abstract concepts in concrete evidence.

This interplay between exposition and execution keeps the scientific reasoning visible throughout the workflow. The notebook becomes both a lesson plan and a lab notebook, capturing not only the final answer but also the steps and logic that lead to it. Students learn to see data analysis as a form of storytelling: one that requires framing a question, selecting evidence, and articulating meaning.

By the end of the semester, their completed notebooks read like miniature research narratives, each demonstrating how computational tools can be used to ask and answer biochemical questions with clarity and purpose.

3.2 Example Data Explorations

Each DE was distributed as a self-contained Jupyter notebook designed for seamless integration into the CyVerse Discovery Environment. The notebooks were version-controlled and organized in a public GitHub repository, which contained four folders corresponding to the four DEs. Each folder included a primary workflow notebook, a short setup notebook, and a data subdirectory containing curated input files (Table 4).

Students were provided a setup.ipynb file that could be uploaded directly into CyVerse. When executed, this file automatically cloned the correct GitHub folder containing the DE and its associated datasets into the student's personal workspace. Once cloned, the DE appeared in their CyVerse directory, allowing students to complete all analyses entirely in the cloud without requiring any local installations. Progress could be saved by downloading the working notebook to a personal computer, and uploaded again later if students needed to relaunch CyVerse. This structure offered a fully reproducible, cloud-hosted workflow while giving students the flexibility to pause and resume their work seamlessly across sessions.

Every notebook followed a similar structure: an introductory overview framed the biological problem, guided code cells demonstrated the analytic steps, and short "try-on-your-own" tasks encouraged students to modify parameters or extend examples. The design intentionally alternated between observation and independent practice, leading students from structured instruction to exploratory inquiry. By the end of each DE, students had produced figures, tables, or structural renderings that linked computational logic to biochemical reasoning (Fig. 5).

3.2.1 Notebook 1: Introduction to Pandas

The first notebook introduced students to the essential syntax and logic of data analysis in Python. Working with small, spreadsheet-style datasets, students learned to import comma-separated value (CSV) files, examine their structure, and perform basic

Table 4
Contents of Data Exploration folders

Contents	Utility
Workflow Notebook (DE#_Workflow.ipynb)	Guides students through activity
Setup Notebook (setup.ipynb)	Verifies that dependencies were correctly installed
/data subfolder	Houses all required datasets

Data Exploration 1

Learning Objective: Introduce Python syntax, DataFrame manipulation, and exploratory data analysis using the pandas library. This first notebook emphasizes computational fluency rather than biochemical complexity, allowing students to focus on the logic of data structures before applying these skills to biological datasets.

Expected Outputs: Mastery of fundamental pandas commands (read_csv, groupby, describe), Increased familiarity with Python data structures (lists, dictionaries, DataFrames), Computation of descriptive statistics (mean, SD, range) for large dataframes, Bar and histogram plots comparing different conditions within the same dataset

Required Libraries:
- pandas
- numpy
- matplotlib

Troubleshooting:
- Ensure filepaths point to the correct dataset
- Use .head() to preview data rather than printing entire DataFrames
- Remind students that syntax and capitalization matter (eg. Movies_df =/= movies_df)

Data Exploration 2

Learning Objective: Visualize, manipulate, and analyze macromolecular structures from the Protein Data Bank (PDB) to explore active-site geometry and ligand binding. Students connect 3-D structure visualization with in-class biochemistry lessons on hemoglobin and extend those principles to clinically relevant systems (BRCA1-BARD1 heteromeric complex. Students connect protein structural features to function by analyzing Multiplex Assay of Variant Effect (MAVE) data.

Expected Outputs: Interactive 3D rendering of hemoglobin and BRCA1-BARD1 complexes, Highlighted ligand representation, Highlighted representation of key residues, Annotated residues identifying different structural components, Ability to connect structural information to in-class instruction and concepts

Required Libraries:
- pandas
- py3Dmol
- nglview

Troubleshooting:
- Encourage students to use Chrome or Firefox for this module
- Clear representations before re-plotting

Data Exploration 3

Learning Objective: Model and visualize metabolic flux distributions within core biochemical pathways to connect quantitative data with metabolic regulation. Learn to connect the structure of data to its biochemical meaning (eg. the number of rows in a DataFrame represent the number of timepoints in an experiment).

Expected Outputs: Ability to interact with an interactive Escher map displaying pathway flux intensity via color gradients, Identification of rate-limiting steps of metabolic branching, Plots showing the effect of different conditions on metabolic flux

Required Libraries:
- pandas
- json
- os
- matplotlib

Troubleshooting:
- Ensure that students understand the structure of a filepath and where data is stored
- Convey the experimental design behind the data (eg. WT vs. engineered conditions)
- Remind students about the importance of consistent labels (eg. lac__D =/= lac_D)

Data Exploration 4

Learning Objective: This module introduces the Escher visualization tool for dynamic mapping of metabolite flow. Following metabolism lectures, students load curated metabolomics data from a published E. coli biofuel production study, execute Flux Balance Analysis, and compare wild-type vs. engineered strains. Using DataFrames, students filter and sort metabolite flux data, then visualize differential pathway activity.

Expected Outputs: Heatmaps showing the effect of different conditions on different metabolites using proteomic and transcriptomic data, Ability to summarize statistics from large-scale datasets (eg. number of proteins in study, number of metabolites in study, etc.)

Required Libraries:
- pandas
- numpy
- os
- matplotlib
- scipy

Troubleshooting:
- Handle missing values by using .dropna()
- Ensure numeric data types for all DataFrames
- Axis misalignment often stems from mismatched gene identifies; ensure students are using correct common keys when mapping

Fig. 5 Overview of the learning objectives, expected outputs, required libraries, and troubleshooting notes for each Data Exploration

manipulations using the pandas library. They practiced selecting columns, slicing rows, computing descriptive statistics such as mean or standard deviation, and generating simple plots with matplotlib (*see* **Note 5**).

Because many students had experience with Excel, this module acted as a conceptual bridge, translating familiar operations into reproducible code. Rather than emphasizing biochemical content, DE1 focused on computational fluency: understanding how data are organized, how commands transform those data, and how visualization can reveal hidden patterns. By the end of the exercise, students could read, summarize, and visualize numerical datasets independently, establishing the foundation for the biological analyses that followed.

3.2.2 Notebook 2: Protein Structure Exploration in 3D

The second notebook extended the same principles of inspection and transformation into three-dimensional biological data. Using py3Dmol [21] and nglview [25], students imported structural files from the Protein Data Bank [22] and visualized them interactively within the notebook. The activity began with hemoglobin, a protein that anchors many classic discussions of structure and function, and then progressed to the BRCA1-BARD1 heteromeric complex, linking the exercise to current biomedical relevance (*see* **Note 5**).

Students manipulated molecular representations, highlighted ligands, and annotated residues, learning to connect visualization with biochemical interpretation. They also integrated a small dataset from a Multiplex Assay of Variant Effect (MAVE) [26, 27], using it to examine how amino-acid substitutions mapped onto structural regions of the BRCA1 protein. Through these exercises, students practiced reasoning spatially about proteins, interpreting data in a tangible, visual form and recognizing how structural features underpin biological function.

3.2.3 Notebook 3: Metabolomic Networks and Metabolomics

The third notebook introduced systems-level reasoning through metabolomics. Here, students analyzed a curated dataset [28] describing metabolic flux under different experimental conditions. Using pandas [29] alongside the Escher visualization library [30], they modeled pathway activity and examined how genetic perturbations altered metabolic flow (*see* **Note 5**).

This DE required students to think quantitatively about biochemical networks. They learned to interpret a pathway map as a dataset, where flux values correspond to reaction rates and color intensity represents metabolic activity. Each student produced visualizations comparing flux distributions between conditions, then interpreted which steps appeared rate-limiting or compensatory. By connecting pathway diagrams with real numerical data, students began to appreciate the scale and complexity of metabolic regulation that underlies cellular behavior.

3.2.4 Notebook 4: Transcriptomics and Proteomics Integration

The final notebook synthesized the skills acquired throughout the semester into a multi-omics analysis. Students combined transcriptomic and proteomic data from a published *E. coli* biofuel-production study [28] to explore relationships between gene expression and protein abundance. Using familiar commands from earlier DEs (e.g., loading files, merging DataFrames, filtering rows, and summarizing statistics), they performed cross-comparisons that revealed how transcript and protein levels align or diverge across conditions (*see* **Notes 5**).

Visual outputs included clustered heatmaps, time-series plots, and summary tables generated with matplotlib, numpy, and scipy. The notebook concluded with a short reflection prompt asking students to interpret one observed pattern in biological terms, reinforcing that computational analysis is most powerful when paired with thoughtful reasoning. By the end of DE4, students could navigate large, multidimensional datasets with confidence and see how quantitative data integrate across molecular scales.

3.3 Assessment and Evaluation

3.3.1 Pre/Post-class Surveys

To evaluate how the integration of coding and omics analysis influenced student learning, confidence, and engagement, we developed a custom survey instrument tailored for large-enrollment biochemistry courses [14]. Existing instruments designed for introductory computer science courses were insufficient for capturing the interdisciplinary context of biochemical data science. Our survey, therefore, aimed to measure both technical self-efficacy and students' evolving perceptions of the relevance of data-driven reasoning to biology.

The survey combined multiple-choice, fill-in-the-blank, and five-point Likert-scale items (1 = strongly disagree to 5 = strongly agree). Items assessed students' comfort with coding, their ability to interpret "omics" data, and their attitudes toward computational tools in scientific research. Additional questions gathered demographic information and prior knowledge and experience to evaluate whether the course structure supported all backgrounds in learning outcomes.

To ensure content validity, the survey was first piloted with 16 students who participated in a one-day data science workshop prior to the semester. This pilot cohort completed abbreviated versions of the first two DEs and provided feedback on question clarity, relevance, and response options. Revisions were incorporated before deploying the survey at scale.

During the semester, all students were invited to complete an anonymous survey at two time points (e.g., once before the first DE and again after the final DE). After the first year of this intervention, a total of 441 students participated in the pre-course survey and 302 completed the post-course version, with 178 matched responses across both. Currently, after 3 years of intervention, we have accumulated 1024 matched responses. Surveys were administered electronically via Qualtrics, and students who completed both were awarded extra credit. All procedures were reviewed and deemed exempt by the Office of Human Research Ethics under 45 CFR 46.104 (Reference ID 357468).

3.3.2 Methods and Validation Summary

Quantitative analyses of the survey results were conducted using IBM SPSS [31]. Because the Likert-scale data were ordinal and non-normally distributed (Shapiro-Wilk $p < 0.001$), nonparametric methods were employed. Wilcoxon signed-rank tests assessed pre/-post changes in student attitudes toward coding, data analysis, and omics literacy. Mann-Whitney U tests compared differences across two groups (e.g., such as demographics or previous programming experience), while Kruskal-Wallis H tests were used for comparisons among multiple socioeconomic categories.

Effect sizes were calculated using the rank biserial correlation (r), with conventional thresholds of 0.2, 0.5, and 0.8 interpreted as small, medium, and large effects, respectively. Dunn-Bonferroni corrections were applied to control for Type I error in multiple comparisons. The use of nonparametric analyses was particularly appropriate for paired pre/post designs in large student populations. This approach allowed us to quantify changes in student confidence and engagement without assuming normal distributions, providing a robust framework for measuring attitudinal shifts across the semester.

3.3.3 Interpreting Learning Outcomes

Interpretation of the survey and quiz data focused on translating statistical results into educational meaning. Items were grouped into two validated subscales: (1) understanding of omics data and (2) research engagement, together, capturing distinct dimensions of learning. Large effect sizes ($r \geq 0.8$) on the understanding subscale reflected strong cognitive gains: students moved from viewing coding as an abstract technical exercise to recognizing it as a practical tool for biochemical reasoning. Smaller but positive changes on the research engagement subscale suggested early growth in curiosity and confidence, indicating that exposure to computational tools sparked interest in research but that sustained engagement may require additional opportunities beyond a single semester.

Effect sizes were emphasized over significance levels, as they offer a more meaningful gauge of practical learning in educational contexts. Patterns of strong improvement in understanding, accompanied by more modest growth in engagement, are consistent with a trajectory in which students first master technical confidence before developing research identity. Visualization tools (e.g., such as paired boxplots or change-score histograms) were used to illustrate these distributional shifts and to highlight the diversity of individual learning experiences within the large cohort.

Post-course convergence of scores across backgrounds indicated that scaffolded, cloud-based instruction can help narrow gaps in computational self-efficacy that stem from deficiencies in prior experience. Where differences persisted, they were viewed as reflections of differences in prior exposure rather than limitations of the instructional design itself.

Finally, qualitative reflections collected from students reinforced the quantitative findings. Many described greater comfort running code, reduced anxiety toward computational tasks, and a clearer sense of how data analysis connects to modern biochemistry. Together, these insights illustrate not only that learning occurred but how and why students' perspectives evolved over the course of the semester.

4 Notes

This section summarizes practical considerations, troubleshooting experiences, and instructional strategies that proved useful when implementing large-scale, cloud-based Jupyter notebook activities. While the technical workflow itself is reproducible across institutions, the effectiveness of the approach depends on anticipating common barriers, designing for inclusivity, managing cognitive load, and maintaining the digital infrastructure over multiple semesters.

1. Common Barriers in Large Classrooms

Large-enrollment courses amplify small logistical issues. The most frequent challenges were related to connectivity, CyVerse performance, and student confidence with coding. Periodic network congestion occasionally caused temporary delays when launching JupyterLab or running heavier rendering tasks; brief refresh intervals and patience generally resolved these interruptions. When CyVerse underwent scheduled maintenance, clear advance communication and flexible deadlines prevented disruptions to student progress.

Login issues may occur (e.g., due to expired institutional credentials or incorrect capitalization in user IDs). These issues were minimized by dedicating the first 5 min of class to verifying access before beginning a DE. Instructors who plan to replicate this model may find it helpful to keep a backup GitHub link available so students can continue locally if the platform experiences downtime.

A subtler but equally common barrier was coding anxiety. Many students entered the course convinced that programming was outside their skill set. Early success proved essential to overcoming this mindset. Simple warm-up cells that generated immediate, visible results (e.g., printing a short message or producing a bar plot) helped students experience progress within minutes. Emphasizing that mistakes are part of the process (and demonstrating how to debug calmly) created a culture of experimentation rather than fear of error.

2. Inclusive Teaching Strategies

Inclusivity in computational instruction involves more than equal access to hardware; it requires thoughtful scaffolding that welcomes students from all backgrounds and learning preferences. Cloud-based delivery ensured that all students, regardless of device or operating system, could participate fully. Within the notebooks, stepwise explanations and plain-language markdown reduced dependence on prior experience.

During live demonstrations, instructors narrated both what they were doing and why, bridging the gap between action and reasoning. Each DE included short "try-on-your-own" prompts immediately after new concepts, giving students the chance to test ideas independently without penalty. Office hours and virtual help sessions were scheduled at predictable times and recorded when possible, ensuring access for students balancing work or caregiving responsibilities.

Peer mentors played an important role in maintaining a supportive environment by circulating during class sessions, offering real-time guidance, and modeling problem-solving approaches. The use of asynchronous quizzes and open-book assessments also supported accessibility by allowing students to demonstrate understanding in low-pressure contexts.

3. Reducing Cognitive Load

Introducing computation to students with no prior coding background requires careful pacing. Cognitive load was managed through repetition, predictability, and clear structure. Each DE reused familiar commands and notebook architecture, allowing students to focus on new biochemical content rather than new syntax.

Short markdown explanations preceded each code block, outlining the goal of the upcoming task so that students understood why they were running a command before executing it. Frequent checkpoints confirmed progress and provided natural pause points for questions. Datasets were intentionally modest in size, large enough to illustrate authentic biological patterns but small enough to load quickly and display clearly within CyVerse.

Breaking lessons into self-contained sections (e.g., introduction, example, and short practice) helped maintain focus. For visual learners, integrated figures and color-coded outputs supported comprehension; for verbal learners, concise narrative summaries clarified meaning. Together, these design choices kept the notebooks approachable and cognitively manageable even for complete beginners.

4. Maintenance and Sustainability Tips

Long-term sustainability depends on reliable version control and efficient grading infrastructure. All course materials were managed through GitHub, which served as a collaborative, version-controlled repository that could be easily updated and shared across semesters. Each semester, a requirements.txt file was frozen from the same CyVerse base image to prevent version drift. Any updates (such as dataset corrections or clarified instructions) were pushed through Git and automatically propagated to student workspaces when the Setup notebook was re-run.

Automating parts of the grading process helped maintain scalability. Concept-check quizzes were delivered through the university's learning management system (e.g., Canvas), where automatic grading handled multiple-choice items and simple code-output checks. This freed instructors and TAs to focus on interpreting open-ended responses and providing qualitative feedback on biological reasoning.

Sustainability also includes human infrastructure. Clear documentation of notebook logic, comments in code cells, and concise change logs allowed new instructors or graduate assistants to take over seamlessly. Each iteration of the course benefited from feedback collected through surveys and informal reflections, ensuring that the notebooks evolved while retaining consistency across semesters.

5. Collected Instructor Tips

Testing and Environment Setup

- Test notebooks from a clean kernel before release. Even small changes to package versions or file paths can produce runtime errors that scale rapidly in large-enrollment settings.
- Create and freeze a requirements.txt file from the same CyVerse base image each semester to maintain consistent software environments.
- Encourage students to terminate analyses and close sessions after each use to free up CyVerse resources.

Introducing Coding and Reducing Anxiety

- Frame coding tasks within biochemical reasoning. For example, equate filtering a DataFrame with selecting specific time points from an experiment to make computation feel familiar and purposeful.
- Begin each notebook with a simple "warm-up" block that verifies the environment, prints a variable, or generates a quick visualization to build early confidence.
- Emphasize that mistakes are normal and valuable; debugging is part of the learning process.

Notebook Design and Flow

- Alternate markdown explanations and executable code cells to maintain a clear conceptual flow from theory to execution to interpretation.
- Keep markdown sections concise and visually readable, using brief paragraphs, bullet lists, or equations as needed to prevent screen clutter.

- Write explanatory text in plain language that describes why a command is being run, not just how to run it.

Pedagogical Structure

- Use "try-on-your-own" prompts after key examples to encourage immediate practice. Have students modify or extend prior code snippets to reinforce comprehension.
- Incorporate checkpoints throughout each notebook to verify correct outputs and give students frequent, low-stakes feedback.
- Maintain consistent structure across all DEs (intro, example, practice, interpretation) so that students know what to expect.

Surveys and Assessment

- Keep pre/post surveys concise and integrated into the course workflow; pairing them with small extra-credit incentives can yield response rates above 80%.
- Emphasize effect sizes over p-values when interpreting educational data to highlight meaningful learning outcomes rather than statistical noise.

Large-Classroom Management

- Allocate the first few minutes of each session to verifying CyVerse login access and resolving credential issues before beginning instruction.
- Maintain a backup GitHub link or downloadable version of each DE in case of temporary CyVerse downtime.
- Communicate early and clearly about any planned platform maintenance or outages.
- Post demo videos to help students understand how to save their work and troubleshoot common issues.

Inclusivity and Accessibility

- Narrate live coding demonstrations, explaining both the technical steps and the underlying reasoning to support diverse learning styles.
- Record office hours or virtual help sessions when possible to provide flexible access for students with scheduling or caregiving constraints.
- Reinforce that coding is learnable by all students, regardless of prior experience, and highlight examples of successful beginners.

Maintenance and Sustainability

- Use GitHub version control as the central hub for distributing, updating, and archiving notebooks.
- Document all changes to notebooks and datasets in short-change logs to maintain transparency for future instructors and teaching assistants.
- Where possible, use automated grading in Canvas (or a similar University-based course system) for concept-check quizzes to streamline feedback and scalability.

Troubleshooting and Technical Reminders

- Remind students to preview data with.head() rather than printing entire DataFrames.
- Ensure file paths point to the correct directories before execution.
- Verify that identifiers (e.g., gene or protein names) match across datasets to prevent axis misalignment in plots.
- Advise using Chrome or Firefox browsers for 3D visualizations, as Safari may cause rendering issues.

5 Future Directions

The success of this work underscores a simple but transformative idea: computational and data-driven learning can be made accessible to all students, regardless of background, resources, or geography. By integrating interactive, cloud-based notebooks into the biochemistry curriculum, we have not only lowered the barrier to learning data science but also demonstrated how these tools can democratize access to authentic scientific exploration.

Future efforts will extend this framework to include more topics on artificial intelligence (AI). As AI becomes increasingly embedded in research and discovery, students must learn not only how to use it but how to use it responsibly and safely. Future iterations of this work will integrate AI-assisted data analysis, predictive modeling, and generative methods directly into the notebooks, pairing technical skill with ethical reflection. Students will learn to question the algorithms that shape their conclusions, to identify bias and uncertainty, and to appreciate that AI is not a substitute for scientific thinking. Teaching these principles early (within the same accessible, hands-on environment that has already proven effective) ensures that students graduate as both computationally fluent and ethically grounded scientists.

This platform also opens the door to vertical learning: a tiered model in which students progress through increasingly advanced modules that mirror their growth as scientists. Introductory notebooks will continue to emphasize basic data manipulation and visualization, while intermediate and advanced materials will introduce machine learning, network modeling, and AI-driven data interpretation. Over time, these layers of instruction will form a seamless educational pathway: one that allows students to "grow up" through the notebooks, from their first lines of Python to independent, AI-assisted research questions. This approach transforms notebooks into more than teaching tools; they become a developmental framework for lifelong scientific learning. This approach offers a foundation upon which the next generation of education can be built with an evolving ecosystem of digital notebooks that grow in complexity and scope, forming an open electronic library of "how-to" resources for scientific inquiry.

Equally important is the accessibility this model provides. Because the notebooks run entirely in the cloud, they can be accessed by any student with an internet connection, whether at a research university, a community college, or a rural institution with limited resources. In this way, the work has implications far beyond a single course or campus. It provides a scalable, equitable model for bringing computational science to primarily undergraduate institutions (PUIs), minority-serving institutions (MSIs), and other communities where students might otherwise never experience authentic data analysis or modeling. Each notebook represents an opportunity: an open window into modern biochemistry that can be freely adapted, shared, and expanded by educators everywhere.

Our vision is to turn education into a globally connected, data-rich ecosystem. Imagine an online library of living, interactive notebooks, continually updated and contributed to by instructors and researchers around the world, each one a digital experiment that students can run, modify, and learn from. In such a system, the boundaries between teaching, research, and discovery blur: every student becomes an active participant in science, every instructor a collaborator in a shared project of understanding. If realized fully, this vision could change the landscape of education. It would make high-quality, hands-on scientific learning available to students everywhere, empowering those in rural or under-resourced settings to engage with the same tools and ideas as those in research-intensive environments. By expanding the framework to include responsible AI education and vertically integrated learning pathways, we can build a system that not only teaches data literacy but cultivates critical thinkers prepared to navigate and shape the future of science and healthcare. Ultimately, this work envisions a world where any student, anywhere, can open a digital notebook, run a cell, and begin discovering how life works at the molecular level.

Acknowledgments

We thank Brian Hogan and Bryant Hutson at the University of North Carolina at Chapel Hill for their support in implementing this curriculum. We also thank Tyson Swetnam, Sarah Roberts, and the entire CyVerse team; this work would not have been possible without their platform and sustained assistance.

Conflicts of Interest The authors have no conflicts of interest to declare that are relevant to the content of this chapter.

References

1. Altaf-Ul-Amin M, Afendi FM, Kiboi SK, Kanaya S (2014) Systems biology in the context of big data and networks. Biomed Res Int 2014:428570. https://doi.org/10.1155/2014/428570
2. Leonelli S (2014) What difference does quantity make? On the epistemology of big data in biology. Big Data Soc 1: 2053951714534395. https://doi.org/10.1177/2053951714534395
3. Marx V (2013) The big challenges of big data. Nature 498:255–260. https://doi.org/10.1038/498255a
4. Woese CR (2004) A new biology for a new century. Microbiol Mol Biol Rev 68:173–186. https://doi.org/10.1128/mmbr.68.2.173-186.2004
5. Biswas N, Chakrabarti S (2020) Artificial intelligence (AI)-based systems biology approaches in multi-omics data analysis of cancer. Front Oncol 10. https://doi.org/10.3389/fonc.2020.588221
6. Gehlenborg N, O'Donoghue SI, Baliga NS et al (2010) Visualization of omics data for systems biology. Nat Methods 7:S56–S68. https://doi.org/10.1038/nmeth.1436
7. Zierer J, Menni C, Kastenmüller G, Spector TD (2015) Integration of 'omics' data in aging research: from biomarkers to systems biology. Aging Cell 14:933–944. https://doi.org/10.1111/acel.12386
8. Attwood TK, Blackford S, Brazas MD et al (2017) A global perspective on evolving bioinformatics and data science training needs. Brief Bioinform 20:398–404. https://doi.org/10.1093/bib/bbx100
9. Drew J, Morgan W, Galindo S et al (2023) Revisiting barriers to implementation of bioinformatics into life sciences education. Front Educ 8. https://doi.org/10.3389/feduc.2023.1317191
10. LaTourrette K, Stengel A, Clarke J (2021) Student-led workshops: filling skills gaps in computational research for life scientists. Nat Sci Educ 50:e20052. https://doi.org/10.1002/nse2.20052
11. Williams JJ, Drew JC, Galindo-Gonzalez S et al (2019) Barriers to integration of bioinformatics into undergraduate life sciences education: a national study of US life sciences faculty uncover significant barriers to integrating bioinformatics into undergraduate instruction. PLoS One 14:e0224288. https://doi.org/10.1371/journal.pone.0224288
12. NIH Strategic Plan for Data Science | Data Science at NIH. https://datascience.nih.gov/nih-strategic-plan-data-science. Accessed 5 Nov 2025
13. Chang H (2021) A comprehensive list of undergraduate bioinformatics programs in the United States
14. Brunk R, Shukla K, Hutson BL et al (2024) Data science for biochemists: integrating and evaluating the use of interactive digital python notebooks in a large-enrollment undergraduate biochemistry course. J Chem Educ 101: 3643–3655. https://doi.org/10.1021/acs.jchemed.4c00167
15. Swetnam TL, Antin PB, Bartelme R et al (2024) CyVerse: cyberinfrastructure for open science. PLoS Comput Biol 20:e1011270. https://doi.org/10.1371/journal.pcbi.1011270
16. Grushow A, Reeves MS (2019) Using computational methods to teach chemical principles: overview. In: Using computational methods to teach chemical principles. American Chemical Society, pp 1–10

17. Yin Stokes G (2021) How faculty with minimal programming experience implemented Jupyter notebooks in physical and general chemistry courses. In: Teaching programming across the chemistry curriculum. American Chemical Society, pp 13–27
18. Weiss CJ (2017) Scientific computing for chemists: an undergraduate course in simulations, data processing, and visualization. J Chem Educ 94:592–597. https://doi.org/10.1021/acs.jchemed.7b00078
19. Bisong E (2019) Google colaboratory. In: Bisong E (ed) Building machine learning and deep learning models on Google cloud platform: a comprehensive guide for beginners. Apress, Berkeley, CA, pp 59–64
20. The center for all your data, analytics, and AI – Amazon SageMaker – AWS. In: Amaz. Web Serv. Inc. https://aws.amazon.com/sagemaker/. Accessed 5 Nov 2025
21. py3Dmol: An IPython interface for embedding 3Dmol.js views in Jupyter notebooks
22. Berman HM, Westbrook J, Feng Z et al (2000) The protein data bank. Nucleic Acids Res 28: 235–242. https://doi.org/10.1093/nar/28.1.235
23. McKinney W (2010) Data structures for statistical computing in python. SciPy 2010. https://doi.org/10.25080/Majora-92bf1922-00a
24. Wilkinson MD, Dumontier M, Aalbersberg IJ et al (2016) The FAIR guiding principles for scientific data management and stewardship. Sci Data 3:160018. https://doi.org/10.1038/sdata.2016.18
25. Nguyen H, Case DA, Rose AS (2018) NGLview–interactive molecular graphics for Jupyter notebooks. Bioinformatics 34:1241–1242. https://doi.org/10.1093/bioinformatics/btx789
26. Esposito D, Weile J, Shendure J et al (2019) MaveDB: an open-source platform to distribute and interpret data from multiplexed assays of variant effect. Genome Biol 20:223. https://doi.org/10.1186/s13059-019-1845-6
27. Rubin AF, Stone J, Bianchi AH et al (2025) MaveDB 2024: a curated community database with over seven million variant effects from multiplexed functional assays. Genome Biol 26:13. https://doi.org/10.1186/s13059-025-03476-y
28. Brunk E, George KW, Alonso-Gutierrez J et al (2016) Characterizing strain variation in engineered E. Coli using a multi-omics-based workflow. Cell Syst 2:335–346. https://doi.org/10.1016/j.cels.2016.04.004
29. team T pandas development (2025) pandas-dev/pandas: Pandas
30. King ZA, Dräger A, Ebrahim A et al (2015) Escher: a web application for building, sharing, and embedding data-rich visualizations of biological pathways. PLoS Comput Biol 11: e1004321. https://doi.org/10.1371/journal.pcbi.1004321
31. Meyers LS, Gamst GC, Guarino AJ (2013) Performing data analysis using IBM SPSS. Wiley

Chapter 3

Analysis of Complex Single-Molecule Kinetics: A Protocol for Single-Molecule Interaction Simulations (SMIS)

Guangjie Yan, Luiz B. Fernandez, and Tai-Yen Chen

Abstract

Single-molecule techniques have transformed molecular biology by enabling direct observation of individual biomolecular events in real time. While single-molecule localization methods have advanced spatial resolution, they lack the temporal resolution needed to capture dynamic interactions. Single-molecule fluorescence resonance energy transfer (smFRET) addresses this gap by revealing transient molecular states often hidden in ensemble measurements. However, extracting reliable kinetic parameters from smFRET data is hindered by noise artifacts, photophysical effects, and the analytical complexity of stochastic state transitions. To address these limitations, Single-Molecule Interaction Simulation (SMIS) provides a robust framework that enables kinetic analysis by simulating molecular transitions and generating interpretable dwell-time distributions. Unlike traditional approaches requiring complex differential equations, SMIS simplifies kinetic modeling and directly addresses challenges in reproducibility. This chapter outlines a comprehensive protocol for implementing SMIS, from defining kinetic schemes to extracting kinetic rate constants by comparing simulation outcomes against experimental data. By integrating SMIS into the analysis workflow, researchers can extract accurate kinetic insights from smFRET experiments, extending their applicability and enhancing the interpretability of single-molecule studies.

Key words Single-molecule interaction simulation (SMIS), smFRET, Kinetic modeling, Dwell-time distribution, Simulated kinetics

1 Introduction

Single-molecule techniques have revolutionized the study of molecular processes by enabling direct observation of individual molecules in real time [1–6]. Although super-resolution methods [7–13] have advanced spatial imaging, they fall short in resolving temporal dynamics critical for studying molecular interactions. Single-molecule fluorescence resonance energy transfer (smFRET) addresses this gap by capturing transitions between molecular states in real time, providing unique insights into protein folding,

Authors Guangjie Yan and Luiz B. Fernandez have equally contributed to this chapter.

Robert E. Dempski (ed.), *Transforming Molecular Biology with Emerging Technologies*, Methods in Molecular Biology, vol. 3043,
https://doi.org/10.1007/978-1-0716-5320-3_3,

DNA-protein interactions, and enzyme-substrate kinetics [11, 14–17]. Unlike ensemble measurements, which average the behaviors of millions of molecules and obscure dynamic heterogeneity, smFRET provides direct access to transient states and rare events. This capacity to resolve events at a single-molecule level has facilitated a deeper understanding of the underlying kinetic processes that drive complex biological systems [18–20].

Despite its power, extracting meaningful kinetic parameters from smFRET experiments often involves solving complex differential equations [21–23]. Data analysis is particularly challenging due to the inherent noise in smFRET measurements, such as camera blurring, shot noise, and photoblinking effects [24–27]. These factors obscure signal interpretation and necessitate sophisticated algorithms to isolate true molecular interactions from artifacts. Additionally, the lack of standardized protocols for smFRET analysis introduces variability across research groups, making reproducibility and cross-study comparisons difficult. Interpretation is further complicated by the dynamic nature of molecular interactions, where rapid transitions between states can lead to conformational averaging [25]. Such analytical challenges limit the broader application of smFRET in deciphering biophysical processes.

Single-molecule interaction simulation (SMIS) complements traditional smFRET analysis by focusing on temporal resolution and kinetic modeling through simulations [28, 29]. Unlike traditional methods that rely on solving complex differential equations, SMIS analyzes stochastic interactions by directly simulating dwell-time distributions and state populations. This approach enhances the usability of smFRET data and broadens accessibility for researchers by providing kinetic insights without the constraints imposed by traditional models.

Here, we use the interactions between ATP:cob(I)alamin adenosyltransferase (MMAB) and its cofactor, 5-′-deoxyadenosylcobalamin (AdoCbl), as a case study to demonstrate how SMIS can overcome this challenge [30]. MMAB is a homotrimeric enzyme critical for vitamin B_{12} metabolism that exhibits negative cooperativity, where binding of the first AdoCbl molecule reduces the affinity for a second one. Kinetically, this cooperativity could originate from a slower association rate for the second binding event or a faster dissociation from the doubly bound state. Distinguishing between these possibilities requires the quantitative determination of each microscopic rate constant. To probe this mechanism, we previously used a FRET-based single-molecule relative fluorescence (SRF) assay, which allowed us to visualize the real-time transitions between MMAB's unbound, singly bound, and doubly bound states.

In this chapter, we demonstrate the use of SMIS for resolving the kinetic mechanism of negative cooperativity in the AdoCbl-MMAB system. This case study highlights how SMIS bypasses analytical intractability by simulating fluorescence-based trajectories and dwell-time distributions from a proposed model and fitting them against experimental data.

2 Materials

2.1 Case Study: Experimental Data Acquisition

The initial step for any kinetic analysis using the SMIS protocol is the acquisition of high-quality, single-molecule experimental data. This section outlines the data acquisition protocol for this chapter's case study: the interaction between MMAB and its cofactors. The following paragraphs detail the methods for protein labeling, the single-molecule microscopy setup, and data acquisition parameters, providing the necessary foundation for understanding the raw data that will be processed in subsequent sections.

To label MMAB with a fluorophore, a Cys-less variant of *Methylobacterium extorquens* MMAB was created by mutating the two non-conserved native cysteines (C64 and C154) to serines. A unique cysteine was then introduced at the solvent-exposed S99 position for labeling. This S99C MMAB variant was selected as it maintained 94% of wild-type catalytic activity. The MMAB mutant was cloned into a pET28 vector, overexpressed in *E. coli* BL21 (DE3), and induced with 1 mM IPTG. Cells were harvested and lysed by sonication in a sodium phosphate buffer (50 mM sodium phosphate, pH 8.0, 0.3 M KCl, 20 mM imidazole, 0.1 mM PMSF, 5% glycerol). The protein was purified from the supernatant using a His-trap column. For labeling, the purified MMAB was first reduced with TCEP overnight at 4 °C. The reduced protein was then incubated with a mixture of Alexa555-Maleimide and Bio-PEG-Maleimide in the dark to facilitate covalent labeling via maleimide chemistry. The final labeled MMAB was purified using a Superdex peptide column followed by a MonoQ anion column. Successful labeling was confirmed by SDS-PAGE analysis.

Experiments were conducted on an Olympus IX71 inverted microscope configured for prism-type total internal reflection fluorescence (TIRF). The excitation source was a 532 nm laser (CrystaLaser), which was passed through a quarter-wave plate to ensure circularly polarized excitation light. The laser power at the sample was approximately 6–7 mW across a 68 × 137 μm^2 area. The fluorescence signal was collected through a 60×, 1.2 NA water-immersion objective (Olympus). The signal was then passed through an HQ550LP filter (Chroma) before being detected by an EMCCD camera (Andor iXon). All imaging was controlled using Andor iQ software.

A homemade flow chamber was constructed using a quartz slide and a coverslip. The quartz surface was functionalized with biotinylated albumin and neutravidin to allow for specific immobilization of the biotinylated, Alexa555-labeled MMAB molecules.

The protein was incubated at a concentration of approximately 30 pM for 5–15 min to achieve a sparse surface density suitable for single-molecule imaging. Once molecules were immobilized, solutions containing AdoCbl or OHCbl at various concentrations (0–80 μM) were introduced into the flow chamber at a velocity of 10 μL/min. To minimize photobleaching and photoblinking of the Alexa555 dye, a photoprotection system consisting of 1 mM Trolox and 10 mM cysteamine was added to the imaging solutions immediately before each experiment. Single-molecule movies were recorded with a typical length of 1–2 min, using an image integration time of 32 ms per frame.

2.2 SMIS APP

The SMIS protocol is implemented through a MATLAB-based graphical user interface (GUI) known as the SMIS_App. This app provides a platform for simulating kinetic trajectories and analyzing dwell-time distributions, as it encapsulates all necessary simulation code and removes the need for command-line operations. A succinct description of the GUI layout and its functions is provided in this subheading. To further guide users, a step-by-step example demonstrating how to generate trajectories and dwell-time distributions for the case study presented here is available on Figshare, where the SMIS_App and all relevant experimental data are also available for download [29].

The SMIS App interface is organized into three main functional panels: a Model Input and Viewer panel, a Status panel, and an Execution panel (Fig. 1a). In the Model Input panel, users construct the kinetic model by specifying the number of fluorescence states and the number of distinct molecular species within each state. The corresponding rate constants for each transition are entered into the "Rate Constants Table," and checkboxes are used to designate which rates are substrate concentration dependent. Additional simulation parameters, such as substrate concentrations, simulation length, time resolution, and data export location, are configured in a separate "Advanced setting" window (Fig. 1b). As parameters are entered, the Model Viewer panel provides a real-time schematic of the kinetic model. In this diagram, molecular species are represented by labeled circles, and the allowed transitions are shown as arrows, allowing for immediate visual verification of the model's structure. The viewer dynamically illustrates the reaction pathways as they are defined: entering a non-zero rate constant draws a single-headed arrow for an irreversible step, and defining the reverse rate converts it to a double-headed arrow for a reversible reaction.

Once the model is finalized, the simulation is initiated using the "Simulation" button in the Execution panel. A status indicator confirms when the computation is complete, at which point the data visualization buttons become active. The "Trajectories" button generates an output window displaying representative

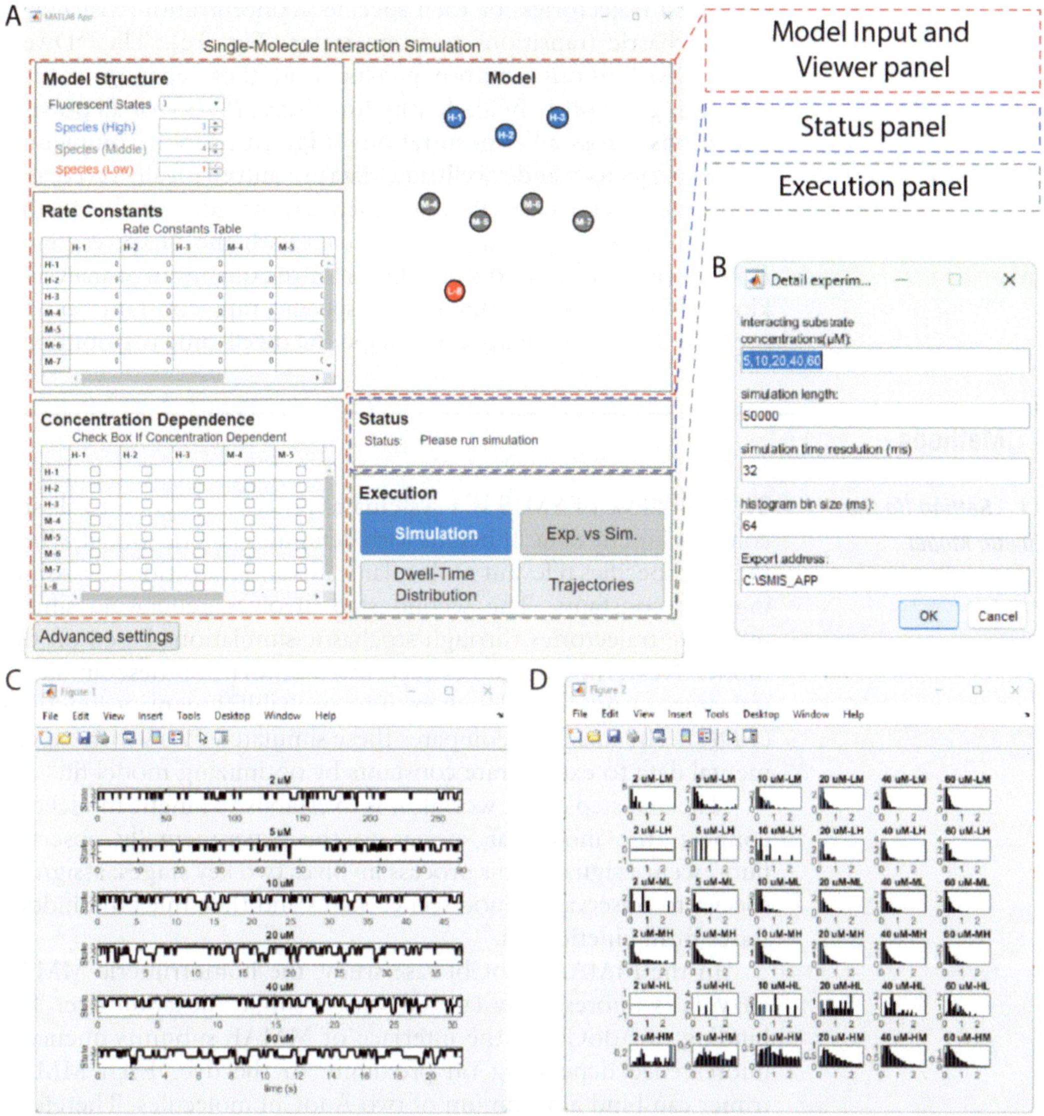

Fig. 1 The SMIS app graphical user interface (GUI) and a representative output window. (**a**) for defining the kinetic model. Users specify the number of fluorescence states and molecular species, input rate constants into the rate constants table, and designate concentration-dependent transitions. A schematic of the kinetic model is automatically generated in the Model Viewer panel for real-time visual verification. (**b**) The "Advanced setting" window, where detailed simulation parameters such as substrate concentrations, simulation length, time resolution, and the data export path are configured. (**c**) The trajectory output window, which displays representative simulated single-molecule trajectories at specified concentrations. Each trajectory visualizes the stochastic transitions between the defined fluorescence states over time. (**d**) The dwell-time distribution output window, which presents the compiled *PDF*s for all possible state transitions. The *PDF*s are shown for each simulated concentration, enabling quantitative kinetic analysis

simulated trajectories for each specified concentration, visualizing the stochastic transitions between states (Fig. 1c). The "Dwell-Time Distributions" button produces another separate window containing the probability density functions (*PDFs*) for all possible transitions across all concentrations (Fig. 1d). Upon completion, the raw trajectory and dwell-time data are automatically saved as ". mat" data files to the user-specified export address for further comparison with the experimental data with the "Exp. vs. Sim." Bottom (Fig. 1a). This output allows for the quantitative analysis of kinetic signatures, revealing how specific rate constants or the presence of intermediate states shape the dwell-time distributions.

3 Methods

3.1 Setting Up the Kinetic Model

The workflow of SMIS is a systematic process designed to dissect complex kinetic data. It begins with defining a kinetic model, where the user specifies relevant molecular states and transitions by assigning rate constants. The second step involves generating single-molecule trajectories through stochastic simulations based on this model. Next, the software extracts dwell times from these simulated trajectories and compiles them into probability density functions (*PDFs*). The final step compares these simulated *PDFs* with experimental data to extract rate constants by optimizing model fit.

The first step of the workflow is to establish a kinetic model that connects the molecular species to the experimentally observed fluorescence signals. This process involves two key stages: assigning species to observable fluorescence states and then inferring hidden states from kinetic data.

In the MMAB-AdoCbl case study, the homotrimeric MMAB enzyme is fluorescently labeled with a single Alexa555 dye. The binding of AdoCbl at the interface of MMAB subunits quenches fluorescence depending on proximity to the dye. Each MMAB trimer can bind a maximum of two AdoCbl molecules. Therefore, the fluorescence intensity of a single labeled MMAB molecule provides a direct readout of the binding stoichiometry (Fig. 2a). The degree of this fluorescence quenching depends on two key factors: the number of AdoCbl molecules bound and their specific location on the trimer relative to the dye's position. This principle allows for the assignment of specific binding configurations to one of three observed fluorescence states: high (F_H), middle (F_M), and low (F_L). Representative single-molecule relative fluorescence (SRF) trajectories at various AdoCbl concentrations show clear transitions between these states (Fig. 2b, c). To quantify the occupancy of these states, hundreds of such trajectories are compiled into *PDFs*, which show three distinct populations corresponding to the F_H, F_M, and F_L states (Fig. 2d). The relative populations of these states shift predictably with increasing ligand concentration

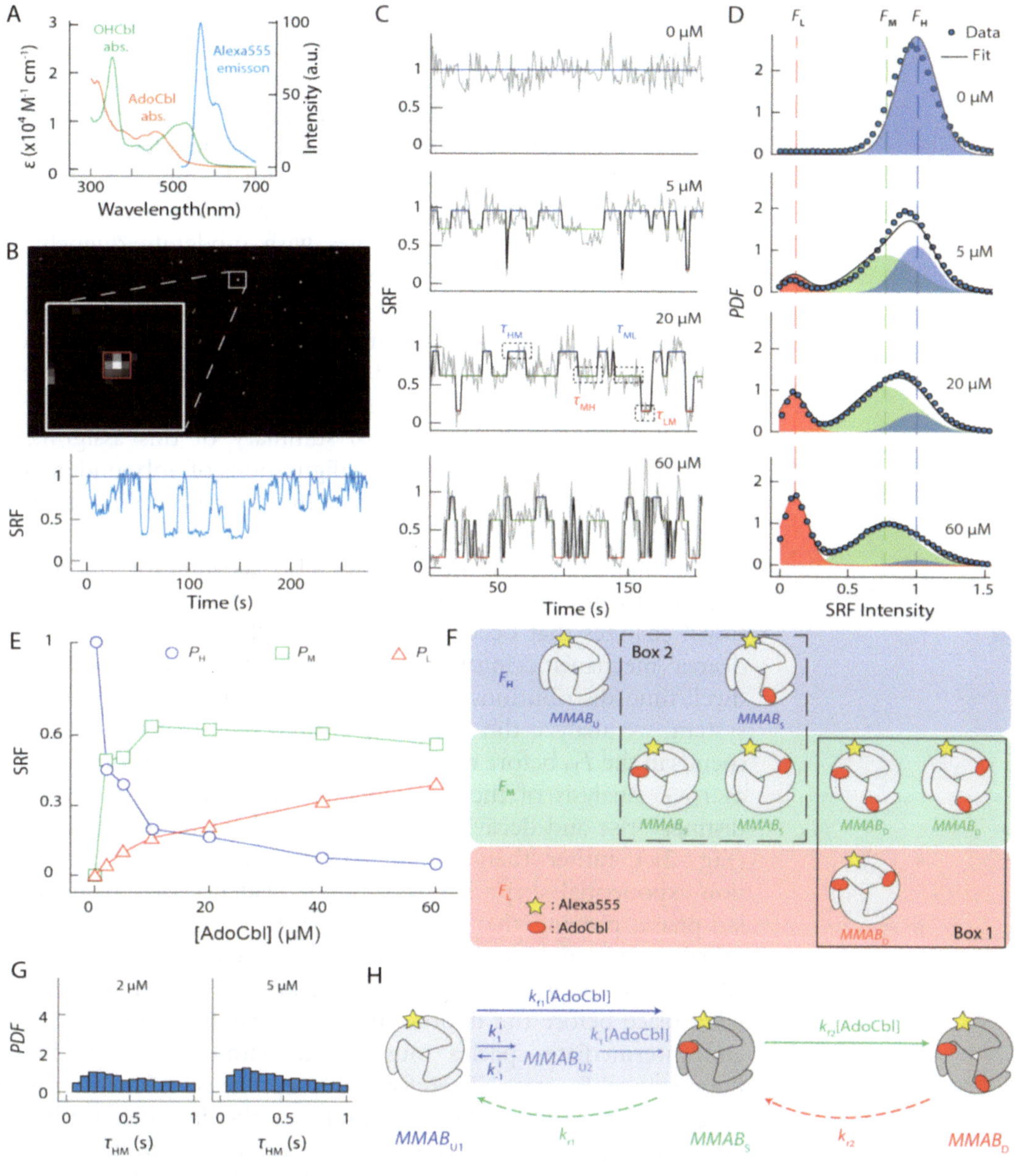

Fig. 2 Experimental workflow for single-molecule analysis of MMAB-AdoCbl interactions. (**a**) Spectral overlap between the emission of the Alexa555 and the absorption of the AdoCbl, demonstrating the physical basis for FRET. (**b**) A representative fluorescence micrograph showing discrete spots, each corresponding to a single immobilized MMAB molecule, for SRF trajectory generation. (**c**) Representative SRF time trajectories at 0, 5, 20, and 60 μM AdoCbl, showing stochastic transitions between F_H, F_M, and F_L states. (**d**) SRF intensity histograms (*PDF*s) were globally fit with a three-state Gaussian model to quantify the three distinct fluorescence populations. (**e**) The relative populations of the F_H (blue circles), F_M (green squares), and F_L (red triangles) states plotted as a function of AdoCbl concentration. (**f**) Assignment of unbound ($MMAB_U$), singly bound ($MMAB_S$), and doubly bound ($MMAB_D$) to the corresponding fluorescence states based on quenching efficiency and population analysis. (**g**) The experimental PDF_{HM} shows a characteristic "rise-and-decay" shape that provides kinetic evidence of a hidden intermediate state. (**h**) The proposed kinetic model for the MMAB-AdoCbl interaction. (Data reproduced from Ref. [30] with permission from Elsevier)

(Fig. 2e). By carefully correlating the relative populations in the high, middle, and low states with the binding configuration, we concluded that F_H corresponds to minimal quenching and is populated by the unbound species ($MMAB_{U1}$) and the hidden conformational intermediate ($MMAB_{U2}$). It also includes the singly bound $MMAB_S$ configuration, where AdoCbl binds to the site farthest from the dye, resulting in insignificant quenching. F_M state contains a mixture of species with moderate quenching, including the two $MMAB_S$ configurations, where AdoCbl is bound closer to the dye and the two remaining $MMAB_D$ configurations. F_L state represents maximum quenching and is exclusively assigned to the doubly bound $MMAB_D$ configuration, where AdoCbl occupies the two binding sites closest to the fluorescent label. Figure 2f provides a visual summary of this assignment scheme, linking specific physical configurations of unbound, singly bound, and doubly bound MMAB to their corresponding fluorescence levels.

A complete kinetic model must also account for transitions invisible to direct fluorescence measurement, such as conformational changes that occur within a single fluorescence state. This deeper mechanistic information is revealed in the shape of the dwell-time distributions for specific transitions. For the MMAB system, we analyze the microscopic dwell time that the molecule spends in the F_H before transitioning to the F_M, a duration denoted as τ_{HM}. Analysis of the experimental τ_{HM} distribution reveals a distinct "rise-and-decay" pattern at low AdoCbl concentrations (Fig. 2g), rather than a simple single-exponential decay. This non-exponential shape is direct kinetic evidence of a mandatory lag phase, proving that the system must first transition from an initial unbound state ($MMAB_{U1}$) to a required intermediate ($MMAB_{U2}$) before binding can occur. Because this entire process takes place before the fluorescence level changes, both the initial and intermediate species must reside within the same high-fluorescence F_H state. Therefore, a comprehensive kinetic model (Fig. 2h) is constructed by combining the fluorescence-based assignment of binding states with the kinetic-based inference of hidden conformational states, establishing the complete reaction pathway that will be used for simulation.

3.2 Simulating Single-Molecule Trajectories

With the comprehensive kinetic model established in Subheading 3.1, the SMIS software is used to generate large ensembles of stochastic single-molecule trajectories based on the kinetic parameters (Table 1). The simulation begins with the system in an initial state (e.g., unbound MMAB at equilibrium) and calculates the dwell time for each molecular species based on the probabilities defined by the assigned rate constants.

These microscopic species transitions are then mapped onto the three observable fluorescence states (F_H, F_M, and F_L) to generate

Table 1
Kinetic parameters for SMIS

Substrate	k_1^i (s^{-1})	k_{-1}^i (s^{-1})	k_1 ($\mu M^{-1}\ s^{-1}$)	k_{f1} ($\mu M^{-1}\ s^{-1}$)	k_{r1} (s^{-1})	k_2 ($\mu M^{-1}\ s^{-1}$)	k_{r2} (s^{-1})
AdoCbl	8.4	2.0	0.57	0.66	3.7	0.12	5.0
OHCbl	2.8	4.0	0.28	0.30	4.5	0.02	3.6

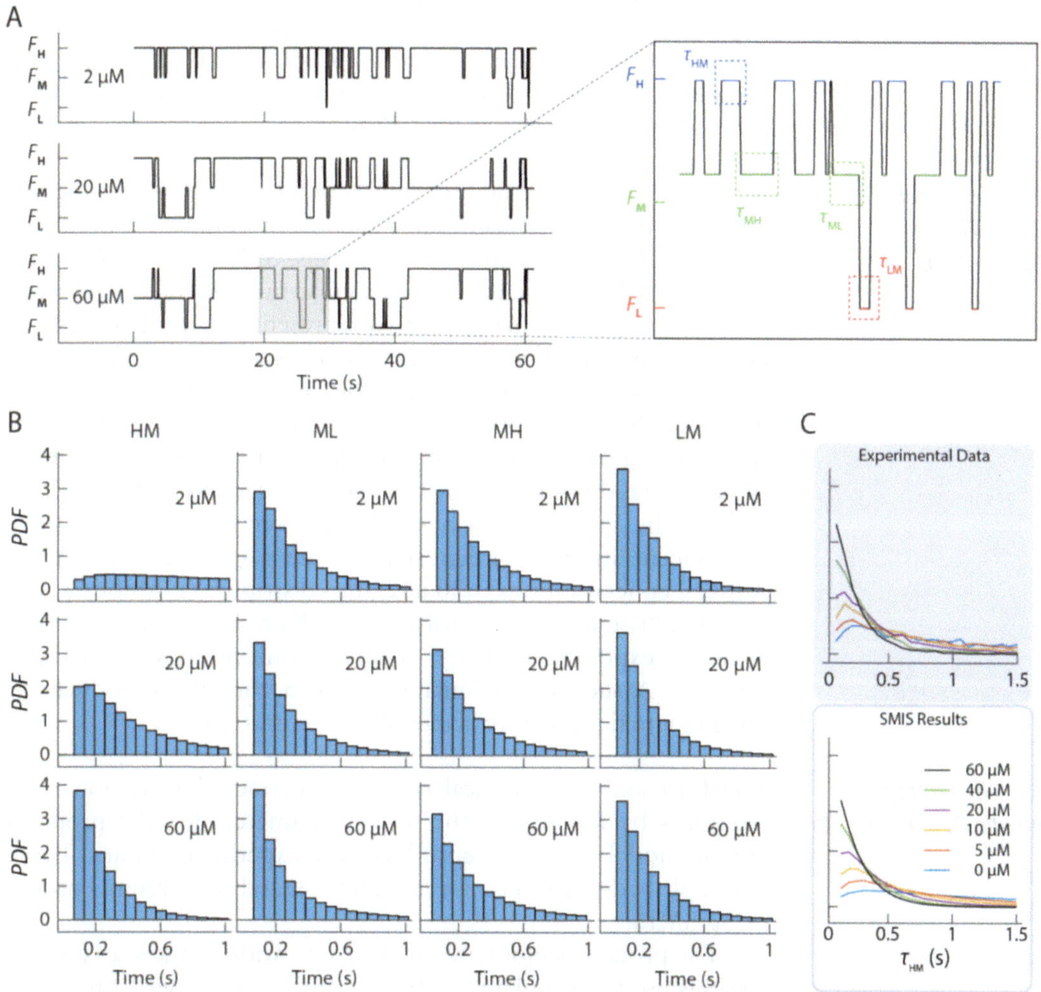

Fig. 3 SMIS-generated trajectories and dwell-time distributions for the MMAB-AdoCbl model. (**a**) Examples of SMIS-generated, noise-free fluorescence trajectories at 2, 20, and 60 μM AdoCbl, showing transitions between the F_H, F_M, and F_L states. The inset defines the notation for dwell times between states (e.g., τ_{HM}, τ_{MH}, τ_{ML}, and τ_{LM}). (**b**) Simulated dwell-time distributions (*PDFs*) for the four transitions (HM, ML, MH, LM) at three different AdoCbl concentrations. Note that the simulated PDF_{HM} at 2 μM correctly reproduces the rise-and-decay pattern observed experimentally. (**c**) The comparison between the experimental τ_{HM} distributions (top) and the best-fit SMIS simulation results (bottom) across a range of AdoCbl concentrations. The excellent agreement demonstrates the extracted parameters from SMIS faithfully capture the kinetic features of experimental results. (Data in (**b**) and (**c**) reproduced from Ref. [30] with permission from Elsevier)

the final fluorescence trajectories (Fig. 3a). To ensure a valid comparison, the time resolution of the simulation can be adjusted to match the experimental imaging timescale (i.e., camera integration time). This entire simulation process is repeated for each substrate concentration that was experimentally tested, for instance, across a range of 0–60 μM AdoCbl.

With proper kinetic rate constants, SMIS will generate a large dataset of simulated trajectories that visually reproduce key experimental features. As shown in the experimental data, the simulated trajectories will exhibit distinct features depending on the ligand concentration; at higher concentrations of AdoCbl, the trajectories show more frequent downward fluctuations from the F_H state as binding events become more common.

3.3 Generating Dwell-Time Distributions

After generating thousands of single-molecule trajectories, the next step is to perform a quantitative analysis by simulating the dwell-time distributions for specific transitions. SMIS extracts the duration the molecule spends in a particular fluorescence state before transitioning to another (e.g., τ_{HM} represents the time spent in F_H before transitioning to F_M, Fig. 3b). These individual dwell times are compiled into histograms and normalized to generate probability density functions for this specific transition (PDF_{HM}). To ensure the resulting distribution is statistically robust, a sufficient number of individual dwell times must be collected (*see* **Note 1**).

For the MMAB-AdoCbl system, the key feature to simulate is the concentration-dependent shape of the τ_{HM} distribution. At low AdoCbl concentrations (<20 μM), this distribution exhibits a characteristic "rise-and-decay" pattern, which gradually transforms into a single exponential decay at high concentrations (>40 μM, Fig. 3c). A successful simulation must accurately reproduce this distinct kinetic signature across all concentrations.

3.4 Extracting Kinetic Rate Constants

The final and most critical step is to extract the unknown rate constants by identifying the set of parameters for the proposed kinetic model that best reproduces the experimental data. This is accomplished through an iterative search-and-refine process using SMIS.

The process begins with an extensive, wide-range search where SMIS is used to simulate kinetic outputs (such as dwell-time *PDF*s and state populations) for thousands of different sets of rate constants, with parameters often spanning several orders of magnitude (e.g., 0.01–1000 s^{-1}). For each simulation, a goodness-of-fit score, such as the average percent residue standard deviation (% RSD), is calculated to quantify the deviation between the simulated and experimental data. Next, the simulations are screened, and only the top-performing sets. Those that provide the best fit to the experimental data (e.g., the top 10% with the lowest % RSD) are selected. The rate constants from this elite group are then used to define a new, much narrower search range for a subsequent round of simulations. This iterative process of simulating, screening, and narrowing the parameter space is repeated until the values for the unknown rate constants converge. This successful convergence validates the kinetic model and yields the most probable values and confidence bounds for the previously inaccessible kinetic parameters.

By systematically adjusting parameters such as k_1^i, k_{-1}^i, and k_1, a unique set of rate constants is identified that allows the simulated *PDFs* to provide the best quantitative fit to the experimental PDFs across all tested concentrations (Fig. 3c). This successful match, demonstrated by the overlay of the simulated and experimental data, validates the proposed conformation-gated binding model and provides the quantitative values for the previously inaccessible kinetic steps. Beyond the analysis of a single interaction, this framework can be extended to dissect more complex biological questions, such as comparing the kinetic mechanisms of a protein interacting with multiple distinct ligands (*see* **Note 2**).

4 Notes

1. To verify the accuracy of SMIS-generated trajectories, we recommend comparing simulated dwell-time distributions with those derived analytically from a simplified two-state kinetic model. This approach ensures that the SMIS platform correctly captures the stochastic transition behavior encoded in the kinetic model.

   In our validation test, we used the Michaelis-Menten enzyme kinetic model (Fig. 4a), a special case of a more general two-state model, where some reverse rate constants are set to zero. By solving the differential equations for the model analytically, we obtained expressions for the dwell-time *PDFs*, average transition rates, and equilibrium state populations. We then used SMIS to simulate the same model with identical rate constants. Figure 4 demonstrates the excellent agreement between the SMIS-generated outputs (blue bars/circles) and the analytical predictions (red curves) for the dwell-time distributions (Fig. 4b), average transition rates (Fig. 4c), and relative state populations (Fig. 4d). This benchmarking process demonstrates that SMIS faithfully reproduces the expected kinetics and can be used confidently to analyze more complex systems where analytical solutions are not available.

   Additionally, we assessed the number of dwell-time events required to achieve statistically robust recovery of kinetic parameters. Simulations revealed that fewer than 250 dwell-time events are typically sufficient to recover transition rates and

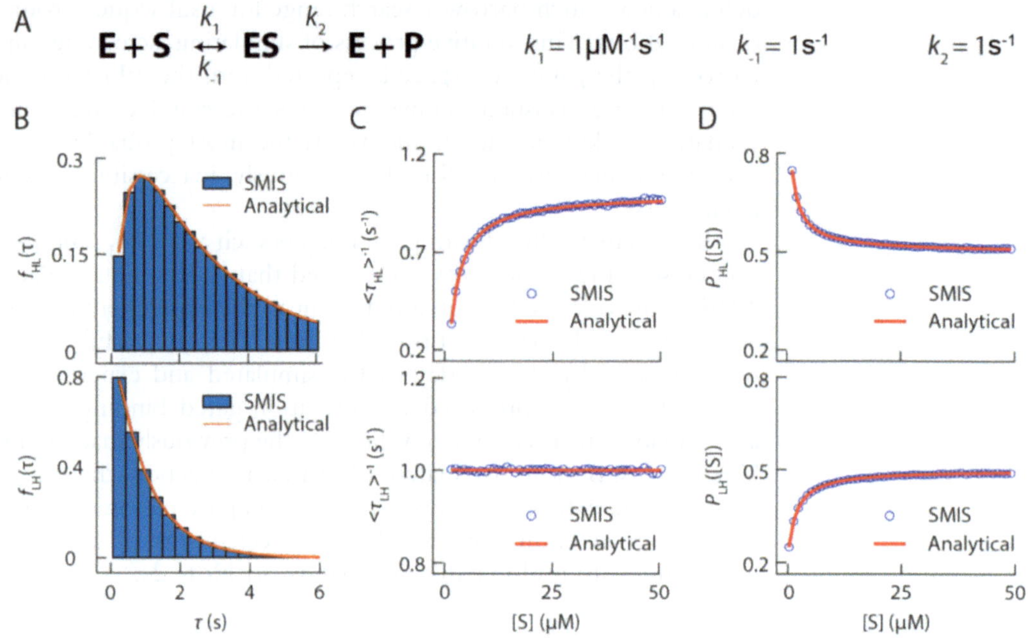

Fig. 4 Benchmarking SMIS accuracy against a known analytical solution. (**a**) The Michaelis-Menten kinetic scheme and kinetic rate constants used as the ground-truth model for the benchmark test. (**b**) Comparison of the dwell-time probability density functions (PDF_{HL} and PDF_{LH}), showing excellent agreement between the SMIS simulation and the analytical function. (**c**) Comparison of the average transition rates ($<\tau_{HL}>^{-1}$ and $<\tau_{LH}>^{-1}$) as a function of substrate concentration. (**d**) Comparison of the relative populations of the high (P_{HL}) and low (P_{LH}) fluorescence states as a function of substrate concentration. (Data reproduced from Ref. [28]. Copyright 2020 American Chemical Society)

PDFs within 10% error, offering practical guidance for experimental design and data collection.

2. The SMIS framework is adaptable to a broad spectrum of biological systems and can be extended beyond analyzing a single interaction to comparing the kinetic mechanisms of a protein with multiple, distinct ligands. For example, in our previous work, SMIS was applied to compare the binding kinetics of MMAB with two different vitamin B_{12} derivatives: the product AdoCbl and a substrate analog, hydroxocobalamin (OHCbl) [30, 31].

   This comparative analysis is highlighted in Fig. 5. The analysis first required a re-evaluation of the fluorescence state assignments, as OHCbl's different quenching efficiency led to a different mapping of species to states compared to AdoCbl (Fig. 5a). Kinetic analysis via SMIS revealed that while OHCbl binding also followed a conformation-gated mechanism, the rate constants for the initial conformational change ($k_1^i$and k_{-1}^i) were distinct from those observed for AdoCbl. Figure 5b shows the experimental dwell-time distributions for the MMAB-OHCbl interaction. This key finding, derived from

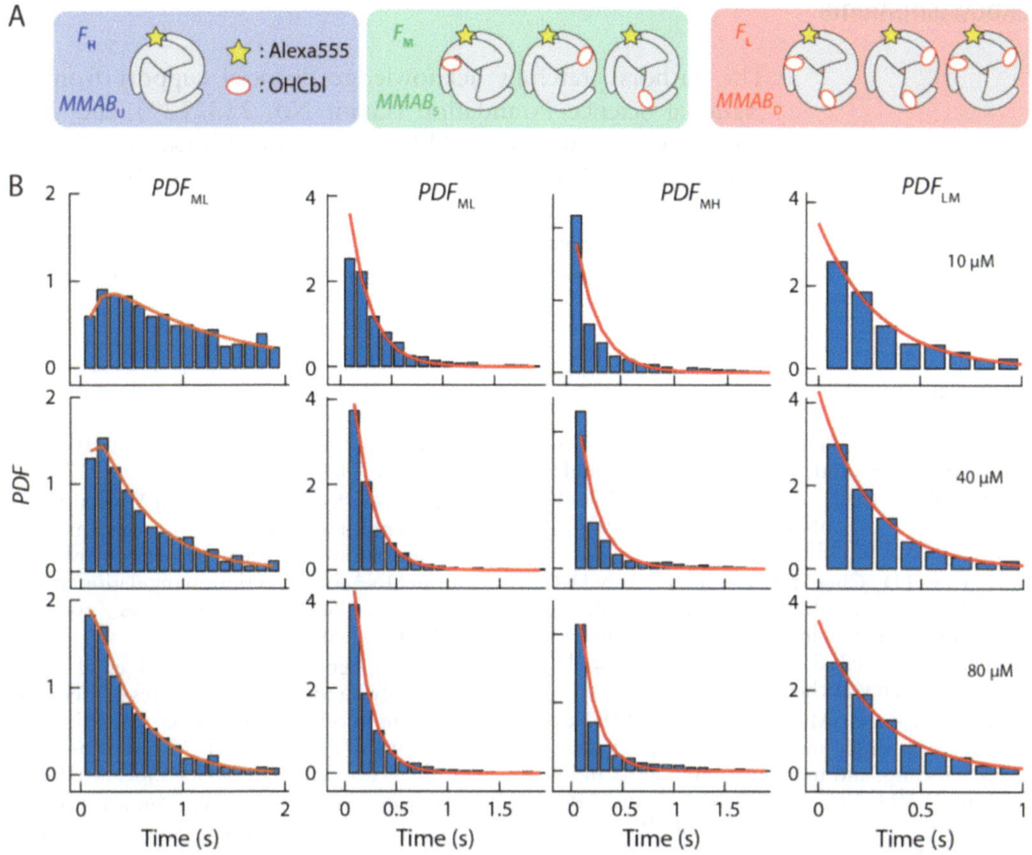

Fig. 5 Advanced application of SMIS for comparative analysis of MMAB-OHCbl interactions. (**a**) The re-evaluated state assignment map for MMAB-OHCbl. Due to OHCbl's higher quenching efficiency, all singly bound (*MMAB*$_S$) and doubly bound (*MMAB*$_D$) species are assigned to the F_M and F_L states, respectively, which differs from the AdoCbl assignment. (**b**) Experimental dwell-time distributions for four key transitions in the MMAB-OHCbl system. Notably, PDF_{ML}, PDF_{MH}, and PDF_{LM} show single-exponential decays that can be fit with analytical solutions (red curves), while PDF_{HM} retains the rise-and-decay feature, indicating a conserved but distinct conformation-gated binding mechanism. (Data reproduced from Ref. [30] with permission from Elsevier)

analyzing such data, suggested that the unbound MMAB enzyme exists in a dynamic equilibrium of multiple conformations and that different ligands preferentially select different preexisting protein conformers for their binding pathways. This example illustrates how SMIS can be used as an advanced tool to dissect complex conformational landscapes and elucidate how ligand-specific adaptations fine-tune molecular recognition and regulatory mechanisms.

Acknowledgments

The authors gratefully acknowledge financial support from the National Science Foundation (Grant No. 2237129), the Welch Foundation (Grant No. E-2229), the National Institutes of Health (Grant No. R35GM161380), and the University of Houston.

Conflict of Interest Statement The authors have no conflicts of interest to declare that are relevant to the content of this chapter.

References

1. Liu Z, Lavis LD, Betzig E (2015) Imaging live-cell dynamics and structure at the single-molecule level. Mol Cell 58:644–659. https://doi.org/10.1016/j.molcel.2015.02.033
2. Nguyen TD, Chen Y-I, Chen LH, Yeh H-C (2023) Recent advances in single-molecule tracking and imaging techniques. Annu Rev Anal Chem 16:253–284. https://doi.org/10.1146/annurev-anchem-091922-073057
3. Ha T, Kaiser C, Myong S et al (2022) Next generation single-molecule techniques: imaging, labeling, and manipulation in vitro and in cellulo. Mol Cell 82:304–314. https://doi.org/10.1016/j.molcel.2021.12.019
4. Chen H, Yan G, Wen M-H et al (2024) Advancements and practical considerations for biophysical research: navigating the challenges and future of super-resolution microscopy. Chem Biomed Imaging 2:331–344. https://doi.org/10.1021/cbmi.4c00019
5. Schönfelder J, De Sancho D, Perez-Jimenez R (2016) The power of force: insights into the protein folding process using single-molecule force spectroscopy. J Mol Biol 428:4245–4257. https://doi.org/10.1016/j.jmb.2016.09.006
6. Yang F, Chen T-Y, Krzemiński Ł et al (2016) Single-molecule dynamics of the molecular chaperone trigger factor in living cells. Mol Microbiol 102:992–1003. https://doi.org/10.1111/mmi.13533
7. Jung W, Chen T-Y, Santiago AG et al (2024) Memory effects of transcription regulator–DNA interactions in bacteria. Proc Natl Acad Sci 121:e2407647121. https://doi.org/10.1073/pnas.2407647121
8. Nguyen D, Yan G, Chen T-Y et al (2023) Variations in intracellular organometallic reaction frequency captured by single-molecule fluorescence microscopy. Angew Chem Int Ed 62:e202300467. https://doi.org/10.1002/anie.202300467
9. Chen T-Y, Santiago AG, Jung W et al (2015) Concentration-and chromosome-organization-dependent regulator unbinding from DNA for transcription regulation in living cells. Nat Commun 6:7445. https://doi.org/10.1038/ncomms8445
10. Xie X, Cheng Y-S, Wen M-H et al (2018) Quantifying the oligomeric states of membrane proteins in cells through super-resolution localizations. J Phys Chem B 122:10496–10504. https://doi.org/10.1021/acs.jpcb.8b10402
11. Santiago AG, Chen T-Y, Genova LA et al (2017) Adaptor protein mediates dynamic pump assembly for bacterial metal efflux. Proc Natl Acad Sci USA 114:6694–6699. https://doi.org/10.1073/pnas.1704729114
12. Rust MJ, Bates M, Zhuang X (2006) Sub-diffraction-limit imaging by stochastic optical reconstruction microscopy (STORM). Nat Methods 3:793–796. https://doi.org/10.1038/nmeth929
13. Betzig E, Patterson GH, Sougrat R et al (2006) Imaging intracellular fluorescent proteins at nanometer resolution. Science 313:1642–1645. https://doi.org/10.1126/science.1127344
14. Schuler B, Lipman EA, Eaton WA (2002) Probing the free-energy surface for protein folding with single-molecule fluorescence spectroscopy. Nature 419:743–747. https://doi.org/10.1038/nature01060
15. Kapanidis AN, Margeat E, Ho SO et al (2006) Initial transcription by RNA polymerase proceeds through a DNA-scrunching mechanism. Science 314:1144–1147. https://doi.org/10.1126/science.1131399
16. Roy R, Hohng S, Ha T (2008) A practical guide to single-molecule FRET. Nat Methods

5:507–516. https://doi.org/10.1038/nmeth.1208

17. Martell DJ, Joshi CP, Gaballa A et al (2015) Metalloregulator CueR biases RNA polymerase's kinetic sampling of dead-end or open complex to repress or activate transcription. Proc Natl Acad Sci USA 112:13467–13472. https://doi.org/10.1073/pnas.151523111
18. Moerner WE, Shechtman Y, Wang Q (2015) Single-molecule spectroscopy and imaging over the decades. Faraday Discuss 184:9–36. https://doi.org/10.1039/C5FD00149H
19. Ha T (2001) Single-molecule fluorescence resonance energy transfer. Methods 25:78–86. https://doi.org/10.1006/meth.2001.1217
20. Lerner E, Cordes T, Ingargiola A et al (2018) Toward dynamic structural biology: two decades of single-molecule Förster resonance energy transfer. Science 359:eaan1133. https://doi.org/10.1126/science.aan1133
21. McKinney SA, Joo C, Ha T (2006) Analysis of single-molecule FRET trajectories using hidden Markov modeling. Biophys J 91:1941–1951. https://doi.org/10.1529/biophysj.106.082487
22. Gopich IV, Chung HS (2022) Theory and analysis of single-molecule FRET experiments. In: Muñoz V (ed) Protein folding: methods and protocols. Springer, New York, pp 247–282
23. Barth A, Opanasyuk O, Peulen T-O et al (2022) Unraveling multi-state molecular dynamics in single-molecule FRET experiments. I. Theory of FRET-lines. J Chem Phys 156:141501. https://doi.org/10.1063/5.0089134
24. Lerner E, Barth A, Hendrix J et al (2021) FRET-based dynamic structural biology: challenges, perspectives and an appeal for open-science practices. eLife 10:e60416. https://doi.org/10.7554/eLife.60416
25. Sasmal DK, Pulido LE, Kasal S et al (2016) Single-molecule fluorescence resonance energy transfer in molecular biology. Nanoscale 8:19928–19944. https://doi.org/10.1039/C6NR06794H
26. Hadzic MCAS, Börner R, König SLB et al (2018) Reliable state identification and state transition detection in fluorescence intensity-based single-molecule Förster resonance energy-transfer data. J Phys Chem B 122:6134–6147. https://doi.org/10.1021/acs.jpcb.7b12483
27. Nir E, Michalet X, Hamadani KM et al (2006) Shot-noise limited single-molecule FRET histograms: comparison between theory and experiments. J Phys Chem B 110:22103–22124. https://doi.org/10.1021/jp063483n
28. Pan M, Zhang Y, Yan G et al (2020) Dissection of interaction kinetics through single-molecule interaction simulation. Anal Chem 92:11582–11589. https://doi.org/10.1021/acs.analchem.0c01014
29. Chen T-Y, Yan G, Fernandez LB (2025) MATLAB software and data for analysis of complex single-molecule kinetics: a protocol for single-molecule interaction simulations (SMIS). Figshare. https://doi.org/10.6084/m9.figshare.29813987.v1
30. Yan G, Pan M, Keller AM et al (2025) Conformation-gated binding underlies kinetic asymmetry and negative cooperativity in ATP:cob(I)alamin adenosyltransferase. Cell Rep Phys Sci 6:102768. https://doi.org/10.1016/j.xcrp.2025.102768
31. Chen T-Y, Yan G (2025) MATLAB software and data.zip for Conformation-gated binding underlies kinetic asymmetry and negative cooperativity in ATP:cob(I)alamin adenosyltransferase. Figshare. https://doi.org/10.6084/m9.figshare.28791878

Chapter 4

Prompt Engineering for Organic Chemistry Education: A Digital Loom-Based Approach

Wilton José Diolindo do Nascimento Júnior, Mayara de Carvalho Santos, Murilo Nícolas Mobelli, Paulo C. M. de Lacerda Miranda, and Gildo Girotto Junior

Abstract

The integration of generative artificial intelligence systems in higher education has created debates about their teaching effectiveness, especially in subjects that need specific scientific formalism, like organic chemistry. This work investigates how different complexity levels in prompt structure influence the quality of generative AI system responses in specific organic chemistry topics. Using the digital loom analogy as a theoretical framework, we compared the performance of two AI systems (ChatGPT 5 and GPT IQQO assistant) in tasks involving the Cannizzaro reaction and structure-reactivity relationship of the guanidine group. Through content analysis of 494 coded segments, we found that both systems showed similar performance (92.8% vs. 93.1% conceptual adequacy), regardless of prompt complexity. The "orbital interactions" category emerged as the most vulnerable area (88.0% adequacy), revealing limitations shared between AIs and students. Computational vision analysis of 117 meaning units showed complementarity between systems, while scientific image generation revealed critical tension between visual appeal and conceptual rigor. Results indicate that user competence in prompt construction is more important than AI architecture, establishing practical guidelines for effective integration of these technologies in chemistry education.

Key words Artificial intelligence, Prompt engineering, Chemistry education, Organic chemistry, Educational technology, Large language models

1 Introduction

The integration of generative artificial intelligence systems in higher education has created debates about their effectiveness and teaching applications. In organic chemistry, representing molecular structures, reaction mechanisms, and abstract concepts like orbital interactions creates recognized teaching challenges [1, 2], making AIs potentially transformative tools.

Robert E. Dempski (ed.), *Transforming Molecular Biology with Emerging Technologies*, Methods in Molecular Biology, vol. 3043,
https://doi.org/10.1007/978-1-0716-5320-3_4,

However, recent studies reveal a critical gap: although these tools show impressive abilities in generating chemical content, their effectiveness depends mainly on user competence in building adequate prompts [3, 4]. This dependence creates an educational paradox: to effectively use AIs as learning support tools, students and teachers need to master a new competence called prompt engineering, which still lacks specific guidelines for chemical education contexts.

Important

The central premise of this work is that, regardless of algorithmic sophistication or the pace of technological advancement, certain human skills will never become obsolete. The first of these fundamental competencies for the AI era is prompt engineering. The ability to communicate effectively with AI systems through structured command construction is not merely a technical skill but represents a new form of literacy that must be cultivated. In chemistry education, this means knowing how to formulate questions that account for the specificities of chemical formalism and the conceptual limitations of artificial systems.

Equally important is the development of critical thinking and conceptual validation. The ability to critically evaluate AI-generated responses becomes even more crucial as these tools become increasingly convincing and employ persuasive language. In the specific chemical context, this involves the capacity to identify inconsistencies in molecular representations, detect conceptual errors in reaction mechanisms, and recognize limitations in explanations of abstract phenomena.

Paradoxically, effective mastery of AI tools requires strengthening human intellectual autonomy. From our perspective, fostering human creativity is a cornerstone of education, as AI should function as an assistant to human thought, not as its replacement. This approach preserves the creative capacity and critical discernment that characterize the authentic educational process, ensuring that technology serves pedagogical purposes without compromising students' intellectual development.

1.1 Research Problem

The central problem investigated in this work is how different complexity levels in prompt structuring influence the quality of generative AI system responses in specific organic chemistry topics and what error patterns emerge in this process.

This question breaks down into specific objectives: (1) compare the performance of two generative AIs in organic chemistry tasks;

(2) identify chemical knowledge categories more vulnerable to AI failures; (3) establish relationships between prompt complexity and conceptual adequacy of responses; and (4) propose specific guidelines for building prompts in organic chemistry.

1.2 Justification

While literature on prompt engineering addresses general principles [5, 6], there is a shortage of studies that specifically investigate their applications in chemistry education. Existing research focuses on broad ChatGPT evaluations in science education [3, 7] or generic AI tool analyses [8], without addressing the particularities of chemical representation and the specific scientific formalism of this discipline.

Current literature on educational GenAI concentrates mainly on general applications or disciplines like medicine and engineering [9, 10], leaving a significant gap in understanding how AI systems respond to the specificities of chemical knowledge. This gap is particularly problematic considering that chemistry requires specific visual languages (molecular structures, orbital diagrams, and reaction mechanisms) that differ substantially from other knowledge areas [11].

Additionally, existing studies on prompt engineering tend to be oriented by technical or commercial applications [12, 13], not considering the specific pedagogical implications of the educational context. The need for teaching supervision [14] and critical validation of AI-generated content [3] remains inadequately addressed in specialized literature.

This work lies in proposing a **structuring analogy** that conceptualizes prompt construction as a digital weaving process, where different elements (instruction, context, input data, output format) function as threads that must be interwoven by the user to produce "conceptual fabrics." This framework not only didactically organizes the prompt engineering process but also offers a theoretical base for understanding why user competence is determining for result quality.

Unlike technical approaches that focus on parameter optimization [4], our digital loom analogy emphasizes the human aspect of AI interaction, positioning the user as a digital craftsperson who must master specific techniques to produce educationally valid results. This humanistic perspective of educational technology represents an original theoretical contribution to the emerging field of educational AI.

2 Theoretical Foundation

Artificial Intelligence, as a computer science field, dedicates itself to developing systems capable of executing tasks typically performed by human intelligence [15–17]. The central concept of AI lies in creating machines that simulate human cognitive processes, such as learning, reasoning, perception, and decision-making.

2.1 Human Cognitive Processes and AI Limitations

Human cognitive processes are characterized by flexibility, abstraction, consciousness, and intentionality [16, 18]. Flexibility represents the capacity to adapt to new situations and learn from experience. Abstraction is a cognitive process of forming concepts and generalizations. Consciousness is the perception of one's own thinking and environment by the sentient individual. Intentionality is evidenced by the capacity to direct oneself toward specific objectives.

These processes involve acquiring, processing, storing, and mobilizing information to interact with the world, requiring a complex mental framework that includes memory, attention, language, and executive functions [15, 18–20]. Neurologically, they involve distributed neural networks in the brain, with specialized but highly interconnected areas [19, 21].

2.2 AI Architectures and Approaches

AI functioning is based on complex algorithms, mathematical models, and instruction sequences that allow machines to analyze large data volumes, identify patterns, learn from experience, and adapt to new situations [15, 16, 22].

The various AI approaches include machine learning, artificial neural networks, natural language processing, and computer vision. Machine learning subdivides into supervised, unsupervised, and reinforcement learning [22, 23]. Artificial neural networks, inspired by the human brain, consist of layers of electronic structures that imitate interconnected neurons [24–26].

AI has equivalents to memory (data storage), attention (focus on specific tasks), and language (natural language processing) [27–30]. However, typically human executive functions—planning, creativity, and complex decision-making—are not adequately mimicked. Fundamentally, AI lacks the intentionality and consciousness that guide human cognition, imitating aspects of cognition without possessing the depth and complexity of the human mind fabric [16, 31, 32].

2.3 Transformative Potential and Educational Applications

AI has transformative potential that extends across various sectors, promising to optimize processes and inaugurate new forms of interaction and work [33]. In education, intelligent tutoring systems can provide individualized feedback, identify areas needing support, automate teaching tasks like evaluation and plagiarism detection [3, 8, 9, 34–36].

In scientific research, artificial intelligence systems assist in scientific writing processes [35], large data volume analysis, optimization, and experiment design, improving the investigative process [37].

2.4 Need for AI Education and Competences

Especially in scientific research and teaching areas, professionals must invest in continuing education focusing on training to use these tools ethically, critically, consciously, and efficiently [34, 38–40], including in Higher Education [10, 41].

Professionals who use AI systems must present critical thinking and data analysis skills to not become prisoners of algorithms [3, 39]. AI education must ensure responsible use of these technologies, avoid excessive dependence, and develop solid critical skills [39].

To make the most of AI, it is essential to develop specific skills. Technological literacy encompasses the capacity to use digital tools, including AI [34, 40–42]. Critical thinking is fundamental for analyzing and interpreting AI results, identifying biases and incorrect information due to the possibility of tool hallucination [43, 44].

2.5 Prompt Engineering: Effective AI Communication

Knowing how to communicate with AI is essential [45, 46]. Although it is an information search tool, it requires a different approach from traditional search engines, requiring mastery of prompt engineering—building commands that optimize AI potential and generate complete and coherent responses.

Tassoti [47] identified that students copied and pasted questions on AI platforms without considering quality, format, and content. This behavior can be attributed to habits in using search mechanisms and a lack of knowledge about AI communication. People believe that communication with intelligent systems is like human communication, which is more capable of perceiving nuances, subjectivities, and implicit contexts [45, 46].

Prompt engineering, concentrating on creating and optimizing instructions for LLMs (large language models), is essential to make the most of AI potential. A well-built prompt directs AI toward precise, relevant, and useful responses, acting as a bridge between human intention and AI capacity [4, 5, 13, 47–49].

The key elements of a prompt are instruction, context, input data, and output data, enriched with demonstrations and examples [5, 6, 49]. To build effective prompts, users must be specific, use clear language, provide context, divide tasks, use examples, and experiment with the tool [50].

Main errors include ambiguity, insufficient context, inadequate ethical considerations, bias reinforcement, conflicting instructions, overfitting, and unrealistic expectations [5, 6, 51, 52].

2.6 Iteration and Refinement Process

Iteration refers to the process of refining and adjusting instructions provided to the tool, based on initial responses obtained. The user submits an initial prompt, evaluates the generated response, and modifies it with details, examples, or restrictions to obtain a more precise response [49, 53]. This repetition allows gradual refinement, exploring different approaches, and adjusting AI direction.

2.7 Specific Challenges in Chemistry and Biochemistry Education

Chemistry education presents intrinsic challenges due to the disciplinary complexity that integrates concepts from multiple areas. This interdisciplinary nature, despite being essential, frequently makes learning difficult, especially in establishing connections between different organizational levels, representation, and used languages [1, 2, 11].

In biochemistry, which unites biology and chemistry, many students find difficulties in using chemical principles to explain biological phenomena. It is important to identify prior knowledge about biology and chemistry to develop more effective teaching strategies [54]. To mitigate this problem, literature suggests content organization, emphasis on key concepts [55], and use of conceptual maps [56].

Research indicates that biochemistry students frequently maintain alternative conceptions [57, 58], even after formal teaching and graduation advancement. These conceptions can be used as a base for developing teaching strategies [59]. Common examples are linked to chemical bond energy and helical protein structure [60]. Difficulties may be related to transferring fundamental general chemistry concepts to biological contexts [58].

Main difficulties are related to understanding metabolic intermediate concentration over time, metabolic pathway interconnection, the intermolecular interaction role in biological systems, macromolecular structure formation thermodynamics, and predicting biochemical reaction spontaneity based on the Gibbs free energy concept [61].

Additional difficulties are associated with a lack of concept contextualization in biochemistry teaching [61] and the use of traditional methodologies [57, 59, 61, 62].

These difficulties are intensified by the need to master chemical formalism—a specific set of conventions for representing molecules, electrons, and chemical transformations [1]. Literature indicates that students frequently maintain alternative conceptions even after formal teaching [57], especially in topics involving abstract representations of molecular phenomena.

2.8 AI in Chemistry Education: Potential and Challenges

Innovative pedagogical and methodological approaches are necessary for chemistry education, and artificial intelligence emerges as a promising tool to overcome teaching and learning challenges. Artificial intelligence systems can answer questions involving high-level reasoning, as required in chemistry studies, showing the potential to support learning complex themes.

Using AI in teaching-learning processes stimulates active learning through dynamic interaction between student and machine [7, 63, 64]. AI can expand access to complex scientific literature, provide concise explanations about chemical processes and concepts, assist in genetic sequence interpretation and disease marker identification [37].

However, pointed-out challenges revolve around the need to validate responses generated by AIs and teaching supervision [7, 14]. Systems demonstrate impressive capabilities but present critical limitations: (a) tendency to "hallucination," generating incorrect but convincing information [43, 44]; (b) difficulty with concepts requiring three-dimensional spatial reasoning; and (c) inconsistency in applying chemical formalism [4].

These limitations make it essential for users to develop specific competencies, particularly in building prompts that maximize precision and minimize conceptual errors. AI should help users organize their thoughts, providing data and structure characteristic of the user, always preserving human autonomy and originality.

3 Methods

3.1 The Digital Loom Analogy as Theoretical Framework

Effective prompt construction can be conceptualized through the analogy of a digital loom, where different essential elements interweave to produce conceptual fabrics. This analogy is not merely illustrative but provides a structuring framework to understand interdependencies between prompt components.

The instruction thread corresponds to warp threads that run vertically in the loom, providing fundamental structure. The instruction defines the prompt purpose and directs AI toward the desired result. Without this thread properly positioned, the entire conceptual fabric loses coherence.

The context thread complements the instruction, adding structural stability. Like secondary warp threads that give body to fabric, context provides necessary prior knowledge for adequate task understanding, creating the conceptual "background."

The input data thread represents the raw material processed by the system. Like cotton or wool, which will transform into fabric, input data is specific information that AI will use to build responses. The quality of this data determines the final product quality.

The output data thread defines the desired pattern and format. Like threads that determine finishing, they specify requirements for result presentation, functioning as structural limits.

The interdependence of these elements is crucial: final fabric quality depends not only on the individual excellence of each thread but also on how they interweave. Inadequate data can compromise a well-developed instruction thread, and ambiguous instruction can lead to unsatisfactory responses even with excellent data.

3.2 Content Selection and Justification

Two specific organic chemistry topics were selected based on empirical criteria: (1) **Cannizzaro reaction**, chosen due to associated chemical formalism complexity, involvement of abstract orbital factors, and documented difficulties for both students and AI systems, and (2) **structure-reactivity relationship of guanidine group**, selected for requiring integrated understanding of electronic effects, resonance stabilization, and acid-base property prediction.

These topics were identified in consultation with a specialist in the organic synthesis area from the State University of Campinas as representative of conceptual and formalism challenges that characterize organic chemistry education.

3.3 Experimental Design

Three complexity levels of prompts were developed for each topic:

1. **Beginner Prompt**: Containing only basic instruction and minimal input data.
2. **Contextualized Prompt**: Including instruction, input data, and specific context.
3. **Complete Prompt**: Incorporating all framework elements (instruction, context, input data, output format specification).

3.4 Evaluated AI Systems

ChatGPT 5: A generative AI system developed by OpenAI, representing a general-purpose architecture with broad access to scientific literature.

GPT IQQO Assistant: The GPT called IQ.QO assistant, built within ChatGPT, was created using a database carefully selected by professionals specializing in organic chemistry. This database comprises 20 academic references, encompassing specialized digital collections and renowned educational platforms. It is complemented by two essential publications: Advanced Organic Chemistry Part A: Structure and Mechanism and Writing Reaction Mechanisms in Organic Chemistry. In addition, there are virtual resources, including the Virtual Textbook of Organic Chemistry and the Libre Texts Chemistry platform.

It is important to highlight that the fundamental system configurations prevent searches on websites that are not part of its pre-established database. In this way, it prevents responses from being formulated based on dubious sources or simply those more accessible on the web, aiming to mitigate a frequent issue in traditional systems, which is the black box related to the sources that the AI uses to construct the response. However, this precaution does not eliminate the possibility of the tool providing conceptually incorrect information to students, as the phenomenon of hallucinations remains a possibility.

3.5 Analysis Protocol

Content analysis based on Bardin's methodology [65] was used, with adaptations for the chemical knowledge evaluation context. Bardin's content analysis involves three phases: (1) pre-analysis, where materials are organized and initial ideas are systematized; (2) exploration of the material, involving coding procedures and categorization; and (3) treatment of results, inference, and interpretation.

This methodology allows for systematic analysis of textual data through categorization of meaning units, enabling both quantitative and qualitative analysis of content. Responses were segmented into meaning units and categorized emergently during the iterative coding process. The categorization was made in MAXQDA software.

Conceptual Adequacy Criteria: For the Cannizzaro reaction: (a) correct use of chemical formalism, (b) adequate mention of orbital interactions, (c) precise description of electronic movement, and (d) correct characterization of reagents, products, and intermediates.

For the guanidine group: (a) adequate application of chemical formalism, (b) correct description of electronic movement, and (c) precise identification of stabilization effects in protonation.

Each criterion was evaluated dichotomously (satisfactory/unsatisfactory) based on a template developed by an organic chemistry specialist. Segments were considered "satisfactory" when they demonstrated adequate conceptual understanding and correct application of chemical formalism.

3.6 Computer Vision and Image Generation Analysis

To evaluate visual interpretation and generation capabilities, two complementary analyses were implemented. Computer vision analysis used a handwritten blackboard image containing comparisons between aldehydes, ketones, and alkenes in the Lewis acid-base theory context, allowing investigation of how AIs "unravel" already-elaborated conceptual fabrics, identifying textual, technical, and conceptual elements in visual representations.

Computer vision tools identify and interpret visual elements in images, converting them to textual information. ChatGPT and GPT IQQO assistant integrate textual processing and computer vision functionalities, including image generation capability via the DALL-E engine. However, for educational purposes in chemistry, we do not recommend using DALL-E for generating scientific images because its outputs lack the necessary rigor of scientific formalism for accurate technical representations.

Image generation analysis compared butadiene molecular orbital representations through two approaches: (a) an integrated DALL-E engine and (b) Python code for programmatic visualization. This comparison investigates the tension between visual appeal and scientific precision in producing educational materials.

Priority was given to requesting graphs, molecular representations, and reaction diagrams via Python code. This approach allows for independence from specific image generation engines and ensures greater precision in scientific visualizations by using programmatic plotting.

4 Results

4.1 Text Generation

Prompts of different complexity levels (beginner, intermediate, and advanced) were specifically developed to answer the following organic chemistry questions. The subsequent analysis evaluated how closely the AI responses matched the expected answer key for each question:

Question i In the Cannizzaro reaction, two equivalents of an aldehyde without -hydrogens, such as benzaldehyde, are transformed into an alcohol and a carboxylic acid in a basic medium. Normally, organic compounds are incapable of transferring a hydride from a C-H bond. However, this phenomenon occurs in the Cannizzaro reaction via a specific intermediate. Explain in detail the mechanism of the Cannizzaro reaction using the appropriate formalism for representing mechanisms in organic chemistry. Using your knowledge of hyperconjugative interactions, draw the reaction intermediate and the orbital interaction that facilitates the release of hydrogen as a "hydride ion."

Question ii The guanidine functional group is a relatively strong organic base, and its conjugate acid has a pKa of 13.6. Guanidine is present in the amino acid arginine as well as various other natural products. The protonation of guanidine occurs on the sp^2 nitrogen instead of the other two nitrogens. Explain this fact using resonance stabilization arguments.

For each question, three prompt levels were constructed following the digital loom framework: beginner prompts (containing only basic instruction and minimal input data), contextualized prompts (including instruction, input data, and specific context), and complete prompts (incorporating all framework elements: instruction, context, input data, and output format specification).

Analysis of 494 coded segments revealed how different conceptual thread combinations interweaved to form chemical knowledge fabrics. Data were distributed equally between systems: GPT 5 with 265 segments (53.6%) and IQQO with 229 segments (46.4%), covering the Cannizzaro reaction (266 segments, 53.8%) and guanidine group (228 segments, 46.2%). Table 1 shows the performance of each system analyzed.

Table 1
Conceptual performance by chemical knowledge category, showing the number of satisfactory segments and adequacy percentages for ChatGPT 5 and IQQO assistant in each analyzed domain

Category	ChatGPT 5			IQQO			Total
	Sat	Unsat	% adequacy	Sat	Unsat	% adequacy	*n*
Chemical formalism	120	8	93.8	104	6	94.5	238
Orbital interactions	22	3	88.0	22	3	88.0	50
Products and intermediates	35	4	89.7	30	3	90.9	72
Electron movement	28	2	93.3	26	2	92.9	58
Stabilization effects	38	2	95.0	34	2	94.4	76
General total	**243**	**19**	**92.8**	**216**	**16**	**93.1**	**494**
Prompt	ChatGPT 5			IQQO			
	Total	Sat	% adequacy	Total	Sat	% adequacy	
Cannizzaro beginner	33	30	90.9	30	28	93.3	
Cannizzaro intermediate	48	45	93.8	36	33	91.7	
Cannizzaro advanced	58	53	91.4	61	57	93.4	
Guanidine beginner	30	28	93.3	26	24	92.3	
Guanidine intermediate	28	26	92.9	22	20	90.9	
Guanidine advanced	68	61	89.7	54	50	92.6	

4.2 Category Distribution and Conceptual Thread Quality

Categorization identified five main domains: **Chemical formalism** emerged as the most representative category with 238 segments (48.2%), confirming its function as a conceptual fabric warp thread. **Stabilization effects** contributed 76 segments (15.4%), **products and intermediates** 72 segments (14.6%), **electron movement** 58 segments (11.7%), and **orbital interactions** 50 segments (10.1%).

This distribution reveals that the **instruction thread**, represented by chemical formalism, truly functions as a conceptual fabric warp, providing structure over which other elements interweave. The analysis showed that categories like **stabilization effects** and **products and intermediates** were major components of the AI's explanation. In contrast, **electron movement** and **orbital interactions** represented more specific, detailed concepts that, while less frequent, were essential for a complete response.

Conceptual adequacy evaluation demonstrated high performance in both systems, with an average adequacy rate of 92.9%. The IQOO system presented 93.1% adequacy (216 of 232 satisfactory segments) compared to 92.8% of GPT 5 (243 of 262 satisfactory segments).

Categories with greater robustness correspond to thicker and more resistant conceptual weaving threads—fundamental consolidated concepts. **Stabilization effects** showed greater consistency (94.5% adequacy), indicating that both systems were adequately based on thermodynamic concepts. **Orbital interactions** presented the greatest conceptual challenge (88.0% adequacy), representing delicate threads requiring refined "digital weaver" expertise.

Contrary to initial expectations, different prompt complexity levels produced equivalent average results: beginner (92.4%), intermediate (92.5%), and advanced (93.0%). These findings indicate that even simple loom configurations—beginner prompts with only instruction and input data—produce adequate conceptual fabrics.

This finding suggests that the user's ability to structure a clear and coherent prompt is more critical to the quality of the output than the technical sophistication of the prompt itself.

Analysis of 35 inadequate segments (7.1% of corpus) revealed specific conceptual weaving failure patterns. In **chemical formalism**, 14 inadequate segments (5.9% of the category) concentrated on incorrect use of curved arrows and ambiguous structural representations, equivalent to warp failures that compromise the entire fabric.

In **orbital interactions**, six inadequate segments (12.0% of the category) evidenced difficulties in applying orbital overlap concepts and molecular frontier theory, problems in delicate thread interweaving, where weaving precision is important.

These results suggest that both GPT 5 and IQQO possess an adequate conceptual foundation for educational applications in organic chemistry, with performance exceeding 90% in most of the categories analyzed. Comparable performance indicates a similar technological maturity between the systems, rendering both viable for implementation in educational contexts. Our data show that, although the overall performance was high, the limitations of AI emerge in topics of greater abstraction. This is in complete accordance with the premise that AI "mimics aspects of human cognition but lacks the depth and complexity of the fabric of the human mind," as stated by Hildt [31] and Teixeira and Gonzales [16]. The difficulty with orbital interactions, a subject that eludes direct representation, materializes this theoretical frontier.

The comparative analysis revealed that both AI models—the one that generates content by freely searching the web and the one that operates with a corpus of predefined sources—were capable of producing specific chemistry material with similar efficacy. The performance equivalence between the GPT 5 and IQQO systems (92.8% and 93.1% adequacy, respectively) is a fundamental result that shifts the focus from the AI's architecture to the user's competence. This finding directly echoes the warnings of Luckin [39] and Cooper [3], who argue that professionals must possess solid critical thinking and data analysis skills to avoid becoming "hostages to the work and outputs of algorithms." Our data suggests that a well-constructed prompt by the user is the true performance equalizer among different systems.

4.3 Computer Vision

This session presents the performance evaluation of two AIs—ChatGPT 5 and GPT IQQO assistant—in interpreting handwritten chemical content through computer vision capabilities. The object of study consisted of the analysis of a whiteboard containing comparisons between aldehydes, ketones, and alkenes, focusing on Lewis acid-base behavior and the interpretation of orbital diagrams. The choice of this specific content is justified by the conceptual complexity involved and the need for precise interpretation of visual elements such as molecular structures, orbital representations, and specialized chemical terminology.

The experimental design included three levels of prompt complexity: beginner, characterized by simple identification instructions; intermediate, incorporating structured context and specific guidelines; and advanced, implementing a multiphase protocol with systematic scanning, functional segmentation, and contextual pedagogical analysis. This progression allowed for the evaluation of not only the basic visual recognition capability but also the AI's response to different levels of instructional structure and complexity.

Visual interpretation capability evaluation investigated how AIs "unravel" already-elaborated conceptual fabrics, identifying elements in handwritten representations (see Fig. 1). 117 meaning units were identified, with 40.2% convergence between systems (47 elements recognized by both).

GPT IQQO demonstrated greater precision in identifying specific technical and conceptual threads (27 and 20 units), acting as a specialized evaluator that quickly recognizes conceptual warp patterns. **ChatGPT 5** showed greater sensitivity to decorative and

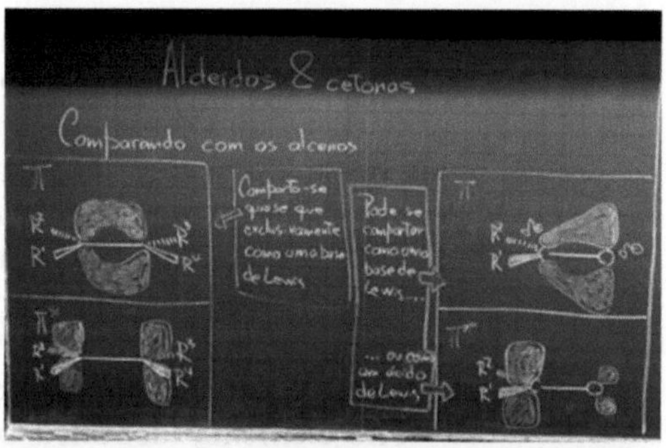

Fig. 1 Handwritten blackboard showing comparison between aldehydes, ketones, and alkenes in Lewis acid-base theory context. The image shows chemical representations drawn by hand with colored chalk, including molecular structures, orbitals, and explanatory notes about acid-base behavior

Table 2
Detailed categorization of meaning units identified in computer vision analysis, demonstrating the distribution of elements recognized by each AI system in different thematic categories

Category	Description	Representative example	US ChatGPT 5	US GPT IQQO	Total
Textual identification	Recognition and interpretation of manuscript texts, titles, specific chemical labels, and nomenclature	"Aldehydes and ketones," "comparison with alcohols"	8	8	16
Technical recognition	Identification of chemical elements: orbitals, groups, functional groups, notations	"Orbitals e *", "C=O group," "substituents R^1, R^2"	12	15	27
Conceptual interpretation	Understanding of fundamental chemical principles and theoretical explanation	"Acid-base behavior of Lewis," "carbonyls reactivity"	9	11	20
Visual elements	Perception of graphic aspects, colors, spatial organization, and visual representations	"Giz colored for orbitals," "vertical vs horizontal blocks"	13	8	21
Methodology	Structuring, sequencing, and protocols for analysis presentation	"Organization in topics," "structured multimodal protocol"	6	9	15
Contextualization	Adequacy to frameworks, references, and pedagogical material	"Supplementary study material"	7	11	18
Total			**55**	**62**	**117**

contextual threads (21 units in Visual Elements), identifying aesthetic nuances of visual composition.

The present descriptive analysis is based on Bardin's thematic content analysis methodology, applied comparatively between ChatGPT 5 (OpenAI) and the GPT IQQO assistant. The study involved a systematic analysis of the responses from the two artificial intelligences when subjected to three levels of prompt complexity: beginner, intermediate, and advanced, focusing on the visual interpretation of aldehydes and ketones in organic chemistry.

The elements identified by the AIs were organized into six emergent thematic categories, each representing specific aspects of visual interpretation in organic chemistry (see Table 2).

The comparative analysis between the GPT IQQO assistant and ChatGPT 5 identified 117 units of meaning while interpreting handwritten visual chemical content. Of the total analyzed, 47 elements were recognized by both platforms, representing a 40.2% convergence in visual interpretation.

In the textual identification category, both AIs exhibited 100% agreement. They identically identified elements such as main titles, subtopics, and structural labels. This result indicates that the capability for handwritten text recognition remains consistent across the different systems evaluated.

The recognition of technical terms showed a numerical difference between the platforms: the IQQO assistant identified 15 units, whereas ChatGPT 5 identified 12. GPT IQQO demonstrated greater precision in identifying specific technical elements, including HOMO/LUMO energy analysis and specialized terminology like “electrophiles” and “nucleophiles.” In conceptual interpretation, GPT IQQO identified 11 units compared to 9 by ChatGPT 5, excelling in explanations of chemical reactivity and theoretical principles.

ChatGPT 5 showed superior performance in the visual elements category, identifying 13 units compared to GPT IQQO’s 8. This difference was manifested through the identification of colors, the physical context of the environment, and the detailing of graphic elements. The result suggests that ChatGPT 5 processes contextual visual information with greater detail, while the IQQO assistant focuses its responses exclusively on the chemical content.

In the methodological dimension, GPT IQQO identified nine units versus six from ChatGPT 5. GPT IQQO applied structured protocols, systematic segmentation, and formal pedagogical approaches. Regarding contextualizing capacity, GPT IQQO presented 11 units against ChatGPT 5’s seven, including academic references, scientific disclaimers, and adaptation to different educational levels.

The observed differences were concentrated in the structuring of the responses, not in the substantive content. The GPT IQQO assistant followed structured protocols, while ChatGPT 5 provided more direct explanations. This difference suggests that the choice of tools may vary according to the specific pedagogical objective.

The analysis demonstrated that ChatGPT 5 and the GPT IQQO assistant operate with complementary approaches to the visual interpretation of chemical content. The GPT IQQO assistant shows greater precision in technical and scientific aspects, whereas ChatGPT 5 offers greater detail in visual and practical contextual aspects.

The 47 elements shared between the platforms establish a common baseline of interpretation, while the 15 elements exclusive to GPT IQQO and the 8 elements exclusive to ChatGPT 5 represent specializations that can be applied according to specific educational needs.

This result indicates that the central issue shifts from selecting a single tool to defining appropriate educational application strategies. The results indicate that the effective integration of AI in chemistry teaching requires the development of educator competencies for the coordinated application of different technological specializations. Educational implementation demands a structured approach that preserves the teacher's role as a mediator of the educational process, utilizing the specializations of each tool according to defined pedagogical objectives.

The capability demonstrated by the AIs in identifying and interpreting visual chemical content establishes these tools as viable auxiliary resources for the educational process. However, effective application requires that teachers and students develop specific competencies: conceptual mastery of chemical content to validate the generated interpretations, the capacity for critical analysis of the responses provided by the AIs, and the ability to select the most suitable tool for the specific educational context. These competencies are fundamental prerequisites, as the goal of technological integration is not to replace the users' critical thinking but to provide support for organizing and structuring their own conceptual elaborations. The development of these competencies is, therefore, a necessary condition for AI technology to function as an amplifier of the teaching and learning process.

4.4 Image Generation

The comparative analysis of the image generation results reveals a significant dichotomy between visual appeal and scientific accuracy in the representation of butadiene's molecular orbitals (Fig. 2). Although the images generated by DALL-E (panels A and C) exhibit greater visual appeal and appear more sophisticated at first glance, they contain a series of fundamental conceptual errors that compromise their scientific validity.

In panel A (simple prompt), produced by DALL-E, inconsistencies are observed in the symmetry of the molecular orbitals and in the proper progression of the number of nodes. Panel C (detailed prompt), also generated by DALL-E, maintains the problems of the previous panel, adding representations of the atomic pz orbitals, which, although correct in their depiction of phases and lobes, do not follow the linear combination rules that govern the formation of molecular orbitals. The numbering of the nodes is present, but the visualization does not clearly demonstrate how these nodes emerge from the constructive and destructive interference of the atomic wave functions.

In contrast, the representations generated in Python (panels B and D), despite their apparent visual simplicity, demonstrate theoretical fidelity. Panel B includes the energy values (= ±1.618 and ± 0.618), which were not requested, demonstrating that the algorithm searched for a source or sources containing this information. The schematic representation of the orbital phases is correct,

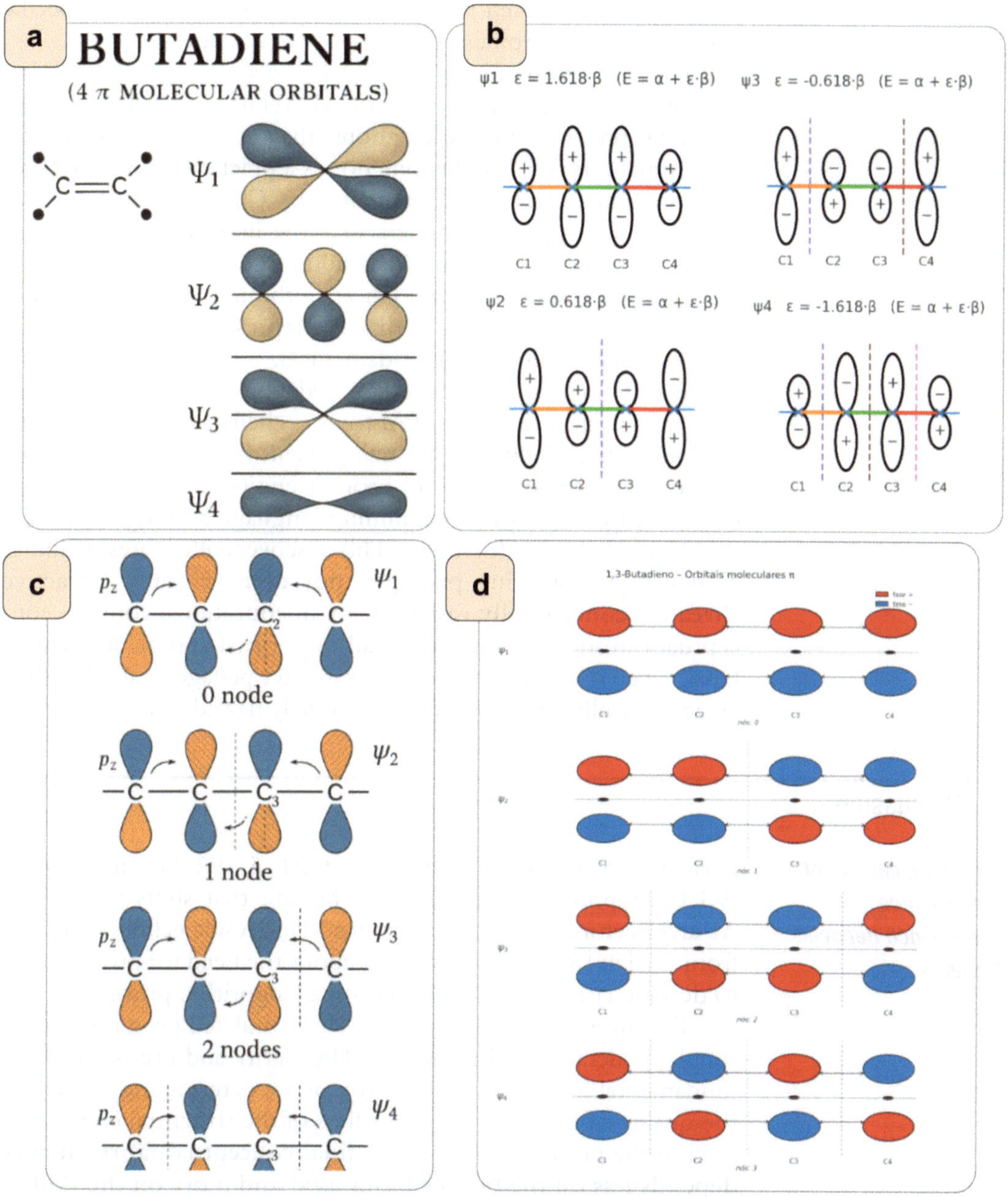

Fig. 2 Comparison between AI approaches for representing butadiene molecular orbitals. (**a** and **c**) Images generated by DALL-E showing visually attractive representations but with conceptual errors. (**b** and **d**) Representations produced in Python demonstrating scientific precision with correct energy values (= ±1.618 and ± 0.618) and adequate representation of orbital phases

showing the appropriate alternation of signs that characterizes each energy level. Panel D offers a systematic and theoretically consistent visualization of the linear combinations of atomic orbitals, correctly representing how each molecular orbital results from the specific overlap of the pz orbitals of the four carbon atoms.

The generation of images of butadiene's molecular orbitals is a flagrant case study on the risks of AI. While DALL-E produced visually sophisticated representations, they contained fundamental conceptual errors. This phenomenon is a practical example of tool hallucination, a problem documented by Chakraborty et al. [43] and Ji et al. [44], and it underscores the importance of the user never uncritically accepting AI results. On the other hand, the programmatic approach in Python, although requiring greater technical knowledge, ensures that the representations are aligned with established theoretical fundamentals.

This discrepancy has important pedagogical implications, suggesting that the integration of AI tools in chemistry education must be done critically, always prioritizing scientific accuracy over visual impact. Furthermore, beyond prompt engineering, some knowledge, even if basic, of programming language and logic can be a recommended asset for users. This discovery illustrates a fundamental digital weaving principle: **not every visually attractive thread is conceptually resistant**. In an educational context, conceptual thread quality must prevail over visual appeal. A scientific fabric, even if visually simple, offers greater pedagogical value than an aesthetically attractive but conceptually flawed weave.

5 Discussion

5.1 Implications of Performance Equivalence Between Systems

Performance equivalence between GPT 5 (92.8%) and IQQO (93.1%) represents a fundamental finding that shifts focus from AI architecture to user competence. This result echoes warnings from Luckin [39] and Cooper [3] about the need for professionals to develop critical skills to not become "algorithm prisoners."

The investigation shows that well-formulated prompts, even if simple, generate good responses. The clarity and precision of the fundamental instruction in the prompt are determining factors for the quality of the response, regardless of the structural complexity level of the prompt. In summary, **final conceptual fabric quality depends less on the type of loom used and more on the skill of who operates the equipment.**

The **orbital interactions** category emerged as the area of greatest vulnerability (88.0% adequacy), revealing theoretical AI limitations that mirror student difficulties. This convergence of limitations between machines and humans suggests that abstract concepts represent fundamental cognitive challenges that transcend the natural-artificial distinction.

Interestingly, this AI difficulty materializes the theoretical frontier proposed by Hildt [31] and Teixeira and Gonzales [16]: AIs imitate aspects of human cognition but do not possess genuinely human abstraction capacity. Concepts that escape direct representation—like orbital interactions—constitute critical tests of this limitation.

Performance stability through different complexity levels (beginner, 92.4%; intermediate, 92.5%; advanced, 93.0%) contradicts expectations that more elaborate prompts necessarily produce superior results. This finding suggests that **clarity and precision of fundamental instruction are more determining than structure sophistication**.

This discovery has important practical implications: beginner users can obtain satisfactory results without mastering advanced prompt engineering techniques, provided they understand basic principles of clear request structuring.

About scientific representation, image generation analysis revealed a critical risk: tools that produce visually attractive representations may contain fundamental conceptual errors. This phenomenon represents a clear manifestation of hallucination documented by Chakraborty et al. [43], where systems generate convincing but factually incorrect content.

Integration of AIs in educational material production must always prioritize scientific precision over visual impact. This discovery has direct implications for pedagogical practices: teachers must develop competencies for the critical validation of AI-generated content, especially complex visual representations. Based on identified patterns, specific guidelines are proposed for building prompts in organic chemistry:

1. **Prioritize fundamental instruction clarity** over structural complexity.
2. **Include specific chemical terminology** in input data to direct AI search.
3. **Explicitly request the correct chemical formalism** application.
4. **Critically validate generated visual representations**, prioritizing conceptual precision.
5. **Use an iterative approach** for refining inadequate responses.

5.2 Study Limitations and Future Research

This study limits itself to two specific organic chemistry topics and two AIs, preventing broad generalizations. Qualitative analysis, although systematic, introduces potential subjectivity in categorization. Future studies should (a) expand analysis to other chemistry domains, (b) include measures of student learning impact, (c) develop quantitative metrics for prompt quality, and (d) investigate temporal effects on AI response consistency.

6 Conclusions

This work demonstrated that generative AIs function as **sophisticated digital looms** capable of helping organize complex conceptual threads in organic chemistry. Final conceptual fabric quality depends fundamentally on **human weaver competence**, that means, the capacity to select adequate threads, structure a clear warp, provide an appropriate context, and define precise output patterns.

Our results shift the focus of the discussion from a mere comparison between AI architectures, which is necessary as a starting point but becomes a superficial debate in the long run, to the imperative need for human training in prompt engineering and critical analysis. This corroborates the view that professionals must master these tools to avoid becoming dependent on the work of algorithms.

The comparative analysis of text generation demonstrated performance equivalence between a specialist chatbot (IQQO) and a generalist one (GPT 5), with adequacy rates exceeding 92% for both. This finding suggests that the equalizing factor lies in the quality of the command provided by the user, rather than in the specifics of the AI model. However, the limits of these tools became evident in domains of high abstraction. The marked difficulty both AIs had with the "orbital interactions" category provides evidence for the theoretical limitations of AI, which lacks the capacity for genuine human abstraction. Interestingly, the machine's difficulty mirrors the learning challenges of students who, as pointed out by Loertscher et al. (2014), also face barriers in understanding complex theoretical concepts.

Image generation emerged as a particularly elucidating case study on the risks of AI. The representations of butadiene's molecular orbitals created by the DALL-E engine, despite being visually attractive, contained fundamental conceptual errors, in contrast to the precision of the images generated via Python code. This phenomenon is a clear manifestation of the tool's hallucination, a risk documented in the literature by authors such as Chakraborty et al. [43] and Ji et al. [44]. This outcome reinforces the premise that the integration of AI tools into chemistry education must be done critically, prioritizing scientific accuracy over visual appeal.

These results, while clarifying, raise new and important questions that deserve future reflection and investigation. The effective integration of AI in education and scientific research compels us to consider the following points:

Question 1 If the human factor is the main differentiator, as our data suggests in line with Luckin [34], and considering that even a beginner prompt was able to generate responses with over 92%

adequacy, where does the true complexity of prompt engineering lie as AIs evolve? Is the essential skill in the clear and direct formulation of the initial instruction or in the iterative and refined process of dialogue with the tool for more complex tasks?

Question 2 We observed that AI manifests difficulties in abstract concepts like those that challenge students. How can we explore this characteristic? Should we view AI as a student whose errors help us diagnose critical points in the curriculum, or can we use it to develop new pedagogical strategies that directly address these conceptual barriers for both humans and machines?

In summary, this work demonstrates that generative AIs are viable and powerful auxiliary resources in the context of chemistry teaching, but they are not substitutes for human intellect and discernment. The priority for the academic and educational community must, therefore, be the development of competencies that allow users—be they teachers, researchers, or students—to interact with these technologies in a critical, ethical, and effective manner. The true revolution is not in the machine that answers, but in the human who learns to ask. The future of chemical education will not be determined by available tool sophistication, but by educators' and students' ability to become **conscious, critical, and competent digital weavers**.

Conflicts of Interest

The authors have no conflicts of interest to declare that are relevant to the content of this chapter

References

1. Johnstone AH (2009) Multiple representations in chemical education. Int J Sci Educ 31:2271–2273
2. Talanquer V (2022) The complexity of reasoning about and with chemical representations. JACS Au 2:2658–2669
3. Cooper G (2023) Examining science education in ChatGPT: an exploratory study of generative artificial intelligence. J Sci Educ Technol 32:444
4. Nascimento Júnior WJD, Morais C, Girotto Júnior G (2024) Enhancing AI responses in chemistry: integrating text generation, image creation, and image interpretation through different levels of prompts. J Chem Educ 101: 3767–3779
5. Giray L (2023) Prompt engineering with ChatGPT: a guide for academic writers. Ann Biomed Eng 51:2629–2633
6. White J, Fu Q, Hays S et al (2023) A prompt pattern catalog to enhance prompt engineering with ChatGPT. https://arxiv.org/abs/2302.11382
7. Luke WAN, Seow Chong L, Ban K et al (2024) Is ChatGPT 'ready' to be a learning tool for medical undergraduates and will it perform equally in different subjects? Comparative study of ChatGPT performance in tutorial and case-based learning questions in physiology and biochemistry. Med Teach 46:1441–1447
8. Zawacki-Richter O, Marín VI, Bond M et al (2019) Systematic review of research on

artificial intelligence applications in higher education—where are the educators? Int J Educ Technol High Educ 16:39

9. Almasri F (2024) Exploring the impact of artificial intelligence in teaching and learning of science: a systematic review of empirical research. Res Sci Educ 54:977–997
10. Southworth J, Migliaccio K, Glover J et al (2023) Developing a model for AI across the curriculum: transforming the higher education landscape via innovation in AI literacy. Comput Educ Artif Intell 4:100127
11. Towns MH, Raker J, Becker N et al (2012) The biochemistry tetrahedron and the development of the taxonomy of biochemistry external representations (TOBER). Chem Educ Res Pract 13:296–306
12. Sahoo P, Singh AK, Saha S et al (2025) A systematic survey of prompt engineering in large language models: techniques and applications. https://arxiv.org/abs/2402.07927
13. Wang L, Chen X, Deng X et al (2024) Prompt engineering in consistency and reliability with the evidence-based guideline for LLMs. NPJ Digit Med 7:1–9
14. Sağın FG, Özkaya AB, Tengiz F et al (2023) Current evaluation and recommendations for the use of artificial intelligence tools in education. Turk J Biochem 48:620–625
15. Boden MA (2009) Computer models of creativity. AI Mag 30:23–34
16. Teixeira JF, Gonzales MEQ (1983) Inteligência artificial e teoria de resolução de problemas. Trans/Form/Ação 6:45–52
17. Turing AM (2009) Computing machinery and intelligence. In: Epstein R, Roberts G, Beber G (eds) Parsing the turing test: philosophical and methodological issues in the quest for the thinking computer. Springer, Dordrecht, pp 23–65
18. Vygotsky LS (1993) Pensamento e linguagem, 1st edn. Martins Fontes, São Paulo
19. Anderson JR, Crawford J (1995) Cognitive psychology and its implications. Worth Publishers, New York
20. Dehaene S (2022) É assim que aprendemos: por que o cérebro funciona melhor do que qualquer máquina (ainda...). Contexto, São Paulo
21. Cosenza RM, Guerra LB (2011) Neurociência e educação, 1st edn. Artmed, Porto Alegre
22. Subasi A (2020) Machine learning techniques. In: Subasi A (ed) Practical machine learning for data analysis using python. Academic Press, Boston, pp 91–202
23. Schmidhuber J (2015) Deep learning in neural networks: an overview. Neural Netw 61:85–117
24. LeCun Y, Bengio Y, Hinton G (2015) Deep learning. Nature 521:436–444
25. Montesinos López OA, Montesinos López A, Crossa J (2022) Fundamentals of artificial neural networks and deep learning. In: Montesinos López OA, Montesinos López A, Crossa J (eds) Multivariate statistical machine learning methods for genomic prediction. Springer, Cham, pp 379–425
26. Shinde PP, Shah S (2018) A review of machine learning and deep learning applications. In: 2018 fourth international conference on computing communication control and automation (ICCUBEA), pp 1–6
27. Hassabis D, Kumaran D, Summerfield C et al (2017) Neuroscience-inspired artificial intelligence. Neuron 95:245–258
28. Lindsay GW (2020) Attention in psychology, neuroscience, and machine learning. Front Comput Neurosci 14:29
29. Van Houdt G, Mosquera C, Nápoles G (2020) A review on the long short-term memory model. Artif Intell Rev 53:5929–5955
30. Veseli B, Singhania S, Razniewski S et al (2023) Evaluating language models for knowledge base completion. In: Pesquita C, Jimenez-Ruiz E, McCusker J et al (eds) The semantic web. Springer, Cham, pp 227–243
31. Hildt E (2023) The prospects of artificial consciousness: ethical dimensions and concerns. AJOB Neurosci 14:58–71
32. Ng GW, Leung WC (2020) Strong artificial intelligence and consciousness. J Artif Intell Conscious 7:63–72
33. Gomes DDS (2010) Inteligência artificial: conceitos e aplicações. Intelig Artif 1:234–246
34. Luckin R, Cukurova M, Kent C et al (2022) Empowering educators to be AI-ready. Comput Educ Artif Intell 3:100076
35. Roumeliotis KI, Tselikas ND (2023) ChatGPT and open-AI models: a preliminary review. Fut Internet 15:192
36. Williamson B, Eynon R (2020) Historical threads, missing links, and future directions in AI in education. Learn Media Technol 45:223–235
37. Rane N (2023) Contribution and challenges of ChatGPT and similar generative artificial intelligence in biochemistry, genetics and molecular biology. SSRN Scholarly Paper 4 603219
38. Chiu TKF, Chai C (2020) Sustainable curriculum planning for artificial intelligence

education: a self-determination theory perspective. Sustainability 12:5568
39. Luckin R (2018) Machine learning and human intelligence: the future of education for the 21st century. Institute of Education Press, London
40. Ng DTK, Leung JKL, Chu SKW et al (2021) Conceptualizing AI literacy: an exploratory review. Comput Educ Artif Intell 2:100041
41. Laupichler MC, Aster A, Schirch J et al (2022) Artificial intelligence literacy in higher and adult education: a scoping literature review. Comput Educ Artif Intell 3:100101
42. Long D, Magerko B (2020) What is AI literacy? Competencies and design considerations. In: Proceedings of the 2020 CHI conference on human factors in computing systems, pp 1–16
43. Chakraborty N, Ornik M, Driggs-Campbell K (2025) Hallucination detection in foundation models for decision-making: a flexible definition and review of the state of the art. ACM Comput Surv 57:188:1–188:35
44. Ji Z, Lee N, Frieske R et al (2023) Survey of hallucination in natural language generation. ACM Comput Surv 55:248:1–248:38
45. Clark L, Pantidi N, Cooney O et al (2019) What makes a good conversation? Challenges in designing truly conversational agents. In: Proceedings of the 2019 CHI conference on human factors in computing systems, pp 1–12
46. Luger E, Sellen A (2016) "Like having a really bad PA": the gulf between user expectation and experience of conversational agents. In: Proceedings of the 2016 CHI conference on human factors in computing systems, pp 5286–5297
47. Tassoti S (2024) Assessment of students use of generative artificial intelligence: prompting strategies and prompt engineering in chemistry education. J Chem Educ 101:2475
48. Jiang Z, Xu FF, Araki J et al (2020) How can we know what language models know? Trans Assoc Comput Linguist 8:423–438
49. Sikha VK, Siramgari D, Korada (2023) Mastering prompt engineering: optimizing interaction with generative AI agents. J Eng Appl Sci Technol 1:1–8
50. Browne GJ, Rogich MB (2001) An empirical investigation of user requirements elicitation: comparing the effectiveness of prompting techniques. J Manag Inf Syst 17:223–249
51. Beg MHR, Mehndiratta V (2024) The ethics of generative AI: analyzing ChatGPT's impact on bias, fairness, privacy, and accountability. In: 2024 international conference on innovative computing, intelligent communication and smart electrical systems (ICSES), pp 1–7
52. Bura C, Myakala PK, Jonnalgadda AK (2025) Ethical prompt engineering: addressing bias, transparency, and fairness. Ethics 12:1
53. Vidhani DV, Mariappan M (2024) Optimizing human–AI collaboration in chemistry: a case study on enhancing generative AI responses through prompt engineering. Chemistry 6: 723–737
54. Villafañe SM, Bailey CP, Loertscher J et al (2011) Development and analysis of an instrument to assess student understanding of foundational concepts before biochemistry coursework. Biochem Mol Biol Educ 39:102–109
55. Voet JG, Bell E, Boyer R et al (2003) Recommended curriculum for a program in biochemistry and molecular biology. Biochem Mol Biol Educ 31:161–162
56. Huang P (2000) The integrative nature of biochemistry: challenges of biochemical education in the USA. Biochem Educ 28:64–70
57. Morton JP, Doran DA, MacLaren DPM (2008) Common student misconceptions in exercise physiology and biochemistry. Adv Physiol Educ 32:142–146
58. Wolfson AJ, Rowland SL, Lawrie GA et al (2014) Student conceptions about energy transformations: progression from general chemistry to biochemistry. Chem Educ Res Pract 15:168–183
59. Fuchs TT, Bonney KM, Arsenault M (2021) Leveraging student misconceptions to improve teaching of biochemistry & cell biology. Am Biol Teach 83:5–11
60. Villafañe SM, Loertscher J, Minderhout V et al (2011) Uncovering students' incorrect ideas about foundational concepts for biochemistry. Chem Educ Res Pract 12:210–218
61. Loertscher J, Green D, Lewis JE et al (2014) Identification of threshold concepts for biochemistry. CBE Life Sci Educ 13:516–528
62. Cicuto CAT, Torres BB (2016) Implementing an active learning environment to influence students' motivation in biochemistry. J Chem Educ 93:1020–1026
63. Ghosh A, Bir A (2023) Evaluating ChatGPT's ability to solve higher-order questions on the competency-based medical education curriculum in medical biochemistry. Biochem Mol Biol Educ

64. Surapaneni KM, Rajajagadeesan A, Goudhaman L et al (2024) Evaluating ChatGPT as a self-learning tool in medical biochemistry: a performance assessment in undergraduate medical university examination. Biochem Mol Biol Educ 52:237–248
65. Bardin L (2011) Content analysis. São Paulo Edições 70(279):978-8562938047

Chapter 5

Biochemistry in 3D: A Virtual Reality Museum Approach to Molecular Learning

Reia Thomas, Joana Thomson, Rahul Kapur, Kevin Park, and Jacqueline Keighron

Abstract

Virtual reality (VR) is rapidly emerging as a powerful educational tool, transforming how students interact with complex scientific topics. This protocol details the use of FrameVR, a browser-based platform, to design an interactive virtual biochemistry museum focused on molecular topics such as protein conformation and ligand binding. Developed collaboratively with undergraduate students, the environment features 3D molecular models, embedded mini-lectures, and gamified assessment tools. This multisensory and immersive design aims to reinforce spatial understanding and cognitive engagement in biochemistry. By integrating pedagogical strategies such as scaffolding, guided exploration, and active learning, the VR space supports student autonomy and conceptual comprehension. Educators benefited from the platform's flexibility, enabling content to be presented in a visually engaging and evidence-based format. Student surveys suggest an improvement in topic understanding and retention, alongside increased engagement and motivation. VR holds immense potential to make abstract topics tangible and accessible, especially in STEM fields. As this medium becomes more widespread, thoughtful integration grounded in educational theory will be essential to ensure its impact is both meaningful and inclusive.

Key words Virtual reality, Biochemistry, Gamification, Instruction, Engagement, STEM, Education

1 Introduction

Virtual reality (VR) is emerging as a powerful educational tool that bridges the long-standing gap between traditional 2D representations and immersive 3D learning environments. In conventional classrooms, complex subjects such as molecular biology, physics, or architecture are often taught using flat images, diagrams, or videos, which can limit students' spatial understanding. VR addresses this challenge by allowing learners to explore and interact with content in three dimensions, leading to improved comprehension and retention. By transforming abstract ideas into tangible, manipulable experiences, VR helps students visualize relationships, processes, and structures in a way that 2D media cannot [1, 2]. This

Robert E. Dempski (ed.), *Transforming Molecular Biology with Emerging Technologies*, Methods in Molecular Biology, vol. 3043,
https://doi.org/10.1007/978-1-0716-5320-3_5,

immersive, multisensory approach not only supports different learning styles but also makes the material more engaging and memorable [3].

The use of VR in educational settings has become increasingly popular due to its cost-effectiveness, accessibility, and ease of implementation, especially in VR laboratory experiments [2]. Studies highlight its role in enabling students to revisit lectures and interactive lessons multiple times, giving them the flexibility to review difficult concepts at their own pace and from different perspectives [3]. For example, in virtual science labs, students can conduct experiments without the constraints of physical materials or safety concerns, deepening their conceptual understanding [4]. Applications that allow users to interact with molecules or simulate historical events bring content to life in ways that static resources cannot [5]. These experiences support experiential learning, a hands-on approach that may increase student engagement and long-term knowledge retention. Moreover, VR can level the playing field by providing equal access to high-quality learning environments, regardless of a school's physical resources [6].

Beyond being a tool for content delivery, VR is also proving to be valuable when used as a platform for content creation. Increasingly, educators are involving students in the design and development of VR environments, allowing them to take an active role in their own learning [7, 8]. This co-creation process fosters critical thinking, creativity, and collaboration, as students must work together to research, design, and build immersive experiences. Multi-user VR environments enable real-time interaction between students and instructors, supporting collaborative problem-solving and peer learning [5]. Such environments cultivate a sense of presence and community, even in remote or hybrid learning scenarios. By combining experiential learning with opportunities for co-creation and collaboration, VR not only enhances how students learn but also how they engage with each other and with the content.

In science education, and especially in disciplines like biochemistry and molecular biology, the application of thoughtful pedagogical design is crucial when incorporating emerging technologies such as virtual reality. Biochemistry poses a distinct challenge for educators, as it involves highly detailed molecular interactions that are both imperceptible to the naked eye and difficult to grasp through conventional teaching tools [9]. In this context, VR can serve as a powerful medium, but only when it is intentionally designed with sound educational principles in mind. Effective VR modules for biochemistry should go beyond simply presenting 3D molecular models; they must be rooted in cognitive science and evidence-based learning strategies [10]. This includes the use of scaffolding to support complex learning, opportunities for learner autonomy, and formative assessments that provide feedback and reinforce understanding. Instructional approaches like guided exploration, where students engage with scenarios or prompts, can further deepen comprehension by encouraging active participation [11]. When designed with pedagogy at the forefront, VR becomes a tool not just for visualizing content but for promoting meaningful, conceptual learning. As research continues to examine how immersive technologies influence cognition and motivation, the need for deliberate, theory-informed design in VR learning environments is increasingly evident [7, 10, 11].

The novelty and interactivity of VR can contribute to increased student motivation, particularly in challenging subjects [8, 12]. Motivation is a critical factor in student persistence and success, especially in STEM disciplines where students often leave classrooms with lasting misconceptions about core concepts such as protein interactions and structural intricacies [9]. VR has been shown to elicit positive emotional responses, including curiosity and enjoyment, which can lead to greater time-on-task and persistence in problem-solving [3]. Importantly, the immersive nature of VR may also foster a growth mindset by encouraging students to engage with difficult material in a low-stakes, exploratory environment. Instead of being penalized for incorrect answers, students can experiment freely and learn from their mistakes in a self-directed manner [3].

While enthusiasm for VR in education continues to grow, it is essential to acknowledge and address potential challenges. Issues such as technology access, device cost, student motion sickness, and equity in implementation must be carefully considered. In some cases, schools may lack the infrastructure or training needed to effectively deploy VR solutions. Limitations in accessing this technology may be improved through partnerships with technology companies, as well as through grants or donors, which may provide discounts, donations, or funding for essential equipment [13]. Health concerns may be mitigated through the gradual implementation of VR, starting with smaller environments and limited headset usage, allowing users to become accustomed to the technology. Furthermore, ongoing research and policy development are needed to ensure that the benefits of VR can be realized across diverse educational contexts. Additionally, collaboration between educators, instructional designers, cognitive scientists, and software developers will be vital to ensure that the pedagogical value of VR is maximized. Meaningful, curriculum-oriented material is presented and enhanced through the use of a captivating and interactive environment.

Furthermore, faculty development and support are critical for the successful adoption of VR in higher education. Instructors may need training not only in operating VR hardware and software but also in rethinking their instructional strategies to leverage immersive environments effectively. Institutions that have allocated specific time to VR training during faculty development days have reported more effective implementation of VR technology in classes as well as collaboration among faculty to further developments in the field [14]. These developments have paved the way for the establishment of pedagogical teaching strategies across institutions and more local funding opportunities, as seen through the extended VR program implementation at the University of Southern Utah [15]. As such, universities and institutions should consider professional development initiatives that encourage faculty to explore and experiment with VR, share best practices, and contribute to a growing community of practice. Institutional investment in this area will help shift VR from a novel curiosity to a sustainable educational strategy [8].

Virtual reality holds tremendous promise as a transformative educational technology. However, the true potential of VR will only be realized through thoughtful integration guided by sound pedagogical principles. As technology continues to evolve, educators have a unique opportunity and responsibility to harness its power not only to teach content more effectively but to reimagine the very nature of learning itself.

2 Materials

The virtual environment was developed using FrameVR, a browser-based platform that enables the creation of immersive, interactive 3D spaces without the need for standalone applications or VR headsets. This platform was selected for its accessibility, cross-platform compatibility, and ability to support diverse multimedia content, including videos, 3D models, and interactive elements, which were essential for supporting visual and experiential learning in science education. It is important to note that VR technology changes at a rapid pace and may become discontinued due to limited support and the dynamic nature of VR design platforms.

The example VR environment demonstrated here was collaboratively designed by a multidisciplinary team of students and professors of digital art and design and biochemistry. This collaboration combined design expertise, user experience principles, and deep subject-matter knowledge to create a scientifically accurate and pedagogically meaningful virtual learning space. Existing course materials, including lab exercises, lecture content, and molecular visualizations, provided the foundation for the virtual exhibits. These were adapted and restructured to promote spatial reasoning, conceptual understanding, and student engagement.

Structured as a virtual biochemistry museum, the environment includes four themed rooms (Fig. 1), each targeting a key biochemical concept or molecule corresponding to a laboratory experiment where students work with a maltose sensor consisting of maltose

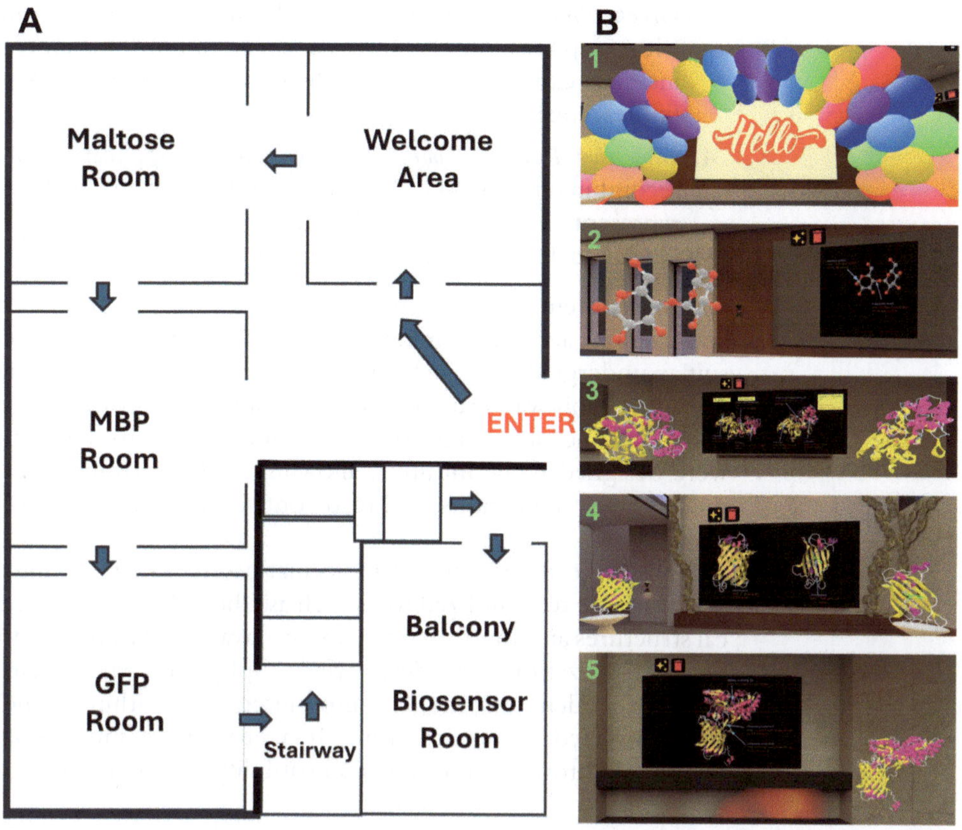

Fig. 1 VR museum design for content accompanying an undergraduate laboratory experiment. (**a**) Floor plan for the environment showing the path VR users follow as they explore the museum contents. (**b**) The five zones of the VR environment corresponding to key concepts of the laboratory experiment and orienting landmarks. (1) Welcome area used to orient users to the VR experience (2) The maltose room with a 3D model of the chemical structure of maltose and accompanying background information (3) The MBP Room with large side by side models of the open and closed conformations of MBP and background information provided through mini-lecture video (4) The GFP room with large VR models of the structure and a cutaway to clearly visualize the chromopore at the center of the barrel (5) The sensor room with the complete structure of the biosensor with accompanying information detailing its use in the laboratory

binding protein circulatory permutated with green fluorescent protein [1]:

1. *Maltose Room*—Features the chemical structure of maltose, contextual background on disaccharides, and explanations of reducing sugar chemistry through both static visuals and interactive components.
2. *Maltose-Binding Protein (MBP) Room*—Displays open and closed conformations of MBP, alongside a video animation demonstrating the conformational change upon maltose binding, reinforcing the concept of induced fit and protein dynamics.

3. *Green Fluorescent Protein (GFP) Room*—Presents a 3D model with emphasis on the internal chromophore structure and includes explanatory content detailing the molecular basis of fluorescence.
4. *Biosensor Protein Room*—Focuses on the application of the recombinant MBP-GFP protein in biosensing, with visual representations of the protein structure and real-world examples of analytical detection technologies.

The content in each room was designed with educational clarity and scientific accuracy in mind. Visual complexity was moderated to suit an undergraduate biochemistry audience, and the spatial layout of each exhibit was structured to promote logical flow, from basic chemical concepts to applied protein functions. Learning elements were designed to be intuitive, allowing students to interact with models, view animations, and read contextual information at their own pace.

To create the visual and molecular assets integrated into the museum, the team utilized tools such as ChemDraw for 2D chemical structures and Jmol (an open-source Java-based tool) for molecular visualization. Models were uploaded via Sketchfab (an independent 3D model editor integrated within FrameVR) and converted as needed using ImagetoSTL (a multi-program conversion browser) to ensure compatibility with FrameVR.

3 Methods

3.1 Setting up FrameVR

1. FrameVR was opened (*see* **Note 1**), and an existing email was used to create a free account.
   (a) The "+ New Frame" option on the Home Screen was selected (Fig. 2a).
   (b) Two options appear:
      (i) The "Name" tab: Must start with a letter and contain only alphanumeric characters or hyphens (Fig. 2b).
      (ii) The "Environment" tab is where the user can pick specific rooms for their environment from a drop-down menu. For this project, the Mid-century was chosen (Fig. 2c) (*see* **Note 2**).

Once the chosen frame was entered, a tutorial regarding how the frame environment can be navigated was followed. The necessary controls for this protocol are described in the following sections.

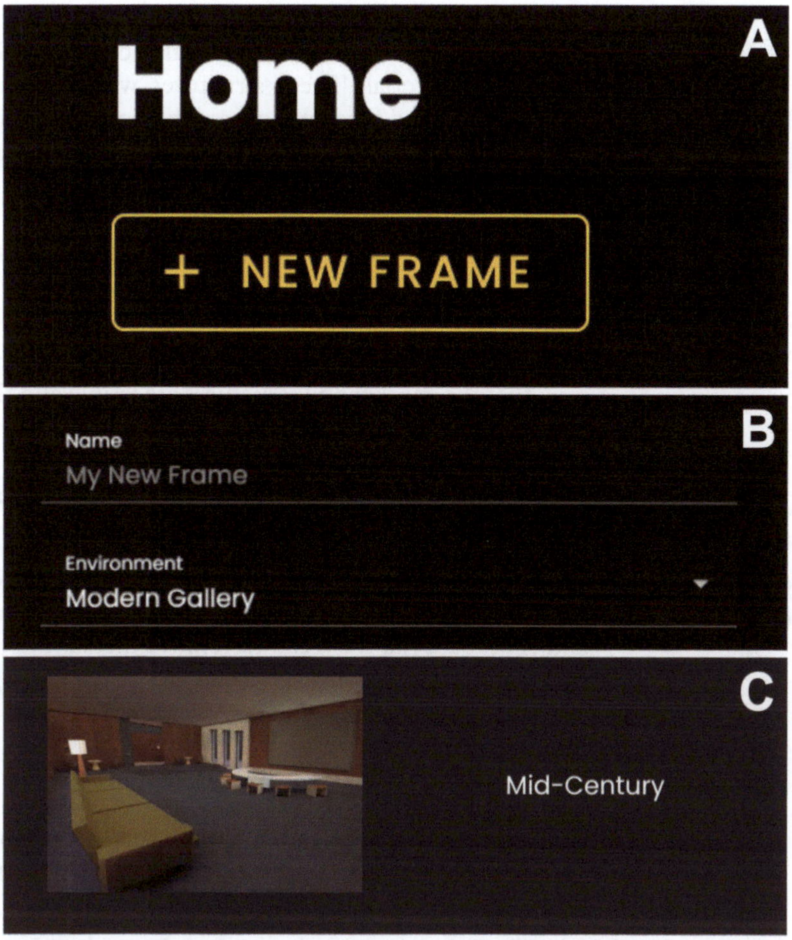

Fig. 2 Getting oriented with FrameVR. (**a**) New frames are added from the home screen; this brings up the menu to (**b**) name the frame and select a template environment; (**c**) thumbnail images are presented for possible options

3.2 Preparation of Accompanying Mini-Lecture Materials

1. The overarching theme for the lectures was determined based on areas that needed supplementation, which can be targeted through student experience surveys and analyses of examination test reports. The main topic was then divided into subtopics, considering all the main points that need to be addressed. For the biochemistry museum, there were five mini-lectures made: an introduction, a description of maltose, a description of MBP, a description of GFP, and the complete biosensor.

   (a) A PowerPoint presentation was created from lecture content. The slides were then reviewed to ensure that the material presented was easy to follow and that the content is condensed into a digestible format.

(b) The required 2D chemical structures were drawn in ChemDraw in accordance with accredited STEM textbook material. These structures were inserted into the PowerPoint presentation on the corresponding slides. A script was prepared for each slide including an introduction of the speaker and topic, summary of the content, and closing statement. Scripts were created to ensure the video did not exceed 90 seconds in time, as this was the agreed length for optimal topic understanding with limited fatigue.

(c) The presentation was recorded on PowerPoint, ensuring that the speaker and structures are visible and that the pace is appropriate. Content was re-recorded if not satisfactory.

(d) The recording was saved and underwent editing through the Microsoft Clipchamp program (*see* **Note 3**).

3.3 Insertion of the "Mini-Lectures" in Frame

1. In the environment, approached sites to insert videos have two buttons, one with (Fig. 3a(1)) yellow sparkles to insert media and (Fig. 3a(2)) a trash can to remove media.

   (a) In the menu that pops up, the "Video" option was selected (Fig. 3a(3)).

   (b) The video was then dragged and dropped into the new tab that appeared.

2. To insert in locations besides these in the environment, the "Hamburger" menu icon in the top left corner of the screen was selected.

   (a) The following panel appeared. First, the "Inventory" tab (Fig. 3b(4)) was selected, then the "Videos" tab was clicked (Fig. 3b(5)) and finally, the "+Add" option was selected (Fig. 3b(6)).

3. The toolbar was accessed at the bottom of the screen, and the pen icon was selected for edit mode (Fig. 3c(8)).

4. A panel appeared on the left side, along with a 3D axis on the video to adjust its position. The arrows and circles can be used to guide the movement in the desired direction (Fig. 3d).

3.4 Creation of 3D Models of Molecules

1. The platform used to create 3D molecules was Jmol, which can be downloaded for free online (*see* **Note 4**). Once the software is opened, the 2D editor option in the "Tools" section is accessed.

2. Once accessed, the desired 2D chemical structure was drawn.

   (a) The default touch setting generated a single, carbon-carbon (ethane) bond.

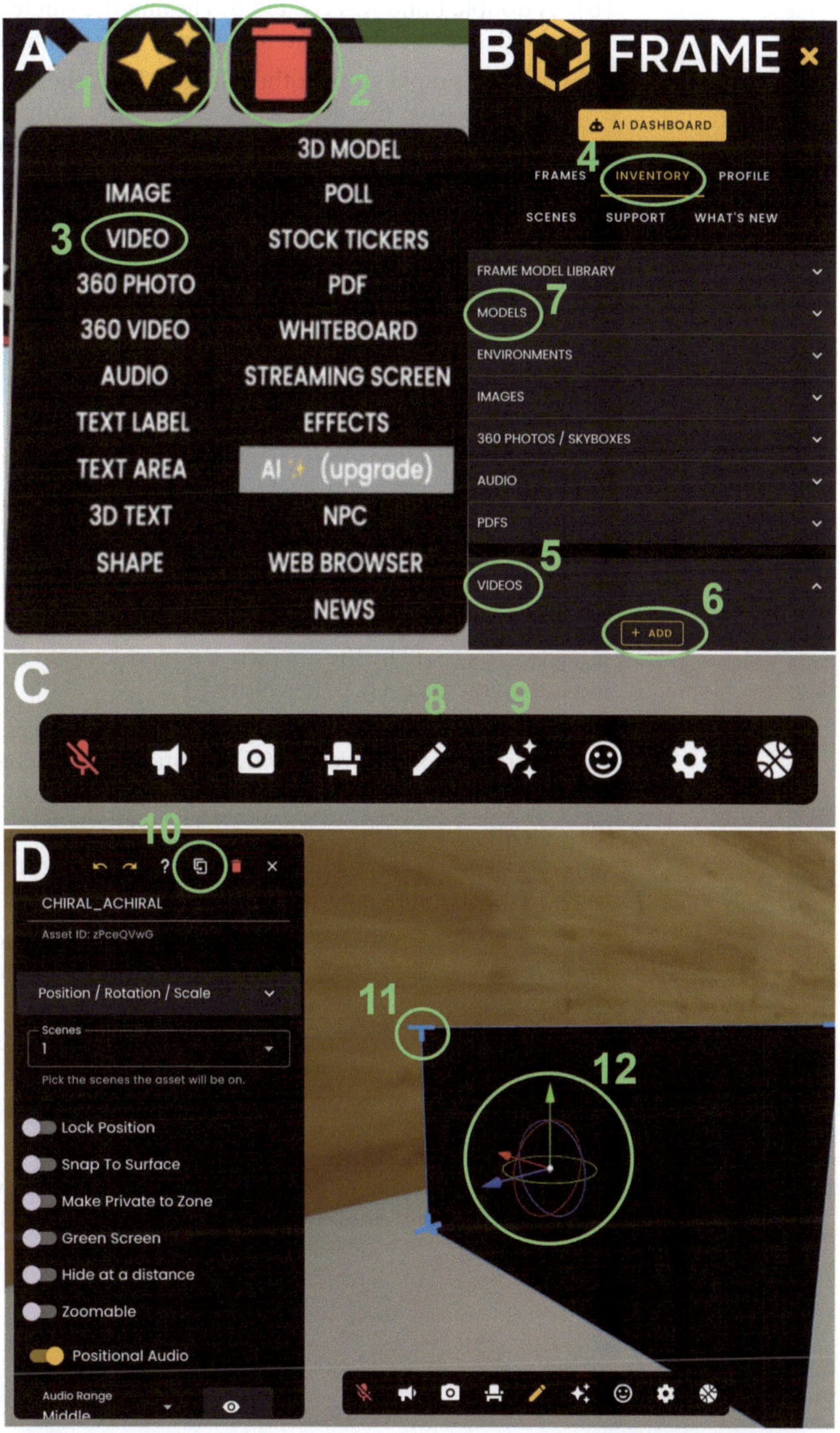

Fig. 3 (**a**) Menu for insertion of mini-lectures into FrameVR showing the following: (1) the icon to insert media, (2) the icon to delete media, and (3) the video insertion button. (**b**) The video menu showing: (4) the inventory tab where videos are stored, (5) the video tab, (6) the button to add a video from a device to inventory, and (7) the model tab. (**c**) Toolbar located on the bottom of the screen showing the following: (8) the pen icon for

(b) A double bond was created by clicking the single bond of the desired double bond interaction a second time.

(c) To add simple atoms other than carbon, the mouse was hovered over the desired location of the atom, and the first letter of the desired atom was typed (H for hydrogen, O for oxygen, F for fluorine, Cl for chlorine, etc.). If not found, the black X at the bottom of the left toolbar was clicked to type or search for the element.

(d) The "clean" button on the bottom left of the screen assisted with polishing structures that were distorted during creation.

3. After the structure was created in the 2D tool, the "add 3D" property option was selected.
4. Once the 3D structure was generated, it's ready to be exported. The molecule was right-clicked, "File" was selected, then "Export," and "Export VRML 3D molecule". A separate folder with all the models needed was created, allowing for facilitated access and design.
5. For the FrameVR environment, the accepted file type for 3D models (which will be the interactive molecules used to supplement the videos) is GLB. (*see* **Note 5**).
6. The file for the molecule was uploaded into the ImagetoSTL website, then the "Convert to GLB" option was selected so the new GLB file could be downloaded (*see* **Note 6**).
7. To download into the computer, the "Download as GLB" option was selected. This file can now be directly input into the frame environment.
8. Navigate to the location where the model will be placed and locate the inventory tab (Fig. 3b(4)) and select the model's subtab (Fig. 3b(7)).
9. The paperclip icon that appears was selected to insert the saved file for the model.
10. The "+" button next to the name of the model was selected to add it to the frame.
11. The position and size of the molecule were adjusted with the resize handles (Fig. 3d(11)) and rotation gizmo (Fig. 3d(12)).

Fig. 3 (continued) editing the position of media and (9) the sparkle icon for the creation menu. (**d**) Panel with (10) duplication button, (11) resize handles, and (12) rotation gizmo

3.5 Creation of 3D Models of Macromolecules

1. RCSB PDB is an open-access database with various 3D macromolecules (*see* **Note 7**). The macromolecule of choice was searched for and downloaded as a PDB file.
   (a) The following macromolecules were used in this VR environment: 1OMP, MBP in the open position [16]; 1ANF, MBP with maltose bound [17]; 3OSQ, the MBP-GFP biosensor [18]; and 1EMA, GFP [19].
2. The PDB file was then inserted into Jmol and modified for visualization in the VR museum using the command console. The Jmol/JSmol interactive scripting documentation (https://chemapps.stolaf.edu/jmol/docs/) [20] can be used to identify specific commands; examples of those used for these models are listed below.
   (a) Water molecules were excluded from the structure with the command "delete water."
   (b) A standard color scheme was used to highlight secondary structure with the command series "select protein" "cartoons on" "color structure."
   (c) The structure of maltose was isolated from PDB 1ANF using the command "delete protein."
   (d) A cut through view of GFP was generated by deleting specific amino acid residues in PDB 1EMA by selecting and deleting specific amino acid residues, for example, the command "delete: A and 5" will delete the fifth amino acid in the: A chain.
3. **Steps 4–11** of Subheading 3.4 were then followed to export the model into the correct file type into FrameVR.

3.6 Developing a Gaming Component

1. For the gaming aspect, which was prepared as a basketball game, the user will attempt to answer the practice question and select a basketball hoop to shoot into that corresponds to an answer choice. They will either be brought to a spot with the next question if correct or prompted to try again if incorrect (*see* **Note 8**).
   (a) The hamburger menu was opened, and the "Frame Settings" tab was selected. Followed by the "General" subtab (Fig. 3b(4)).
   (b) The "Basketball enabled" option was located and turned on (Fig. 4a).
   (c) The basketball icon that appears in the toolbar was selected (Fig. 4b).
   (d) A hoop and basketball appeared (Fig. 4c). The hoop was selected in edit mode, and the position and location of the hoop were adjusted to be close to the practice question.

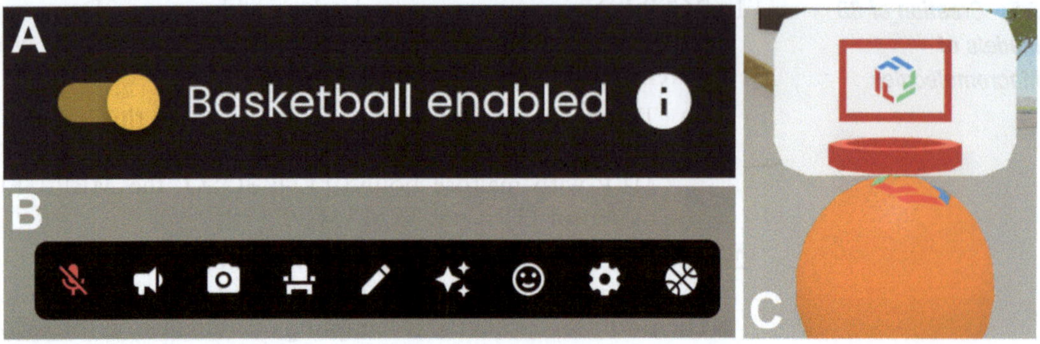

Fig. 4 Addition of basketball for gamification (**a**) slide menu enabling basketball in the environment, (**b**) basketball icon in toolbar, and (**c**) basketball hoop and ball that appear in the environment

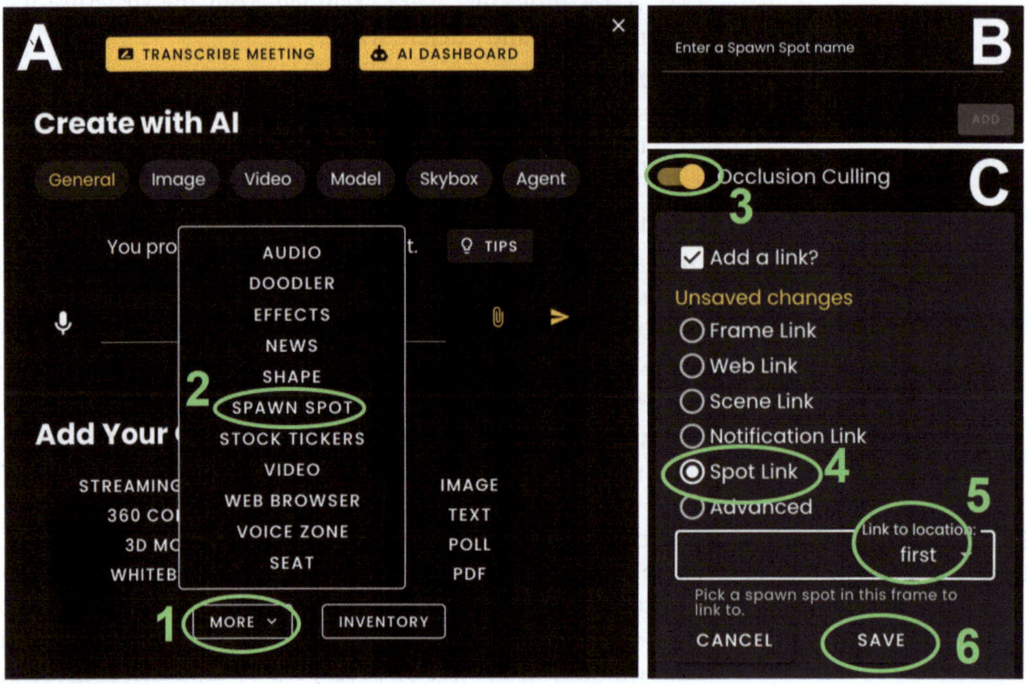

Fig. 5 Creating spawn spots. (**a**) Sparkle creation menu icon. (1) Additional menu options. (2) Spawn spot option. (**b**) Naming the spot. (**c**) (3) Enabling occlusion culling. (4) Setting up a spot link. (5) Selecting the location that was named in the previous step. (6) Saving the new spot

The hoops were duplicated to coincide with the number of answer choices for the question (Fig. 3d(10)). Images of letters above the hoops were added to indicate the corresponding answer choice.

(e) Relocation of the next question occurred in order to create a connection from the correct basket of the previous question to the next. In the toolbar, the sparkles for the creation menu were selected. (Fig. 3c(9)).

(f) In the screen that appears, the "More" button was selected (Fig. 5a(1)) followed by the "Spawn Spots" (Fig. 5a(2)) option in the list that appears.

(g) In the tab that appears, a name that corresponds to the question number was entered (Fig. 5b). The "Add" button underneath was then selected.

(h) Relocation to the correct answer basket of the previous question occurred. The basket was selected in edit mode, and occlusion culling was turned on (Fig. 5c(3)). The "Spot Link" option was then selected (Fig. 5c(4)), followed by the "Link to location" box, and the name of the spawn spot appeared in a drop-down menu and was selected (Fig. 5c(5)), followed by the "Save" option (Fig. 5c(6)).

3.7 Assessment of Learning

1. A common method to quantify learning outcomes from VR content is with pre- and posttesting and engagement surveys. For the purposes of this protocol, the assessment used in Calma-Roddin et al. 2023 [1] will be used as an example. The questions used to assess learning and engagement are provided in the supplementary material. Assessments should conform to local Institutional Review Board (IRB) protocols to prevent any violation of the student or participant's privacy and willingness to participate.

(a) The first step in this process is to identify the learning outcomes that can be assessed from the VR content provided to students (*see* **Note 9**).

(b) For each learning outcome, create pairs of questions that assess the same content but use different specific examples or ask the question from a different perspective (*see* **Note 10**).

(c) Basic statistics can be compared for the entire pre- vs. posttest or for individual sections of related questions. Mean scores and standard deviations can be gathered, for example, in the VR environment, the Wilcoxon signed-rank test was used to determine significance after first assessing that a non-normal sample distribution was collected.

(d) Including answer choices such as "I can't tell" or "I don't know" in multiple-choice questions may allow researchers to assess the confidence of students in their answer choice and prevent random guessing. In the example VR environment, the result showed a decrease in selecting this answer choice and an increase in selecting the correct answer after interacting with the VR environment.

2. Surveys using a defined number scale or ranking can also be used as an indicator of learning outcomes, interest, and engagement.
   (a) Questions should be clear and concise, such as "On a scale from 1 (strongly disagree) to 5 (strongly agree) I feel like I learned or understood something better after interacting with the VR content?" Survey questions can be used to determine if the student or participant enjoyed the experience, felt more confident in their understanding of the presented material, or would prefer VR to other educational platforms. Results can be presented as histograms or percentages, selecting a particular ranking to assess these more qualitative measures.
3. Free response comments can also be a useful indicator of student learning, providing an open-ended question or interviewing students or participants can provide information that the researcher has not expected to see or qualify results that were not predicted by the researcher's hypothesis.

3.8 VR Setup

1. The Meta Quest 2 VR headset setup and safety protocols included in the kit were followed (*see* **Note 11**). Alternatively, students who experience difficulties wearing VR headsets or with motion sickness can explore the VR museum using a computer or a larger projector screen for better immersion.
2. Access to the museum is accomplished by loading the url provided in a web browser.

4 Notes

1. The link to open FrameVR: https://learn.framevr.io/.
2. FrameVR only allows 3 Trial Frames for free. Trial Frames can have up to 8 editors and 15 visitors at the same time.
3. The program Microsoft Clipchamp can be accessed through this link: https://clipchamp.com/en/windows-video-editor/. It offers various generic editing functions (crop, trim, etc.) and new AI features. The software can also incorporate tools such as transitions, filters, and audio modulations. Online tutorials were followed to perform necessary editing. The program Microsoft Clip Champ offers various generic editing functions, as well as new AI functions.
4. Link to download Jmol: https://jmol.sourceforge.net/
5. Some limitations when creating the environment include inadequate presentation of molecules from one media format to another and discontinuity of certain features in FrameVR.

Models created in other design software (Jlab, Blender) might experience difficulties in translating to Frame's network.

6. The website used during this experiment is Image to STL (https://imagetostl.com/convert/file/vrml/to/glb#google_vignette).
7. The link to access RCSB PDB: https://www.rcsb.org/
8. Due to a lack of funding and continual changes made to VR platforms in general, it's important to note that some features (the basketball game feature mentioned above, for example) might only exist for a short period of time. This requires the instructor who is creating the environment to explore other gaming options and adapt them to the material presented.
9. Pre- and posttests can be given in the VR environment directly or provided on a computer or paper outside of the experience. There are pros and cons of assessing learning outcomes in the VR environment. For example, it allows students to answer questions directly after interacting with the related material, while also increasing the time students spend in VR, which can lead to fatigue from extended use of the VR equipment.
10. Using the same question on both the pre- and posttest can complicate the evaluation of learning outcomes; if students or participants can remember the question, it may produce a bias when answering the posttest.
11. It's important for the instructor to reinforce the practice of safely using goggles. The user of the headset will have a limited awareness of their external environment; therefore, the instructor must ensure that the headset user is away from all tripping hazards (chairs, loose/lay items on the floor) and protruding surfaces. Additionally, many users might experience dizziness, headaches, and vision fatigue after prolonged use of the headset, so it's vital to allow users to take breaks in between uses.

Acknowledgments

The authors would like to thank Dr. Nicole Calma-Roddin, from the Department of Psychology and Counseling, and Nouman Wajid, a recent M.A. graduate from the Department of UI/UX Design and Development, for their valuable guidance and insightful perspectives, which significantly contributed to the development of this work.

Conflict of Interest Statement The authors have no conflicts of interest to declare that are relevant to the content of this chapter.

References

1. Calma-Roddin N, Park K, Keighron J (2023) Exploring the structure of proteins and other biomolecules with a VR Museum: lessons in classroom integration. J Chem Educ 100(7): 2574–2582
2. Dunnagan CL, Dannenberg DA, Cuales MP, Earnest AD, Gurnsey RM, Gallardo-Williams MT (2019) Production and evaluation of a realistic immersive virtual reality organic chemistry laboratory experience: infrared spectroscopy. J Chem Educ 97(1):258–262
3. Makransky G, Lilleholt L (2018) A structural equation modeling investigation of the emotional value of immersive virtual reality in education. Educ Technol Res Dev 66(5): 1141–1164
4. Cheng K-H, Tsai C-C (2019) A case study of immersive virtual field trips in an elementary classroom: students' learning experience and teacher-student interaction behaviors. Comput Educ 140:103600
5. Theisen KE (2022) Two active learning models of protein dynamics for use in undergraduate biochemistry courses. J Chem Educ 99(6): 2245–2251
6. Jensen L, Konradsen F (2018) A review of the use of virtual reality head-mounted displays in education and training. Educ Inf Technol 23(4):1515–1529
7. Liu J, Li R, Yang W, Lei H, Wang L, Zhang Y, Wang Y, Liu W, Deng J, Hu L, Jiang Y, Wang Z, Tang J, Yang L (2023) Project-based teaching model in pharmaceutical integrated experiment course for undergraduates implementing the case study: design, synthesis, and biological evaluation of potential SIRT5 inhibitors. J Chem Educ 100(11):4414–4422
8. Seco-González A, Porto C, Piñeiro Á, Garcia-Fandino R (2023) Corona VRus coaster: the virtual reality roller coaster of the proteins of SARS-CoV-2. J Mol Model 31(9):233
9. Howell ME, Booth CS, Sikich SM, Helikar T, Roston RL, Couch BA, Dijk K (2019) Student understanding of DNA structure–function relationships improves from using 3D learning modules with dynamic 3D printed models. Biochem Mol Biol Educ 47(3):303–317
10. Pinger C, Geiger M, Spence DM (2020) Applications of 3D-printing for improving chemistry education. J Chem Educ 97(1):112–117
11. Radianti J, Majchrzak TA, Fromm J, Wohlgenannt I (2020) A systematic review of immersive virtual reality applications for higher education: design elements, lessons learned, and research agenda. Comput Educ 147:103778
12. Bejjarapu DS, Chen Y, Xu J, Shaffer E, Garg N (2024) Enhancing learning outcomes for materials crystallography via virtual reality: role of user comfort and game design. J Chem Educ 101:973–983
13. Samala AD, Rawas S, Rahmadika S, Criollo-C S, Fikri R, Sandra RP (2025) Virtual reality in education: global trends, challenges, and impacts—game changer or passing trend? Discov Educ 4:229
14. Peisachovich E, Appel L, Sinclair D, Luchnikov V, Da Silva C (2021) CVRriculum program faculty development workshop: outcomes and suggestions for improving the way we guide instructors to embed virtual reality into course curriculum. Cureus 13(3):e13692
15. Frahm A (2025) Transforming higher education through virtual reality: implementation, faculty engagement, and pedagogical impact. In: Practitioner proceedings of the 11th international conference of the immersive learning research network, pp 206–210
16. Sharff AJ, Rodseth LE, Spurlino JC, Quiocho FA (1992) Crystallographic evidence of a large ligand-induced hinge-twist motion between the two domains of the maltodextrin binding protein involved in active transport and chemotaxis. Biochemistry 31:10657–10663
17. Quiocho FA, Spulino JC, Rodseth LE (1997) Extensive features of tight oligosaccharide binding revealed in high-resolution structures of the maltodextrin transport/chemosensory receptor. Structure 5:997–1015
18. Marvin JS, Schreiter ER, Echevarria IM, Looger LL (2011) A genetically encoded, high-signal-to-noise maltose sensor. Proteins 79: 3025–3036
19. Ormo M, Cubitt AB, Kallio K, Gross LA, Tsien RY, Remington SJ (1996) Science 273:1392–1395
20. Jmol/JSmol interactive scripting documentation (2025). https://chemapps.stolaf.edu/jmol/docs/. Accessed 25 Aug 2025

Chapter 6

Molecular Modeling and Simulations of Biologics

Dariya Baizhigitova and Pin-Kuang Lai

Abstract

Molecular dynamics (MD) simulations are widely used in computational and structural biology to study the time-dependent behavior of biological macromolecules at atomic resolution. Over the years, they have become an indispensable tool in the study of biomolecular systems and have been applied to investigate protein folding, conformational change, ligand interaction, protein structure determination, and refinement of X-ray crystallographic models. This chapter presents a practical, end-to-end protocol for performing MD simulations on large biomolecules, using a full-length antibody as a case study. It also introduces foundational software tools commonly used in molecular modeling and MD simulations, including PyMOL, visual molecular dynamics (VMD), nanoscale molecular dynamics (NAMD), MDAnalysis, and others. Additionally, the protocol covers the modeling of antibody-excipient interactions and identifying binding sites from MD trajectories. Together, this accessible and reproducible framework aims to equip researchers with the computational tools and foundational knowledge needed to study the dynamic behavior of complex biomolecular systems and integrate MD simulations into their own biological research.

Key words Molecular dynamics simulations, Biologics, Antibodies, Excipients

1 Introduction

Since their first introduction in the mid-twentieth century, molecular modeling and computational simulation methods have evolved dramatically, driven by advances in algorithms, computational power, and structural databases. Among them, molecular dynamics (MD) simulations have emerged as particularly powerful, offering a unique ability to study the time-dependent behavior of biological macromolecules at atomic resolution [1–3]. Unlike other molecular modeling and simulation methods, MD simulations allow us to study the fluctuations in the relative positions of atoms in proteins and other large biomolecules over time, capturing atomic positions with femtosecond temporal resolution [1].

A key strength of MD simulations lies in their ability to characterize a wide array of biophysical properties of large biomolecules,

Robert E. Dempski (ed.), *Transforming Molecular Biology with Emerging Technologies*, Methods in Molecular Biology, vol. 3043,
https://doi.org/10.1007/978-1-0716-5320-3_6,

including protein folding, conformational change, and their interaction with various ligands [1]. Moreover, MD has been applied to determine protein structures from nuclear magnetic resonance (NMR), refine the X-ray structures of proteins, and calculate the free energy changes associated with mutations [2]. Taken together, the capabilities make MD simulations especially valuable for studying large, complex proteins such as monoclonal antibodies (mAbs), which have demonstrated significant therapeutic potential in recent years [3]. Given the high cost and time intensity of experimental methods, MD simulations offer a cost-effective approach for guiding the formulation design of mAbs and other large biologics, providing insight into their structure, stability, and interactions with small molecules such as excipients [3].

This chapter presents a practical, end-to-end protocol for performing MD simulations on large biomolecules, using a full-length antibody model of Infliximab as a case study. Moreover, the protocol provides step-by-step instructions on the study of small molecule (excipient) interactions with mAbs, and an Infliximab-caffeine system is used here as an example. While the chapter focuses on conducting MD simulations on mAbs, the methods outlined are widely applicable to a range of biomolecules, including enzymes, receptors, and protein-ligand complexes. The workflow covers: (1) assembly of the initial structure using PyMOL [4], Alphafold3 [5], and ABodyBuilder2 [6]; (2) structure validation and refinement with visual molecular dynamics (VMD) [7]; (3) assignment of charges and force fields using PDB2PQR [8, 9] software tool and CHARMM36m [10, 11] parameters; (4) energy minimization, solvation, and MD simulations of the antibody structure with nanoscale molecular dynamics (NAMD) [12, 13]; (5) study of antibody-excipient interactions; (6) trajectory and binding site analysis with MDAnalysis [14, 15]. Installation guidelines for all utilized software can be found online through the cited sources, and most of the custom scripts are provided in the Supplementary Materials section of our earlier publication [3]. Together, this chapter shall provide a reproducible and accessible framework for researchers interested in studying the dynamic behavior of complex biomolecular systems and integrating MD simulations into their own biological research.

2 Methods

2.1 Constructing a Full-Length Antibody Model

1. The Fab crystal structure of IgG1 can be downloaded from the PDB Database (PDB code: 4G3Y). Otherwise, the heavy and light chain antibody sequences can be used to generate the Fab structure using AlphaFold3 [16] or ABodyBuilder2 [17] (*see* **Note 1**).

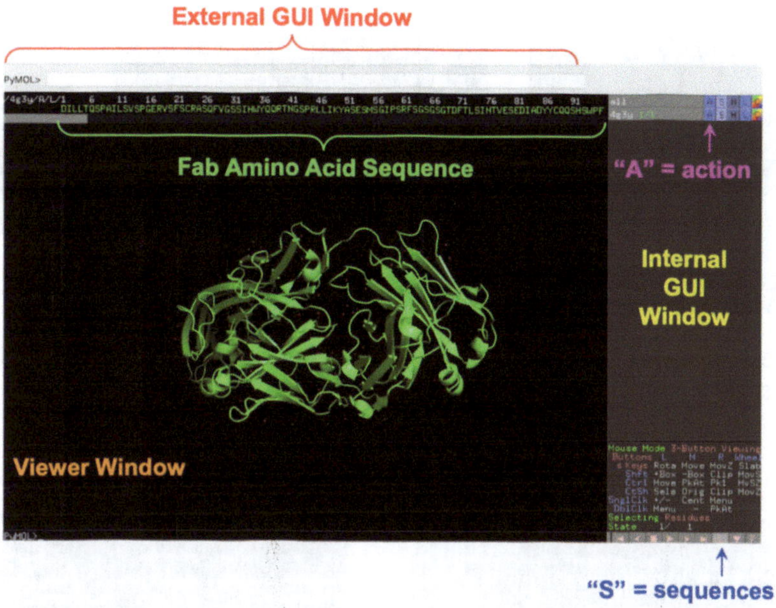

Fig. 1 PyMOL's three windows, names panel, and movie controls. Infliximab Fab structure (PDB code: 4G3Y) is loaded in the viewer window

2. To preprocess the antibody structure, a molecular visualization system PyMOL is used [4]. Installation guidelines for the open-source version are available online (https://github.com/schrodinger/pymol-open-source).
3. Upload the Fab PDB structure to PyMOL. The structure will be displayed on the Viewer Window (Fig. 1). The Internal GUI is displayed on the right and allows performing actions on specific objects and specific object selections. Click on "A" → "remove waters". Alternatively, type in the following command in the External GUI (top) window:

```
remove solvent
```

4. Navigate to Movie Controls in the Internal GUI window and click on "S" (bottom right). This will show the sequences of the template and Fab structures (Fig. 2).
5. Select chain "C" by dragging the cursor to the left or right of the sequence. A selection "(sele)" will appear in the Internal GUI. Click on "C" → "delete selection" (Fig. 2).
6. The clean Fab structure needs to be saved twice, as there are two Fab regions in a full-length antibody structure. To do so, navigate to "File" → "Export Molecule" → "PDB Options" → "Save" in PyMOL, ensuring to choose PDB file format and saving the files as "Infliximab_Fab.pdb" and "Infliximab_Fab2.pdb".

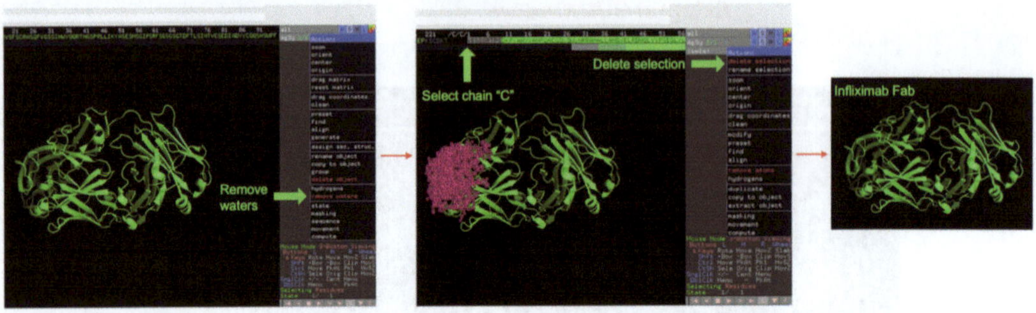

Fig. 2 Preprocessing process for the Infliximab Fab structure (left to right): remove waters, delete chain "C", clean Fab structure

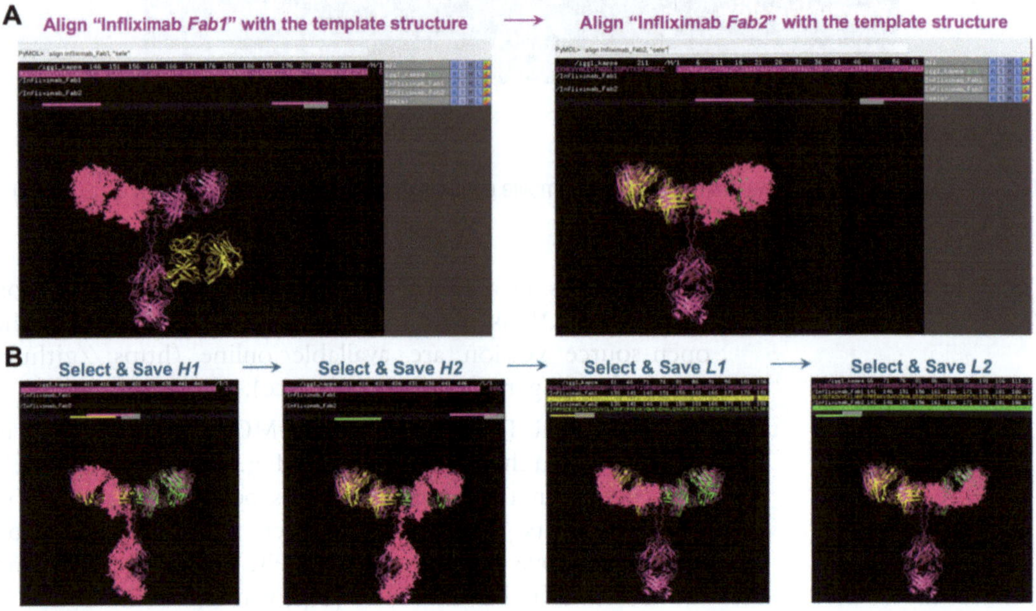

Fig. 3 (**a**) Alignment of Infliximab_Fab1 and Infliximab_Fab2 structures with the KOL/Padlan Template. (**b**) Extraction of the two heavy chains (chains H and I) and two light chains (chains L and M) sequences

7. Create a new window in PyMOL by navigating to "File" → "New PyMOL Window".
8. Upload the template KOL/Padlan structure [18, 19] "igg1_kappa.pdb" (available in our earlier publication) [3] and the "Infliximab_Fab.pdb" and "Infliximab_Fab2.pdb" structures saved earlier (Fig. 3).
9. Navigate to Movie Controls in the Internal GUI window and click on "S" (bottom right). This will show the sequences of the template and Fab structures.
10. Select one heavy chain (marked as "H" in the sequence window) of the Fab region of the template structure (here, ends

with KKV at residue number 218) by dragging the mouse to the right or left of the sequence (Fig. 3). Select the corresponding light chain of the template structure (marked as "L" in the sequence window). Type in the following command in the External GUI (top) window:

```
align Infliximab_Fab1, "sele"
```

11. Infliximab_Fab1 is now aligned with the Fab region of the template structure (Fig. 3a). Repeat the same process for Infliximab_Fab2 by selecting the second heavy chain (marked as "I" in the sequence window) and light chain (marked as "M" in the sequence window) of the template structure and typing the same command but with Infliximab_Fab2 in the External GUI window:

```
align Infliximab_Fab2, "sele"
```

12. The two Fab structures are now correctly aligned to the template structure (Fig. 3a).
13. Select the Fab region of one heavy chain (Infliximab_Fab1) and the corresponding heavy chain Fc region on the template structure (starting from EPK at residue number 219). Save the selection as "Infliximab_H1.pdb" by navigating to "File" → "Export Molecule" → "PDB Options" → "Save" and choosing "sele" as the "Selection". Repeat the same process for the second Fab region (Infliximab_Fab2) and save as "Infliximab_H2.pdb" (Fig. 3b).
14. Save the two light chains from Infliximab_Fab1 and Infliximab_Fab2 structures as "Infliximab_L1.pdb" and "Infliximab_L2.pdb" accordingly (Fig. 3b).

2.2 Postprocessing of a Full-Length Antibody Model

1. Sequence renumbering: The heavy and light chain PDB files are renumbered using a python program (reindex_pdb.py) with the command below. If the code does not get executed, "python3" command can be used instead of "python". Before running the command, the PDB files must have consistent chain names (*see* **Note 2**). The command is similar for all four chains, i.e., H1 can be simply replaced with H2, L1, L2.

```
python reindex_pdb.py 1 Infliximab_H1.pdb Infliximab_H1_renumber.pdb
```

2. Sequence extraction: To ensure that the sequence of the full-length antibody model matches the target sequence, we first extract the antibody sequences with the command below. For the three remaining PDB files, replace

"Infliximab_H1_renumber.pdb" with H2, L1, and L2, respectively. For the two light chains, replace "H" with "L" (*see* **Note 3**).

```
cat Infliximab_H1_renumber.pdb | awk '/ATOM/ && $3 == "CA" &&
$5 == "H" {print $4}' | tr '\n' ' ' | sed 's/ALA/A/g;s/CYS/C/g;
s/ASP/D/g;s/GLU/E/g;s/PHE/F/g;s/GLY/G/g;
s/HIS/H/g;s/HSD/H/g;s/HSE/H/g;s/HSP/H/g;s/ILE/I/g;s/LYS/K/g;
s/LEU/L/g;s/MET/M/g;
s/ASN/N/g;s/PRO/P/g;s/GLN/Q/g;s/ARG/R/g;s/SER/S/g;s/THR/T/g;
s/VAL/V/g;s/TRP/W/g;
s/TYR/Y/g' | sed 's/ //g' | fold -w 60
```

3. Identifying incorrect and missing amino acid residues: To identify the missing or incorrect amino acid sequences, EMBOSS Needle Pairwise Alignment Tool (https://www.ebi.ac.uk/jdispatcher/psa/emboss_needle) can be used to align the sequences from the PDB files to the target structures (*see* **Note 4**, Fig. 4). For reference, the target structure here is Infliximab Remicade [3]. Any differences should be noted down.
4. Mutating the sequences: Based on the differences between the PDB and target structures, structure modification is performed using the Tool Command Language (TCL) available in VMD

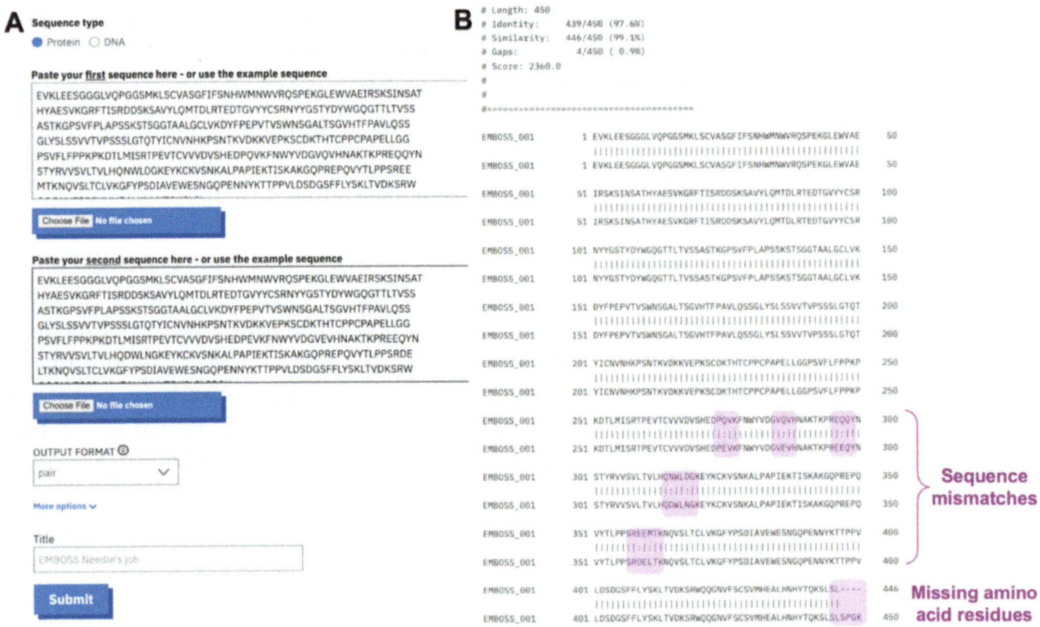

Fig. 4 (**a**) Alignment of sequences from the PDB files (first sequence) to the target structures (second sequence) using EMBOSS Needle Pairwise Alignment Tool. (**b**) EMBOSS Needle results. Incorrect and missing amino acid residues are highlighted and labelled in purple

[7]. Our previously published TCL script (*Infliximab.tcl*) is used to mutate the protein residues, connect the antibody chains, and add glycan structures [3]. The script can be used as a reference to modify the structures of other antibodies and proteins.

5. After aligning the sequences using EMBOSS Needle Pairwise Alignment Tool, differences should only be present for the heavy chains. Alignment results for light chains in the PDB and target structures should indicate that the two are identical. Thus, in the TCL script, mutations of the following residues for the heavy chains are made and listed under "segment H" and "segment I" in the *Infliximab.tcl* script.

```
mutate 275 GLU
mutate 286 GLU
mutate 297 GLU
mutate 315 ASP
mutate 318 ASN
mutate 359 ASP
mutate 361 LEU
```

6. Alignment results should also show that four residues are missing from the heavy chains of the PDB structures. They are indicated as follows in the *Infliximab.tcl* script.

```
residue 447 SER
residue 448 PRO
residue 449 GLY
residue 450 LYS
```

7. Load structures and the TCL file to one directory, i.e., "Infliximab_H1_renumber.pdb", "Infliximab_H2_renumber.pdb", "Infliximab_L1_renumber.pdb", "Infliximab_L2_renumber.pdb", "Infliximab_G1_renumber.pdb", "Infliximab_G2_renumber.pdb", and *Infliximab.tcl.* Execute the TCL script by running the following command in the terminal window:

```
vmd -dispdev text -e Infliximab.tcl
```

8. Two files of the antibody structure will be generated after running the script: "Infliximab_mod.pdb" and "Infliximab_mod.psf". The PDB file contains all antibody chains (H, I, L, M) with the correct sequences. PSF stands for protein structure file and contains information for bonds, angles, and dihedral angles.

2.3 Assigning Charge States for Titratable Residues

1. Next, it is important to account for the charge states of titratable amino acids, which have side chains that can gain or lose protons (H^+) depending on their environment. The different protonation states can significantly affect electrostatic interactions, hydrogen bonding, and protein conformational dynamics, all of which are critical for obtaining reliable structure-based predictions [20]. For example, histidine can exist in protonated HSP (+1) or neutral HSD (+0)/HSE (+0) states at biologically relevant pH 6.0, which is commonly used in antibody formulation. To incorporate these effects, the pdb2pqr software [8, 9] is used to assign protonation states based on pH and local environment, and CHARMM36m force fields [10] are employed to parameterize the resulting structures for molecular simulations.
2. Copy "Infliximab_mod.pdb" and save as "Infliximab_pqr.pdb" in the command window.

```
cp Infliximab_mod.pdb Infliximab_pqr.pdb
```

3. Next, the residue names of different histidine titration states need to be changed to HIS. The *sed -i* command in Linux can be used to make changes.

```
sed -i "s@HSP@HIS@g" Infliximab_pqr.pdb
sed -i "s@HSD@HIS@g" Infliximab_pqr.pdb
sed -i "s@HSE@HIS@g" Infliximab_pqr.pdb
```

4. Execute the command below to determine the residue titration states at a given pH (here, pH 6.0 is used). "Infliximab_pqr.pqr" file is generated after running the command.

```
pdb2pqr --with-ph=6.0 --ph-calc-method=propka --apbs-input --
ff=charmm --ffout=charmm --verbose Infliximab_pqr.pdb Inflix-
imab_pqr.pqr
```

5. Using the generated "Infliximab_pqr.pqr" file, titration states for all histidine residues were determined (*see* **Note 5**, Fig. 5).
6. After examining the "Infliximab_pqr.pqr" file, it is noted that histidine residues 32, 61, 171 on the heavy chain are predicted to be in their protonated form HSD (+1), and residues 207, 432 on the heavy chain and 34, 198 on the light chain are predicted to be HSE (+0). We modify the *Infliximab.tcl* script accordingly to account for these mutations by adding the lines below in "segment H", "segment I" and "segment L", and "segment M" for the heavy and light chains, respectively (Fig. 5).

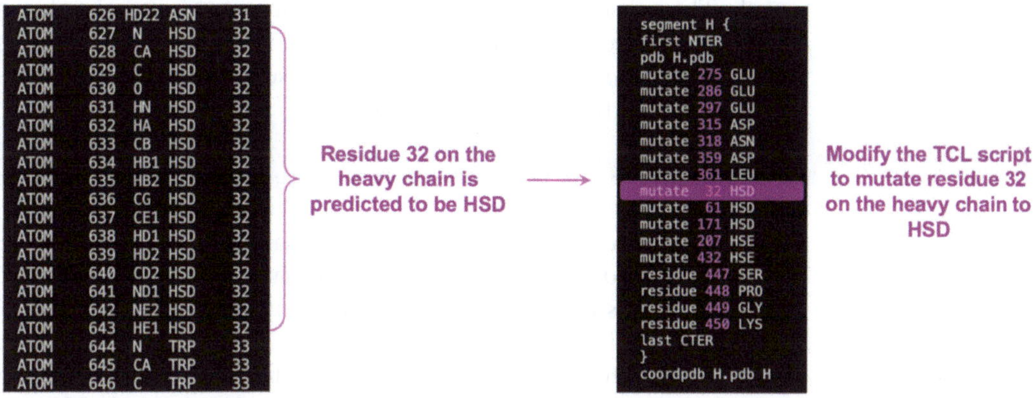

Fig. 5 Left: Predicted titration state of amino acid residue number 32 on the heavy chain by pdb2pqr software. Right: Section of the TCL script with the corresponding mutation

For heavy chains:

```
mutate 32 HSD
mutate 61 HSD
mutate 171 HSD
mutate 207 HSE
mutate 432 HSE
```

For light chains:

```
mutate 34 HSE
mutate 198 HSE
```

7. Rerun the "Infliximab.tcl" script to generate the final PDB and PSF files—"Infliximab_mod.pdb" and "Infliximab_mod.psf".

```
vmd -dispdev text -e Infliximab.tcl
```

2.4 Minimization and Solvation of Input Structures

1. Prior to MD simulations, minimization and solvation of the generated input structures ("Infliximab_mod.pdb" and "Infliximab_mod.psf") must be performed. Energy minimization is performed for 5000 steps using the NAMD program [12, 13], while the solvation is conducted with VMD [7]. Note that the exact execution commands depend on the utilized computing systems. Thus, only the general commands are given below.
2. Readers may choose to create 3 directories in the working folder: "minimization", "solvation", and "namd".
3. Move the previously published *Infliximab_min.conf* file [3] to the "minimization" directory and execute the code below:

```
namd3 Infliximab_min.conf
```

4. The solvation step is done to add water and ion molecules to neutralize the initial structures. Move the previously published *Infliximab_solv.tcl* file [3] to the "solvation" directory and execute the following command:

```
vmd -dispdev text -e Infliximab_solv.tcl
```

5. "Infliximab_ion.pdb" and "Infliximab_ion.psf" files will be generated after executing the command. These files are used as input structures for MD simulations.

2.5 Molecular Dynamics Simulations of Antibodies

1. MD simulations described here follow 4 major steps: (1) Energy minimization on the solvated system; (2) Heating of the system from 100 K to 300 K with position restraints on the antibody non-hydrogen atoms; (3) Equilibrium of the system for 10 ns; (4) Production run performed for 50 ns at 300 K and 1 atm.
2. Move the *Infliximab_ion_min.conf*, *Infliximab_ion_heat.conf*, *gen_restraints_heat.tcl*, *Infliximab_ion_eq.conf*, *Infliximab_ion_prod.conf* files published previously [3] to the "namd" directory.
3. Execute the minimization step by running the following command:

```
namd3 Infliximab_ion_min.conf
```

4. Generate the position restraint file for antibody non-hydrogen atoms by running the following:

```
vmd -dispdev text -e gen_restraints_heat.tcl
```

5. Perform heating from 100 K to 300 K by running the command below.

```
namd3 Infliximab_ion_heat.conf
```

6. Equilibrium and production steps are executed by running the commands below.

```
namd3 Infliximab_ion_eq.conf
namd3 Infliximab_ion_prod.conf
```

2.6 Generating Initial Configuration of Antibody-Excipient Systems

1. First, the input excipient molecule must be obtained—in this case, caffeine. Navigate to PubChem (https://pubchem.ncbi.nlm.nih.gov/) and type in "caffeine" in the search bar. Download the structure in Structure-Data Format (SDF) and open it

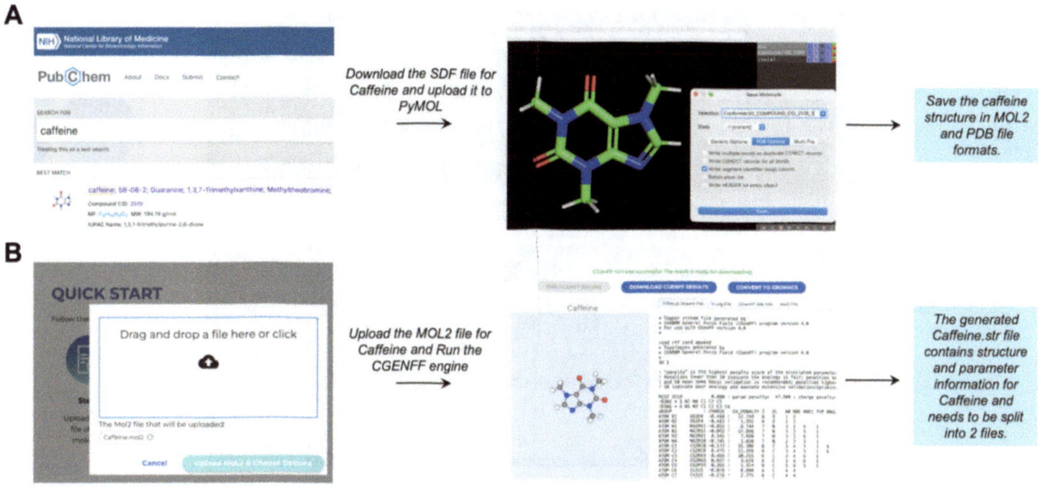

Fig. 6 (**a**) The workflow for obtaining the SDF, MOL2, and PDB files for caffeine from PubChem. (**b**) Generation of structure and parameter files for caffeine by using CGenFF

in PyMOL (Fig. 6a). Save in PDB and MOL2 file formats. Thus, 3 structures should be saved: "Caffeine.sdf", "Caffeine.mol2", and "Caffeine.pdb".

2. Generating excipient force fields: To generate the excipient force fields, tools like CGenFF by SILCSBIO can be used (https://cgenff.com/). After navigating to CGenFF's web server, login to your account or create a new one. Next, upload the structure in its mol2 file format—"Caffeine.mol2". Click on "Process input" → "Run CGenFF Engine" → "Download CGenFF results" (Fig. 6b). This will generate a "Caffeine.zip" file.
3. After extracting the file contents, open the "Caffeine.str" file, which contains information for caffeine structure and parameters. Split the structure and parameter information into "Caffeine.str" and "Caffeine.prm" files, respectively (*see* **Note 6**). The "Caffeine.str" file is comprehended by the VMD program and is used to generate the initial configuration of the antibody-excipient system, while the "Caffeine.prm" file is the parameter file needed for performing MD simulations using the NAMD program.
4. Next, a TCL script is used to generate an input file for Packmol, which is a software tool that creates the initial antibody-excipient configuration for MD simulations [21] by adding caffeine molecules to the antibody system in desired concentrations. Execute the *solvate_PKL.tcl* script published previously [3] in the command window to generate the Packmol input file "packmol_input.inp" (*see* **Notes 7** and **8**):

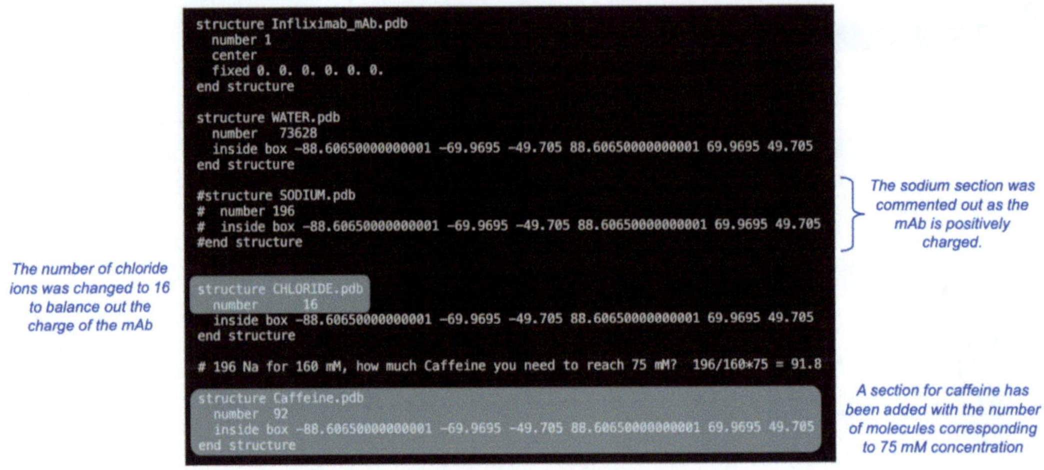

Fig. 7 Modifications made in the generated Packmol file, "packmol_input.inp" to account for the +16 charge of the infliximab mAb and caffeine molecules at 75 mM concentration

```
./solvate_PKL.tcl Infliximab_mAb.pdb -shell 10. -charge +16
-density 1.0 -o packmol_solvated.pdb
```

5. The generated "packmol_input.inp" file automatically adds water molecules as well as 160 mM concentration of sodium and chloride ions. As the Infliximab mAb carries a charge of +16, the sodium lines in the "packmol_input.inp" file should be commented out and the number of chloride ions should be edited to 16 to balance out the charge on the mAb (Fig. 7). Next, add a section for caffeine molecules with the same box size (Fig. 7). The number of caffeine molecules to be added can be calculated using Eq. 1.

$$\#\text{Excipient Molecules} = \frac{\#\text{Sodium Ions}}{160\ \text{mM}} \times \text{Desired Excipient Concentration} \quad (1)$$

6. The initial antibody-excipient configuration file is generated by executing the script below:

```
packmol < packmol_input.inp
```

7. Finally, the PDB and PSF input files needed for MD simulation are generated using VMD. Run the previously published *Infliximab_EXP.tcl* script by executing the command below [3]. This will generate "Infliximab_EXC_mod.pdb" and "Infliximab_EXC_mod.psf" files (Fig. 8).

```
vmd -dispdev text -e Infliximab_EXP.tcl
```

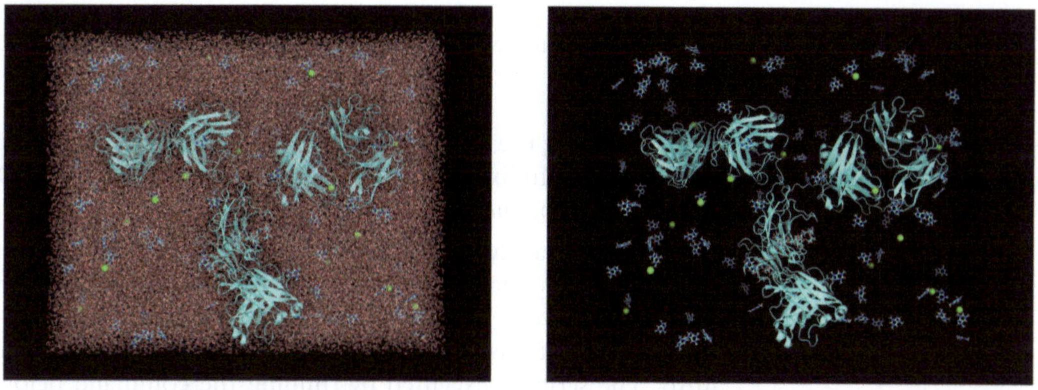

Fig. 8 Input structures for MD simulations with caffeine molecules. Left: Water molecules are shown. Right: Water molecules were removed from the image. The mAb is highlighted in cyan, chloride ions are in green, and the surrounding structures are caffeine molecules with carbon atoms colored in cyan

2.7 Molecular Dynamics Simulations of Antibody-Excipient Systems

1. The protocol for running MD simulations on antibody-excipient systems is like that of antibody-only system. Similarly, readers may choose to create a subfolder called "namd".
2. The configuration files and TCL scripts for minimization (*Infliximab_EXC_min.conf*), heat (*Infliximab_EXC_heat.conf* and *gen_restraints_heat.tcl*), equilibrium (*Infliximab_EXC_eq.conf*), and production (*Infliximab_EXC_prod.conf*) have been published earlier [3]. Move these to the "namd" folder. Notice that unlike for the antibody-only system, the configuration files for the antibody-excipient system also include the caffeine parameter file "Caffeine.prm".
3. Perform MD simulations by executing the commands below:

```
namd3 Infliximab_EXC_min.conf
vmd -dispdev text -e gen_restraints_heat.tcl
namd3 Infliximab_EXC_heat.conf
namd3 Infliximab_EXC_eq.conf
namd3 Infliximab_EXC_prod.conf
```

2.8 Analyzing Antibody-Excipient Binding Sites

1. Following the completion of MD simulations of the antibody-excipient structure, the number of binding events between caffeine molecules and infliximab mAb can be calculated by using MDAnalysis—an open-source Python library for analyzing MD trajectories [14, 15].
2. Prior to running the MDAnalysis Python script, the generated DCD file from the production run "Infliximab_EXC_prod.dcd" that contains atom trajectories needs to be edited to remove all water and ion molecules. This is done by executing a TCL script—*extract_chains.tcl* by running the command below (*see* **Note 9** for the code):

```
vmd -dispdev text -e extract_chains.tcl -args Infliximab_EXC_-
mod.psf Infliximab_EXC_prod.dcd Infliximab_EXC_prod_edited.
dcd
```

3. After executing the *extract_chains.tcl* script, an edited DCD file is generated "Infliximab_EXC_prod_edited.dcd". This DCD file does not contain any water and ion molecules.
4. Next, the MD analysis script is used to calculate the number of binding events that occurred between infliximab mAb and caffeine molecules. The binding event is defined as any excipient (caffeine) atoms within 4.0 Å from any residue atoms of the mAb. The script is executed by running the command below (*see* **Note 10** for the MDAnalysis code):

```
python3 binding_events_analysis.py contact_hist.txt
```

5. The MDAnalysis code will generate a histogram file in TXT file format containing the chain name, residue number, and the number of binding events occurred for each caffeine residue. These results can be plotted and analyzed using Excel in Microsoft Office Suite or any other tool preferred by the reader (Fig. 9).

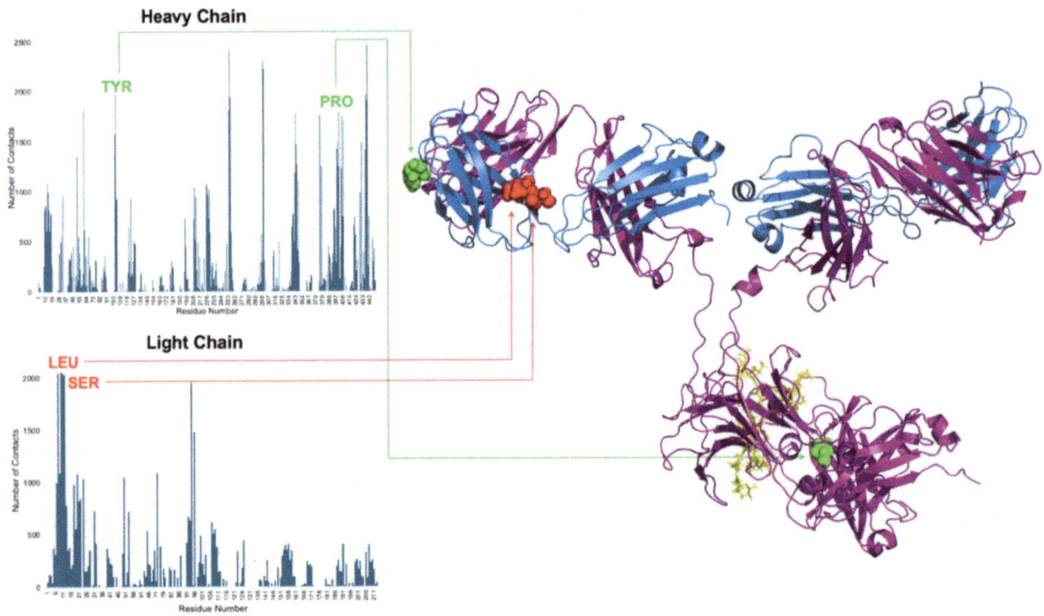

Fig. 9 Histograms of the number of contacts between caffeine molecules and Infliximab mAb on the heavy and light chains from 50 ns MD simulations. The light chains are colored in blue, heavy chains are in purple, and the glycan chains are in yellow. Two residues on each of the heavy and light chain with high number of caffeine contacts are showcased as spheres and colored in green and red, respectively

3 Notes

1. AlphaFold3 is an artificial intelligence (AI) system that predicts a protein's 3D structure from its amino acid sequence (https://alphafoldserver.com/welcome) [5]. To generate the structure, open the FASTA sequence file [3] and select the heavy chain sequence up until amino acid number 218 (ending with amino acids KKV, this is where the Fab region ends). Copy the sequence and paste to AlphaFold under the "Protein" selection (Fig. 10). Make another "Protein" entry in AlphaFold, select the entire light chain sequence, and paste it into the new entry. Specify the job name and submit. The resulting structure can be downloaded in Crystallographic Information File (CIF) format from AlphaFold3 and visualized in PyMOL (Fig. 10). After opening the structure in PyMOL, save it in PDB file format. ABodyBuilder2 is an automated antibody modeling pipeline that can be used to produce models of the antibody Fv regions (https://opig.stats.ox.ac.uk/webapps/sabdab-sabpred/sabpred/abodybuilder2/) [6]. The process is similar to the one described above for AlphaFold3: after navigating to the web server, paste the heavy and light chain sequences, specify the job name, and submit. Download the resulting structure in PDB file format (Fig. 10).
2. The chain names must be consistent, i.e., H for heavy chain and L for light chain. This can be done by renaming the chains

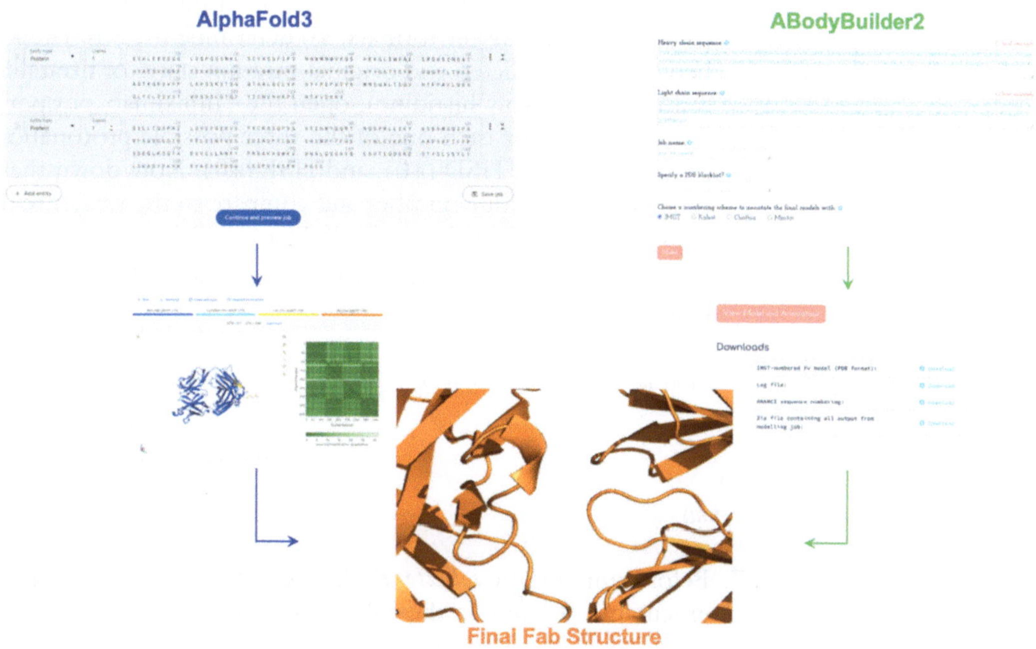

Fig. 10 Workflow for obtaining the antibody variable region using AlphaFold3 (left) and ABodyBuilder2 (right)

using a Linux command below, which changes all instances of item "A" present in "filename" into item "B". Note that Linux accounts for the spaces before and after "A" and "B".

```
sed -i "s@A @B @g" filename
```

3. When extracting the sequences of the light chain PDB files, "H" needs to be replaced with "L". See the command below (chain name "L" is highlighted).

```
cat Infliximab_L1_renumber.pdb | awk '/ATOM/ && $3 == "CA" &&
$5 == "L" {print $4}' | tr '\n' ' ' | sed 's/ALA/A/g;s/CYS/C/g;
s/ASP/D/g;s/GLU/E/g;s/PHE/F/g;s/GLY/G/g;
s/HIS/H/g;s/HSD/H/g;s/HSE/H/g;s/HSP/H/g;s/ILE/I/g;s/LYS/K/g;
s/LEU/L/g;s/MET/M/g;
s/ASN/N/g;s/PRO/P/g;s/GLN/Q/g;s/ARG/R/g;s/SER/S/g;s/THR/T/g;
s/VAL/V/g;s/TRP/W/g;
s/TYR/Y/g' | sed 's/ //g' | fold -w 60
```

4. To perform sequence alignment in EMBOSS Needle Sequence Alignment Tool, paste the sequence from the PDB file as the "first sequence" and the sequence of the target structure as the "second sequence" in the browser. Choose "pair" option for the "output format". The misaligned or incorrect sequences will be marked with dots in the browser.
5. The "Infliximab_pqr.pqr" file generated with the pdb2pqr software contains structural information of the antibody, similar to PDB and PSF file formats. To determine the mismatches between the predicted and original titration states of titratable residues (here, only histidine), open the "Infliximab_pqr.pqr" file and search for each one of the three possible protonation states: HSP (+1), HSD (+0), and HSE (+0). Note down their chain and the residue number and compare to the structure in the original "Infliximab_mod.pdb" file.
6. The parameter information starts and ends with the following lines:

```
read param card flex append
.....
....
END
RETURN
```

7. Before running the *solvate_PKL.tcl* script, it needs to be made executable by running the command below:

```
chmod +x solvate_PKL.tcl
```

8. The charge of the mAb is determined from the *Infliximab_EXP.tcl* script executed earlier. The charge of the infliximab mAb is +16.
9. The TCL script below is used to remove water and ion molecules from the antibody-excipient system.

```
# extract_chains.tcl
package require pbctools
# Arguments
# 0: PSF file
 # 1: input DCD file
# 2: output DCD file
if { [llength $argv] != 3 } {
puts stderr "Usage: vmd -dispdev text -e extract_chains.tcl
-args your.psf in.dcd out.dcd"
exit 1
}
set psffile [lindex $argv 0]
set dcd_in [lindex $argv 1]
set dcd_out [lindex $argv 2]
# Load files
mol new $psffile
mol addfile $dcd_in waitfor all
# Get the number of frames
set nframes [molinfo top get numframes]
# Chain selection (protein, excipients molecules, and ions) -
change the chain list here if you ever want a different subset
set sel [atomselect top "chain H I L M G J A"]
# Write the edited trajectory
animate write dcd $dcd_out sel $sel waitfor all beg 0 end [expr
{$nframes-1}]
# Clean up
$sel delete
exit
```

10. The MDAnalysis script below is used to calculate the number of binding events between the antibody and the excipient.

```
#!/usr/bin/env python
import sys
import numpy as np
import MDAnalysis as mda
from MDAnalysis.analysis.distances import distance_array
if len(sys.argv) != 2:
print("Usage: python binding_events_analysis_fixed.py out-
put_histogram.txt")
sys.exit(1)
```

```
out_file = sys.argv[1]
# Load topology and trajectory files
u = mda.Universe("Infliximab_EXC.psf", " Infliximab_EXC_pro-
d_edited.dcd")
# Select antibody chains (H, I, L, M)
mAb = u.select_atoms("segid H I L M")
if len(mAb) == 0:
raise RuntimeError("No mAb atoms found! Check your segid
names.")
print(f"Found {len(mAb)} mAb atoms.")
# Build a unique list of (segid, resid) for each residue
residues = list(mAb.residues)
n_residues = len(residues)
print(f"Detected {n_residues} total mAb residues.")
# Create a map: (segid, resid) → index in [0..n_residues-1]
residue_keys = [(res.segid, res.resid) for res in residues]
resid2idx = { key: idx for idx, key in enumerate(residue_keys)
}
# For each atom in mAb, store its residue-index
atom2res = np.array([resid2idx[(atom.segid, atom.resid)]
for atom in mAb.atoms], dtype=int)
# Select the excipient molecules (resname LIG)
ligand = u.select_atoms("resname LIG")
if len(ligand) == 0:
raise RuntimeError("No ligand atoms found! Check resname
LIG.")
print(f"Found {len(ligand)} ligand atoms.")
# Create a binding site histogram (cutoff 4 Å
hist = np.zeros(n_residues, dtype=int)
cutoff = 4.0 # Å
# Loop over frames using the numpy "where" function
print(f"Processing {len(u.trajectory)} frames ...")
for ts in u.trajectory:
D = distance_array(mAb.positions,
ligand.positions,
box=u.dimensions)
# find the atom-pairs within cutoff
mab_idxs, _ = np.where(D <= cutoff)
if mab_idxs.size == 0:
continue
contacted_res = np.unique(atom2res[mab_idxs])
hist[contacted_res] += 1
# Record the binding events
with open(out_file, "w") as fp:
fp.write("segid\tresid\tcount_frames\n")
for (segid, resid), count in zip(residue_keys, hist):
fp.write(f"{segid}\t{resid}\t{count}\n")
print(f"Wrote contact histogram ({n_residues} residues) to
{out_file}")
```

Acknowledgments

This work was financially supported by start-up funds from Stevens Institute of Technology. We thank the Advanced Cyberinfrastructure Coordination Ecosystem: Services & Support (CHM210016) and the Oak Ridge Leadership Computing Facility (BIP232) for supporting computing resources.

Conflict of Interest Statement The authors have no conflicts of interest to declare that are relevant to the content of this chapter.

References

1. Hollingsworth SA, Dror RO (2018) Molecular dynamics simulation for all. Neuron 99:1129–1143. https://doi.org/10.1016/j.neuron.2018.08.011
2. Karplus M, Petsko GA (1990) Molecular dynamics simulations in biology. Nature 347:631–639. https://doi.org/10.1038/347631a0
3. Phillips A, Srinivas A, Prentoska I et al (2024) Teaching biologics design using molecular modeling and simulations. Biochem Mol Biol Educ 52:299–310. https://doi.org/10.1002/bmb.21813
4. Schrödinger LLC (2015) The PyMOL molecular graphics system, Version 1.8
5. Abramson J, Adler J, Dunger J et al (2024) Accurate structure prediction of biomolecular interactions with AlphaFold 3. Nature 630:493–500. https://doi.org/10.1038/s41586-024-07487-w
6. Abanades B, Wong WK, Boyles F et al (2023) ImmuneBuilder: deep-learning models for predicting the structures of immune proteins. Commun Biol 6:575. https://doi.org/10.1038/s42003-023-04927-7
7. Humphrey W, Dalke A, Schulten K (1996) VMD: visual molecular dynamics. J Mol Graph 14:33–38. https://doi.org/10.1016/0263-7855(96)00018-5
8. Dolinsky TJ, Czodrowski P, Li H et al (2007) PDB2PQR: expanding and upgrading automated preparation of biomolecular structures for molecular simulations. Nucleic Acids Res 35:W522–W525. https://doi.org/10.1093/nar/gkm276
9. Dolinsky TJ, Nielsen JE, McCammon JA, Baker NA (2004) PDB2PQR: an automated pipeline for the setup of Poisson–Boltzmann electrostatics calculations. Nucleic Acids Res 32:W665–W667. https://doi.org/10.1093/nar/gkh381
10. Huang J, Rauscher S, Nawrocki G et al (2017) CHARMM36m: an improved force field for folded and intrinsically disordered proteins. Nat Methods 14:71–73. https://doi.org/10.1038/nmeth.4067
11. Vanommeslaeghe K, Hatcher E, Acharya C et al (2010) CHARMM general force field: a force field for drug-like molecules compatible with the CHARMM all-atom additive biological force fields. J Comput Chem 31:671–690. https://doi.org/10.1002/jcc.21367
12. Phillips JC, Braun R, Wang W et al (2005) Scalable molecular dynamics with NAMD. J Comput Chem 26:1781–1802. https://doi.org/10.1002/jcc.20289
13. Phillips JC, Hardy DJ, Maia JDC et al (2020) Scalable molecular dynamics on CPU and GPU architectures with NAMD. J Chem Phys 153:044130. https://doi.org/10.1063/5.0014475
14. Gowers RJ, Linke M, Barnoud J et al (2016) MDAnalysis: a python package for the rapid analysis of molecular dynamics simulations. In: Benthall S, Rostrup S (eds) Proceedings of the 15th python in science conference, pp 98–105
15. Michaud-Agrawal N, Denning EJ, Woolf TB, Beckstein O (2011) MDAnalysis: a toolkit for the analysis of molecular dynamics simulations. J Comput Chem 32:2319–2327. https://doi.org/10.1002/jcc.21787
16. Jumper J, Evans R, Pritzel A et al (2021) Highly accurate protein structure prediction with AlphaFold. Nature 596:583–589. https://doi.org/10.1038/s41586-021-03819-2

17. Leem J, Dunbar J, Georges G et al (2016) ABodyBuilder: automated antibody structure prediction with data–driven accuracy estimation. MAbs 8:1259–1268. https://doi.org/10.1080/19420862.2016.1205773
18. Padlan EA (1994) Anatomy of the antibody molecule. Mol Immunol 31:169–217. https://doi.org/10.1016/0161-5890(94)90001-9
19. Boehm MK, Woof JM, Kerr MA, Perkins SJ (1999) The fab and fc fragments of IgA1 exhibit a different arrangement from that in IgG: a study by X-ray and neutron solution scattering and homology modelling. J Mol Biol 286:1421–1447. https://doi.org/10.1006/jmbi.1998.2556
20. Lasham J, Djurabekova A, Zickermann V et al (2024) Role of protonation states in the stability of molecular dynamics simulations of high-resolution membrane protein structures. J Phys Chem B 128:2304–2316. https://doi.org/10.1021/acs.jpcb.3c07421
21. Martínez L, Andrade R, Birgin EG, Martínez JM (2009) PACKMOL: a package for building initial configurations for molecular dynamics simulations. J Comput Chem 30:2157–2164. https://doi.org/10.1002/jcc.21224

Chapter 7

Generative AI Agents as Generalized Data Visualization Tools

Morgan A. Vincent and Benjamin J. Lear

Abstract

Every practicing scientist will need to produce data visualizations. At present, the most common way to accomplish this is with dedicated software tools, be they GUI-based (i.e., Origin or Excel) or text-based (i.e., Python or R). In both cases, one must have intimate knowledge of the structure of graphs, from the perspective of the tool, to control the design of the plot. However, newly emerged LLM-based generative AI chatbots can both write and run code that will produce data visualizations. Such chatbots effectively serve as a generalized front-end for text-based plotting tools that remove the need to know all the specific terminology of the tool. Herein, we show how one can use these tools to process data and plot it, all using a natural language interface.

Key words Data visualization, Artificial intelligence, Large language models, Data analysis, File formats, Analysis workflow

1 Introduction

Data visualizations are the backbone of scientific communication. They can be found in all forms of scientific communication, from formal discourse (in the form of presentations and publications) to public outreach (in the form of press releases and popular science articles) to informal discussions (i.e., the proverbial sketch on a napkin).

Since the invention of science, the workflows for creating data visualizations have changed drastically, and so too have the tools that enable them. Early data visualizations were drawn by hand, often by the very scientists who generated the data. This was the workflow from the earliest products that resemble modern data visualizations in the tenth century, [1] to at least the middle of the seventeenth century [2]. At the time, there was little demand for reproduction (especially compared to today), and so such

Robert E. Dempski (ed.), *Transforming Molecular Biology with Emerging Technologies*, Methods in Molecular Biology, vol. 3043, https://doi.org/10.1007/978-1-0716-5320-3_7,

visualizations were also copied by hand. Many of these visualizations appear crudely designed to our modern eye.

As the public became more scientifically and numerically literate, a demand emerged for wider discrimination and data visualizations were reproduced via engravings, produced by specialists who worked with those who were collecting and analyzing data [3]. This close partnership between tradesman and scientist during the production of data visualizations remained until the late nineteenth century. While technology changed—moving from engravings [3] to photographs to early specialized computer-based tools [4]—the involvement of an expert trained specifically in the production of these data visualizations generally resulted in high-quality visualizations that were well-thought-out and employed sound design choices. This era might reasonably be considered the "golden age" of data visualizations [5]. It was during the early to middle of this era that many of the data visualizations that we are familiar with today were created, including the bar chart, [6] line chart, [6] and pie chart [7].

In the late nineteenth century, the popularization of the personal computer and the emergence of accessible tools for plotting data democratized the production of data visualizations. It was after this time that many new data visualizations—harnessing ever-increasing computational power—were invented [5]. However, it is also true that many scientific departments stopped employing drafts people to produce these visualizations. While remnants of this partnership remain in the form of publicity offices that produce highly stylized schematics of scientific ideas, the burden of producing standard data visualizations has largely shifted back to the practicing scientist [5]. The end result is that modern scientists are not only expected to be experts in the design of experiments, collection of data, and analysis of data, but in the production of data visualizations as well. We might reasonably term this era the "modern age" of data visualization, marked by an explosion in the raw number of data visualizations—as more scientists learn to use the tools required to produce data visualizations. At the same time, one might argue that, as the number of data visualizations has increased, the average quality of the graphics has decreased. The decrease in quality seems likely a result of the fact that scientists are not well-versed in graphic design. Recognition of this knowledge gap has spawned a series of resources that are aimed at helping practicing scientists learn to produce as well as design data visualizations. This manuscript might be classified as the former, while there are books, [8–10] journals, [11–13] and websites [14] that could be classified as the latter.

At present, we are firmly entrenched in this modern age of data visualizations, but it is possible that we are entering a new era, one defined by the emergence of a new tool: generative artificial intelligence (AI) based on large language models (LLMs). In this new era, scientists can again find themselves with a ready partner, in the form of an AI agent, that can help them think through the design, perform the creation of, and even refine data visualizations. This chapter presents one example of a workflow that uses a generative AI agent to produce a publication-quality data visualization. Though the workflow is presented using a specific tool (Open AI's ChatGPT 5), the approach is expected to be generalizable to other current and future tools.

2 Materials

The workflow shown here relies on the following materials:

1. Data that is to be visualized in a format readable by the AI agent. The data we have presented here is fabricated. It is also stored in a .csv file. Other file formats are amenable to the workflow, as long as the AI agent can process the information. For this reason, structured formats that are plain text will often be the best. Common formats that meet these requirements are .txt, .csv, .tsv, .json, and .xlsx. The latter of these is the Excel file format, which many AI agents can process. However, it is worth noting that many other proprietary formats are not easily processable by AI agents. Thus, it is quite probable that raw data taken from an instrument would not be amenable to the approach demonstrated below.
2. Access to an AI agent that can work with data files, produce and run code, and then provide access to final files for download. Each of these components is necessary for the workflow shown below. Typically, if an AI agent can both produce and run code, the other two capabilities will also be present. At the time of this writing, there were a myriad of such AI agents, including ChatGPT 4.o, ChatGPT 5, Claude Opus 4.1, and Gemini 2.5. In the workflow demonstrated below, we use ChatGPT 5, but similar results are expected from other agents that possess the required capabilities.
3. A program for editing scalable vector graphics. One of the authors uses Adobe Illustrator for this, while the other uses Affinity Designer 2. These programs are not strictly necessary for creating data visualizations, but in our experience, it can often be time-consuming to have the AI agent accomplish every small-scale change that you desire. For instance, if you are trying to ensure proper placement of a feature, like an annotation, it can be faster to open up a .svg of the image for this final modification. An example of the sort of transformations where this approach is useful is also illustrated in the method below.

3 Methods

1. Begin by compiling the raw data into a structured file format, preferably a comma-separated values (.csv) file or an Excel (.xlsx) spreadsheet. Ensure that the dataset contains informative column names, consistent data types, and that each row represents a single observation (*see* **Note 1**). These guidelines are observable in the data shown in Fig. 1. Note that, though this data is fabricated, it reflects the sort of data one might expect to encounter in the biological sciences. Save the file in an accessible location for upload to the AI agent.
2. Upload the prepared dataset into the ChatGPT interface (*see* **Notes 2** and **4**). With the AI agent open, this can be done by simply dragging the file into the prompt window (*see* **Note 3**). Once the file has been successfully uploaded, the system will display the file name, storage path, and/or a data preview. The file should now be available for subsequent reading and analysis. Figure 2 shows the interface that indicates the file "Sample Data.csv" has been uploaded.

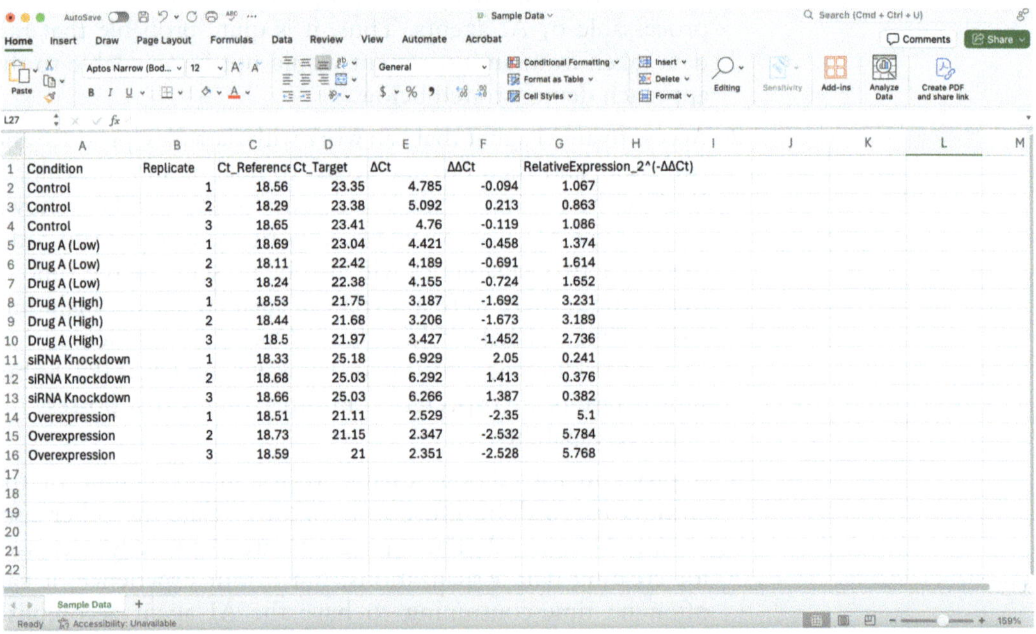

Condition	Replicate	Ct_Reference	Ct_Target	ΔCt	ΔΔCt	RelativeExpression_2^(-ΔΔCt)
Control	1	18.56	23.35	4.785	-0.094	1.067
Control	2	18.29	23.38	5.092	0.213	0.863
Control	3	18.65	23.41	4.76	-0.119	1.086
Drug A (Low)	1	18.69	23.04	4.421	-0.458	1.374
Drug A (Low)	2	18.11	22.42	4.189	-0.691	1.614
Drug A (Low)	3	18.24	22.38	4.155	-0.724	1.652
Drug A (High)	1	18.53	21.75	3.187	-1.692	3.231
Drug A (High)	2	18.44	21.68	3.206	-1.673	3.189
Drug A (High)	3	18.5	21.97	3.427	-1.452	2.736
siRNA Knockdown	1	18.33	25.18	6.929	2.05	0.241
siRNA Knockdown	2	18.68	25.03	6.292	1.413	0.376
siRNA Knockdown	3	18.66	25.03	6.266	1.387	0.382
Overexpression	1	18.51	21.11	2.529	-2.35	5.1
Overexpression	2	18.73	21.15	2.347	-2.532	5.784
Overexpression	3	18.59	21	2.351	-2.528	5.768

Fig. 1 Example of a (fabricated) dataset formatted in Excel. Each column contains clearly labeled variables (e.g., condition, replicate, Ct values, calculated ΔCt and $\Delta\Delta Ct$, relative expression), and each row corresponds to a single observation. While the data shown are not real experimental results, they illustrate the type of structured file needed for subsequent upload and visualization steps

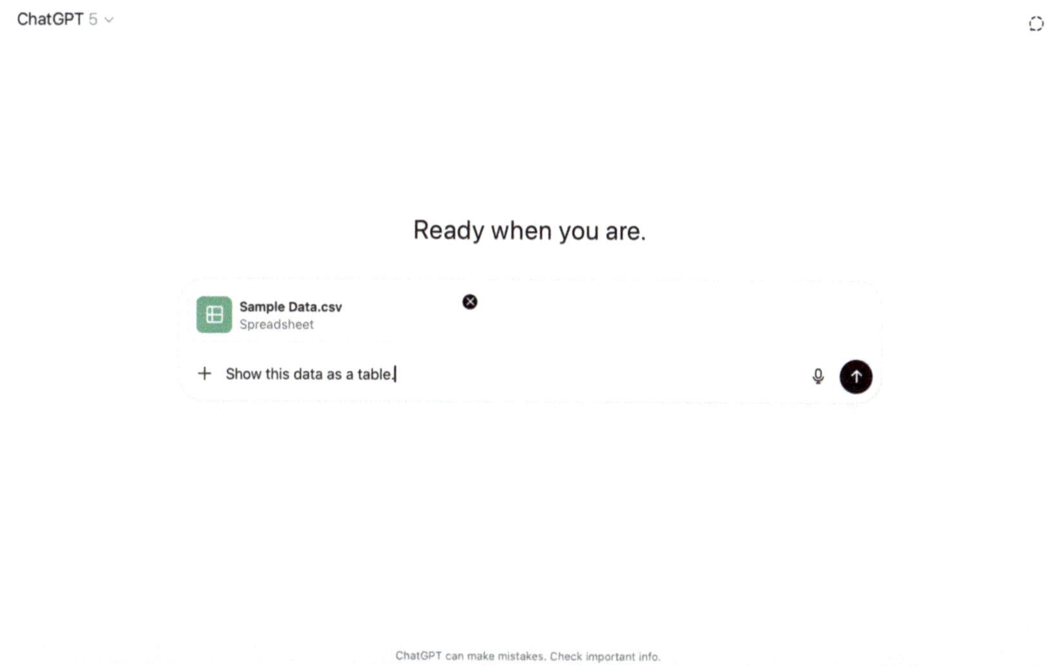

Fig. 2 Screenshot of uploading a dataset into the ChatGPT interface. The file Sample Data.csv appears in the prompt window, confirming that the upload was successful and that the dataset is now accessible for subsequent reading and analysis. If the upload fails, no file name will appear, and the process should be repeated to ensure the data is properly loaded

3. Instruct the ChatGPT agent to display the data as a table. Review the output to confirm that the columns have been correctly interpreted and that no data corruption has occurred during upload and reading by the AI agent. At this stage, also inspect for missing values or inconsistent entries that may require cleaning (*see* **Note 5**). Figure 2 contains a prompt asking for such a display, and Fig. 3 shows the response to this prompt. The output contains a scrollable table contained within a sub-window and a textual explanation confirming that the table has been made.
4. (Optional) Define the transformations necessary to prepare the dataset for visualization. This may include converting raw measurements into relative units, normalizing values to a reference condition, performing background subtraction, etc. (*see* **Note 6**).
5. (Optional) Request that ChatGPT compute summary statistics for each group of interest (*see* **Note 6**). Typical measures include the mean and standard deviation. The resulting summary table provides the values used for error bars and contextual interpretation of the data. Figure 4 shows the prompt asking for the mean and standard deviation for the conditions,

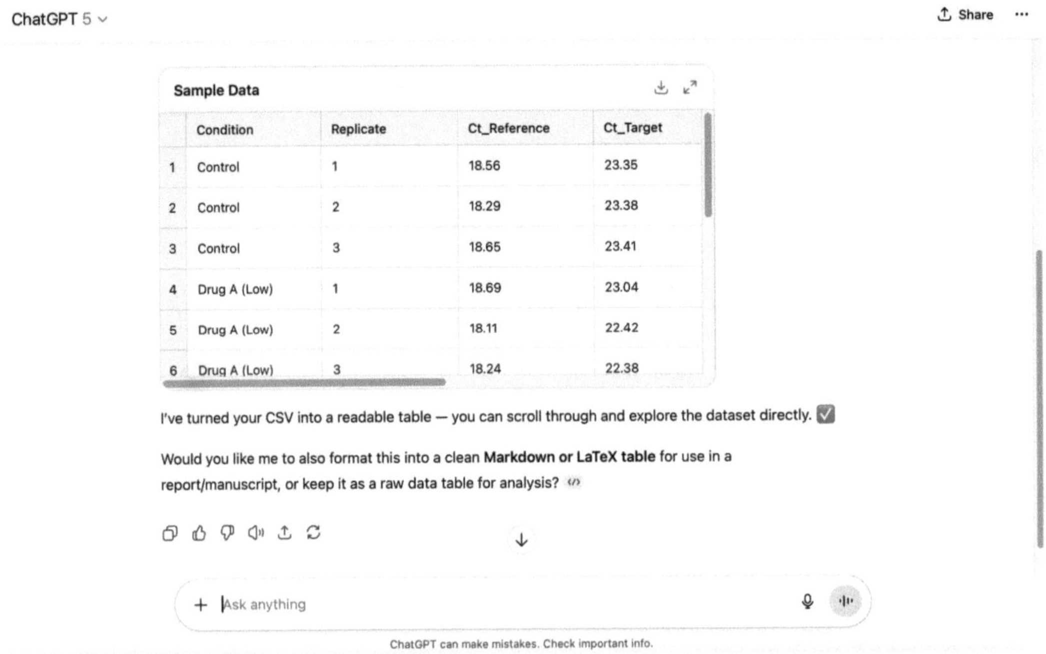

Fig. 3 ChatGPT output displaying the uploaded dataset as a scrollable table. Each column and row has been correctly interpreted (compare to Fig. 1), and the interface provides a textual confirmation that the data have been rendered in table format

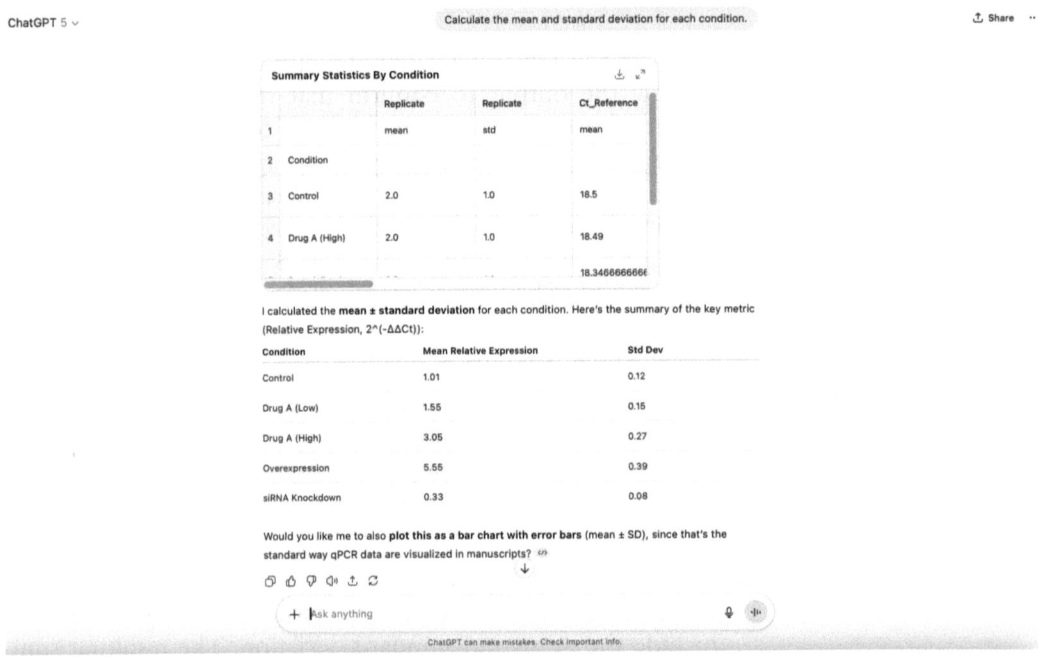

Fig. 4 ChatGPT output showing summary statistics (mean and standard deviation) for each condition, presented in both tabular form and a concise text summary

Fig. 6 ChatGPT output in response to a prompt requesting a bar chart of the dataset. The preliminary figure displays mean values with error bars (standard error of the mean) and overlays individual replicates as multi-colored "×" markers. This first draft is used to confirm that categories, axis labels, scales, and summary statistics are mapped correctly before any styling or refinement. A short textual explanation is also provided beneath the figure to confirm how the data are represented

as well as the output, again structured in a table, this time directly in the chat.

6. (Optional) Before generating the first draft of a figure, it can be useful to ask ChatGPT for recommendations on different ways to visualize or analyze the data. Suggestions may include new types of visualization, as well as more advanced analysis approaches. Either way, this step encourages consideration of multiple perspectives on the dataset and may reveal patterns or relationships that would not be apparent in the originally planned visualization. Figure 5 shows the prompt, "Are there any ideas to refine the data workup before plotting?" and the responses—in this case, a full page of bulleted things to consider. The user is, of course, responsible for vetting these ideas and ensuring that any implementation of them is appropriate.

7. Generate a preliminary visualization using ChatGPT's plotting capabilities (*see* **Note** 7). Common options include bar charts with error bars, scatter plots, or line graphs, depending on the structure of the data. Many other options are possible, depending on the underlying capabilities (*see* **Note** 7). At this stage,

more robust and manuscript-ready. Here are some ideas:

🔍 Data Quality & Pre-Processing

1. **Check replicates for outliers**
   - Run a simple outlier test (e.g., Grubbs' test, Dixon's Q, or visualize boxplots of ΔCt).
   - In qPCR, even a single bad pipetting event can throw off the averages — removing or flagging those increases confidence.
2. **Normalize to the reference gene consistently**
   - Confirm your ΔCt was calculated as *Ct_Target – Ct_Reference* in all cases.
   - Sometimes people flip it by mistake, which inverts fold-change.
3. **Set control as the baseline**
   - By definition, the control's ΔΔCt mean should be zero and fold-change ~1.
   - Double-check that this normalization holds, since all other conditions are relative to it.

📊 Statistical Refinement

4. **Work with log2 fold-change**
   - Relative expression (2^(-ΔΔCt)) can be skewed. Taking the log2 makes effects symmetric around zero (up vs. down regulation).
   - This also makes error bars more interpretable in publications.
5. **Report variability from ΔCt stage**
   - Instead of only propagating variability at the expression stage, you can calculate mean ± SD at the ΔCt level, then convert.
   - This avoids inflating errors after exponentiation.
6. **Consider biological vs. technical replicates**
   - If your replicates are only technical, SD may underestimate biological variation.
   - Flag clearly in figure captions whether error bars represent technical or biological replicates.

📐 Visualization Prep

7. **Decide on scale**
   - If relative expression spans wide values (like your Overexpression ~5.5 vs Knockdown ~0.3), a log scale y-axis avoids squashing small bars.
   - Alternatively, present both linear and log views.
8. **Annotate significance**
   - Plan your statistical testing (e.g., t-tests vs Control, or ANOVA with post-hoc tests) so you can add stars or brackets above bars.
   - Readers expect this in molecular biology plots.
9. **Use consistent bar coloring**
   - Grayscale for conditions or a subtle color palette helps readability.
   - Consider leaving the control as neutral (e.g., gray) and treatments in color.

✅ A streamlined pipeline would be:
Raw Ct → ΔCt (check replicates & remove outliers) → ΔΔCt (normalize to control) → log2 fold-change (for stats/plots) → plot bar chart with mean ± SD and significance annotations.

+ Ask anything

ChatGPT can make mistakes. Check important info.

Fig. 5 ChatGPT's response to a prompt requesting ideas prior to plotting. The output includes a structured set of considerations for data quality, statistical refinement, and visualization preparation, presented as a bulleted list

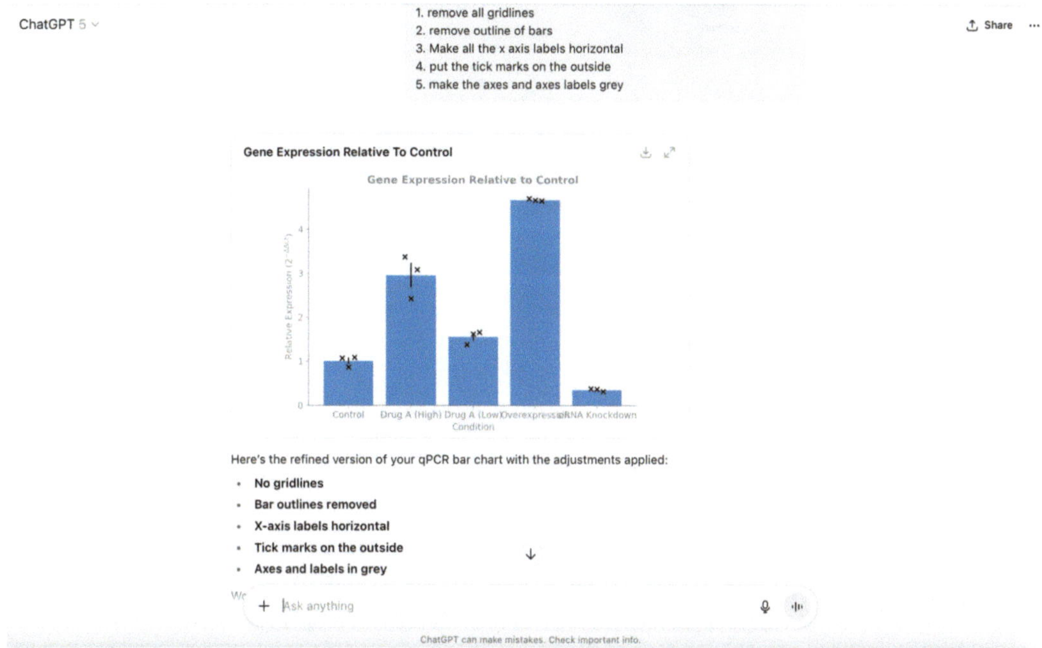

Fig. 7 Refined version of the preliminary bar chart generated by ChatGPT after applying specific formatting instructions. The updated figure removes gridlines and bar outlines as well as sets axis labels and ticks in gray. The changes demonstrate how prompt-based adjustments can bring the figure closer to conventional visualization presentation standards

the emphasis is on verifying that the figure correctly represents the data and includes all necessary elements. Figure 6 shows the output from ChatGPT when asked to make a bar chart of means, with error bars representing the standard deviation (*see* **Note 8**).

8. Iteratively refine the visualization to match expected conventions and design standards (*see* **Note 8**). For example, in this case, we wished to adjust the order of categories on the axis, modify font sizes and labels, remove extraneous grid lines, and set appropriate axis limits (*see* **Note 9**). During this stage, it is important to apply principles of prompt engineering (*see* **Note 10**) by giving ChatGPT specific and detailed instructions (e.g., "make the x-axis labels horizontal and gray," or "limit the y-axis to 5 units", *see* **Note 11**). Though not shown here, one may need to repeat directions across prompts, as discussed in **Note 12**. This refinement process may be repeated as many times as needed to arrive at a satisfactory design (*see* **Note 13**). In this workflow, we perform two refinements. Figure 7 shows a plot that provides a list of desired changes, with the output of the refined plot based on these instructions. Figure 8 shows even more refinements that follow, this time asking for bars to

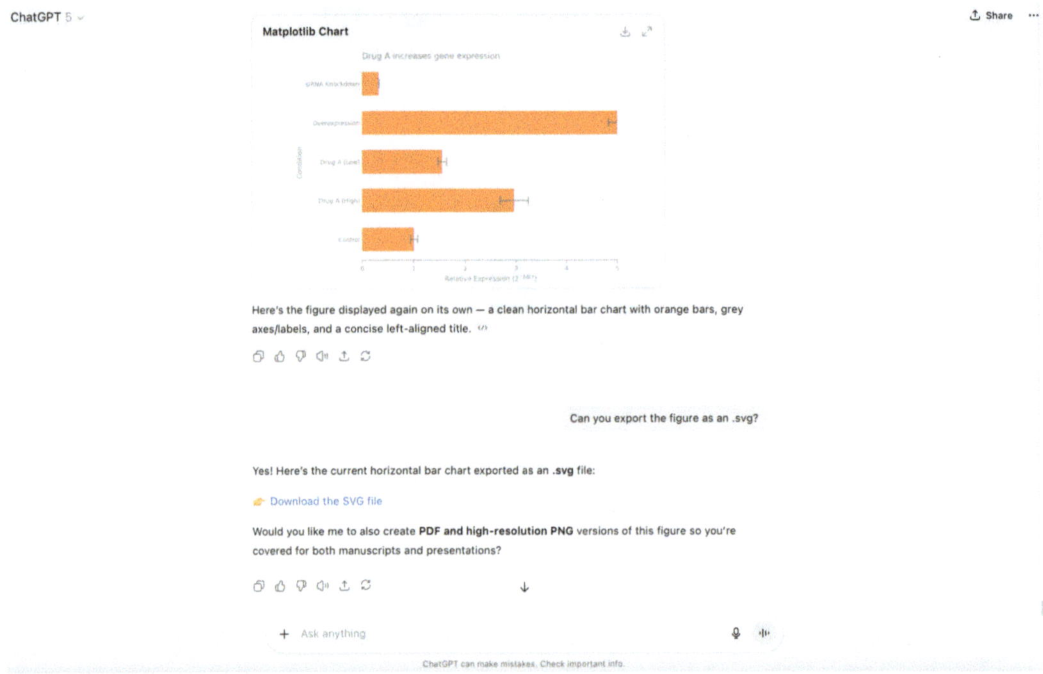

Fig. 8 Another refinement of the data visualization, this time changing the colors of the bars and their orientation. Additionally, there is an example of exporting a refined figure from ChatGPT. In this case, the finalized horizontal bar chart is provided with a clickable link to download the file in a svg format

be recolored orange, the title to make a claim (*see* **Note 14**), and the bars to run horizontally, rather than vertically.

9. Once you have refined the figure as much as you wish in ChatGPT, export it in one or more publication-quality formats. Scalable vector graphics (.svg) and portable document format (.pdf) are often the best choice, though bitmaps (such as a .png) may be required by certain publications. Ensure that all exported files are stored with clear filenames and version control. Figure 8 has a prompt below the chart, requesting the image to be exported as a .svg file. ChatGPT then produces a clickable link that will allow the user to download the file (*see* **Note 15**).

10. Save both the processed dataset and the summary statistics used to generate the figure (*see* **Note 15**). This provides a record that helps with transparency, facilitates peer review, and ensures reproducibility of the visualization. You can request these files from ChatGPT, and it will provide a link to download the files. This works similarly to what is shown in Fig. 8 for the .svg file but will provide a link to a data file. You can specify the format, but if you do not, it will likely produce a .csv file.

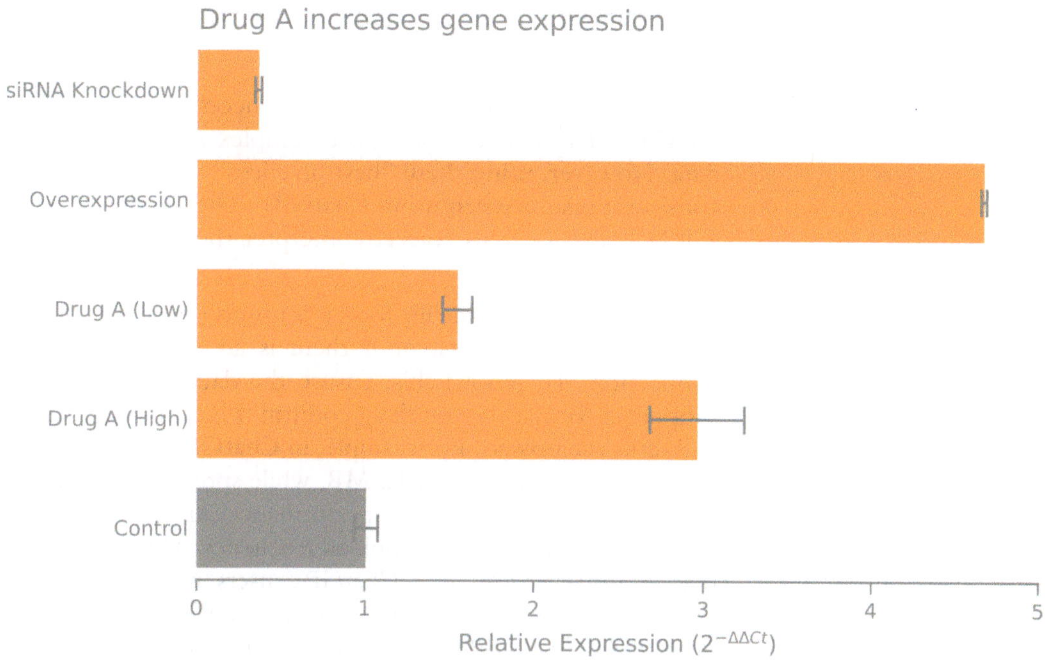

Fig. 9 Finalized version of the bar chart after adjustments in Adobe Illustrator. Compared to the final ChatGPT-generated figure, refinements include removing redundant axis labeling, modifying font and line styling, and recoloring the control bar to improve contrast. These edits illustrate how external graphics software can be used to polish figures for publication or presentation

11. Although ChatGPT can generate high-quality figures, certain refinements are more efficiently carried out in a dedicated graphics program (e.g., Adobe Illustrator, Affinity Designer 2, Boxey, etc.). These refinements may include adjusting font sizes, standardizing line weights, and ensuring color palettes meet accessibility requirements (such as color-blind accessible schemes, *see* **Note 16**). Additional annotations, labels, or graphical elements that are difficult to produce directly in ChatGPT can also be added at this stage. Once finalized, the figure should be re-exported in the format required by the target journal or presentation medium. Figure 9 shows the bar chart after some refinements in Adobe Illustrator (*see* **Note 17**). The major changes were that the redundant title on the y-axis was removed, and the bar associated with the control condition had its color changed. Though the figure produced entirely within ChatGPT (Fig. 8) was probably sufficient for publication, these small changes (working from the figure produced by ChatGPT) have improved the design and communication of the message (*see* **Note 9**).

4 Notes

1. This structure for the data is not strictly necessary. It is possible for ChatGPT to figure out more complex structures for the data. However, using these "best practices" will provide more consistent results when using ChatGPT. That is, the AI agent will be most likely to correctly interpret the structure of the data on the first try.
2. An upload may fail if the file format is unsupported, the file size exceeds platform limits, or if there is an unstable Internet connection. To resolve this, ensure the dataset is saved in a supported format (csv or xlsx), confirm it is within size limits, and retry the upload. For example, in ChatGPT 5, most document types can be up to 512 MB, while spreadsheets are best kept under 50 MB for reliable performance. Another reason for upload failure could be the upload frequency. This depends on the account version: free ChatGPT users may only upload about three files per day, whereas ChatGPT Plus users can upload as many as 80 files in a three-hour rolling window.
3. Though we are illustrating using ChatGPT in a "conventional" chat window, there are other options for interfacing with the AI agent. For instance, ChatGPT has a "canvas" feature, in which code can be written collaboratively with the AI agent. Also, ChatGPT can be included within integrated development environments (IDEs) such as VS Code, Positron, Sypder, and Marimo notebooks. Other AI agents can also be used in similar manners.
4. The uploaded data can expire. Each time you start a new session (or restart an old one), you will need to upload the data again. If you made progress in cleaning and/or analyzing the data, you may choose to upload that file instead of, or in addition to, the original. There are also limits on the amount of information that can be uploaded (*see* **Note 2**).
5. It is important to realize that the AI agent's responses are probabilistic in nature. That is to say that, given the exact same user prompt, they almost always produce different outputs. This is worth keeping in mind when asking ChatGPT to perform analysis or transformations of data. Though there could be multiple ways in which to *correctly* perform analysis, it is also possible that incorrect analysis was performed (also the case when done by humans). Thus, we strongly recommend looking at the code that is produced by the program and ensuring that it is working as intended. If you are unable to understand the code on your own, then the AI agent can serve as an interpreter as well. Though it may seem odd to ask an AI agent to check its own work, it can be quite successful.

However, be aware that the system is often very agreeable when you suggest corrections, and it may confirm changes even if they are not technically accurate or appropriate. As with *any* technical analysis, caution is required.

6. Analysis can be done either in the AI chat or outside of it (using conventional approaches). You can mix and match these approaches, ensuring that (if you have modified the data outside the chat window), you update the uploaded file to have the most recent form of the data to make a visualization.
7. Even when the AI agent simply produces a data visualization, it is important to realize that the way this is done is by writing and then executing computer code. At present, this is always the case. For ChatGPT 5, the AI agent is producing Python code and then executing this and displaying the output to the chat window. The reason this is important to realize is that you can access the code produced by ChatGPT 5, which you may wish to do to verify that the processing of your data was done correctly, or to copy to a different environment for use. In addition, having access to the underlying code allows you to create a polished visualization that can be easily reused with other datasets, ensuring consistent formatting and results. The authors recommend maintaining commonly used code in a shared GitHub repository (https://github.com/), a platform for storing, organizing, and version-controlling code so it can be easily accessed, updated, and reused by others in the lab. This approach is especially valuable for laboratories that frequently generate the same types of plots, as it saves time and enforces standardization across figures.
8. If you have a particular style of plotting or a particular approach to analysis that you anticipate using repeatedly, then you can try to create your own custom AI agent for these tasks. For instance, in ChatGPT, you can create a "custom GPT" where you provide overall context on these preferences, and this can streamline the process of working with the AI agent.
9. A detailed discussion of the theory behind the design of data visualizations is beyond the scope of this chapter, which is focused on the usage of AI agents. However, there are many good resources for this. Good starting places include the website www.when-data-meets-design.com as well as the book "Effective Data Visualization" [8] and "The Data Visualization Handbook" [9].
10. The AI agent may not always produce the result you had in mind. There are several reasons for this. For one, the AI agents are probabilistic in nature (*see* **Note 5**), and so their responses have some chance of being quite different than what would typically be expected. However, it could also be the case that

the prompt was unclear. For instance, someone might colloquially ask for the "average" of a data set and then the AI agent could (correctly) return either the mean or the median. So, being clear about what is desired can be critical. To help alleviate this confusion, one can use "prompt engineering" where the prompt is designed to make the desired outcome clearer. While part of this involves using precise language, another might be providing context in the prompt. For instance, you might add "I am a second-year graduate student in a molecular biology laboratory..." into your prompts, which will help focus the AI agent to produce responses that are more appropriate for such cases.

11. Understanding the code underlying the data visualizations can help tailor your requests. For instance, ChatGPT 5 uses the Matplotlib library of Python as the basis of its plotting. Thus, it will be very challenging to ask ChatGPT to directly produce data visualizations that are not supported by Matplotlib. Additionally, Matplotlib has a particular conception of how plots are constructed, which biases some of the design language. Though the AI agent most often functions as an effective translator of user input to the language of the plotting library, there may be times when using the precise language of this library will be required to attain the desired results. In this case, online documentation for the library can be quite useful (e.g., Matplotlib Library: https://matplotlib.org/stable/users/index), or you can ask the AI agent to give the precise language for the changes you are attempting to implement.

12. LLM-based AI agents are trained on large sets of textual data, and the outputs that are produced reflect the most likely output, given the text it was trained on. This means that there can be aspects of plotting that are extremely common in the training set, and these will then be very likely to be produced by the AI agent. For instance, a very large number of Matplotlib examples on the Internet use grid lines. Since ChatGPT 5 uses Matplotlib as its plotting library, it will draw from these examples and almost always default to having grid lines in the plot. The tendency can be extremely strong, and it can be very hard to "convince" ChatGPT 5 not to use grid lines. In such cases, it might be more efficient to accept the plot with the grid lines and then remove them in svg editing software later (*see* **Note 14**). This holds for many other aspects of plots, depending on the training set used for the specific AI agent you are using.

13. While it is true that one can indefinitely continue to refine a chart using ChatGPT, each chat session is limited to a certain number of "tokens," which are the chunks that language is broken into for the AI agent. Each time a prompt is given,

ChatGPT reads the entire history of the session, which functions as the "context window" for the prompt. At the time of this writing, the context window limit for ChatGPT 5 was 400,000 tokens. When a session reaches this length, new prompts can no longer be submitted. Instead, one would need to open a new session to continue the work. Doing so can have mixed results, as the context for new prompts is no longer present. Thus, it is best to attempt to complete figure design within a single session. Though there is no strict correspondence between words and tokens, a rough estimate is that 400,000 tokens $\approx$ 300,000 words (a lengthy novel). Thus, in practice, this limit is not likely to be met during a normal data visualization workflow, but it is worth being aware of.

14. A useful strategy when working with AI agents for data visualizations is to prompt for titles that make a clear, interpretable claim about the data. Framing the title as a claim can help ensure that the visualization communicates a specific takeaway rather than simply describing variables or axes. However, not all visualizations require a title. Instead, the central claim can be conveyed through other elements, such as the figure caption, accompanying text, or the surrounding narrative (e.g., an oral presentation) in which the figure is embedded. In these cases, one might directly prompt the agent to generate a draft caption or narrative that articulates the intended takeaway. These outputs should be treated as starting points to be refined and verified by the creator of the visualization. As such, prompting an AI agent to generate claim-focused language can still be valuable.

15. The downloadable links provided by ChatGPT can expire. Essentially, each time a new session starts (or an old session restarts), any existing links will no longer be valid. This is similar to how uploaded data can expire, as discussed in **Note 4**.

16. Visualizations should be designed with accessibility in mind, particularly for individuals with color vision deficiencies. Color should not be relied upon as the only means of conveying information; it is most effective when paired with other encodings such as position, shape, or direct labeling. Therefore, when using AI agents to generate or refine visualizations, it may be useful to prompt for accessible or colorblind-friendly palettes. These outputs should be treated as drafts and evaluated by the user. Tools such as the Coolors contrast checker (https://coolors.co/contrast-checker) can be used to assess whether color combinations provide sufficient contrast, while the Coblis color blindness simulator (https://www.colorblindness.com/coblis-color-blindness-simulator/) allows designers to preview how a visualization may appear to

individuals with different types of color vision deficiencies. Although these tools support more deliberate design decisions, the responsibility for verifying accessibility and making appropriate revisions remains with the individual creating the visualization.

17. The edits shown here were performed in Adobe Illustrator, but there are several software tools that can edit svg files. One of the authors of this manuscript uses Adobe Illustrator as their primary tool, while the other uses Affinity Designer 2. There are also other tools such as Boxy, which are free-to-use online .svg editors.

Acknowledgments

BJL thanks NSF (grant number 2518899) for financial support of this work.

Conflict of Interest Statement This study was funded by the National Science Foundation award Number 2518899.

References

1. Funkhouser HG (1936) A note on a tenth century graph. Osiris 1:260–262
2. Tufte ER (2001) The visual display of quantitative information. Graphics Press
3. Wallis HM, Robinson AH (1987) Cartographical innovations, an international handbook of mapping terms to 1900. Map Collector Publications
4. Tukey JW (1962) The future of data analysis. Ann Math Stat 33:1–67
5. Friendly M (2008) A brief history of data visualization. In: Handbook of data visualization. Springer Berlin Heidelberg, Berlin/Heidelberg, pp 15–56
6. Playfair W (1801) The commercial and political atlas representing, by means of stained copper-plate charts, the progress of the commerce, revenues, expenditure, and debts of England, during the whole of the eighteenth century
7. Playfair W The statistical breviary: shewing the resources of every state and Kingdom in Europe
8. Evergreen SDH (2016) Effective data visualization: the right chart for the right data. SAGE Publications
9. Koponen JHJ (2019) Data visualization handbook. Aalto University, Espoo
10. Few S (2012) Show me the numbers: designing tables and graphs to enlighten. Analytics Press, Burlingame
11. Forrest J Nightingale – home. In: Nightingale https://nightingaledvs.com/. Accessed 29 Aug 2025
12. International Conference on Visualisation, VIS – conference Table of Contents | IEEE Xplore. https://ieeexplore.ieee.org/xpl/conhome/1001944/all-proceedings. Accessed 2 Sept 2025
13. Market Cafe Magazine. In: Mark. Cafe Mag. https://www.marketcafemag.com. Accessed 2 Sept 2025
14. When Data Meets Design. http://www.data-meets-design.com/wiki/

Chapter 8

Interactive Dashboards for Modeling-Based Education: Methods and Best Practices with Python and Plotly Dash

Renato Lombardo

Abstract

The integration of interactive digital tools into science education has become increasingly important for helping students grasp abstract and multiparameter concepts that are often difficult to convey with static figures. This chapter presents a reproducible method for building interactive dashboards with Python and the Plotly Dash framework, illustrated through the case of ligand binding equilibria. The approach combines classical binding models (single site, multiple independent sites, heterogeneous sites, and cooperative Hill binding) with a browser-based interface that allows users to manipulate parameters and immediately visualize the corresponding curves. From a pedagogical perspective, dashboards offer a powerful way to bridge symbolic formalism and intuitive understanding. By experimenting directly with constants, numbers of sites, and cooperativity factors, learners can explore how these parameters affect the shape of saturation curves, compare models side by side, and develop a deeper conceptual grasp of molecular recognition and cooperativity. The method also highlights the value of dashboards as teaching aids that promote active learning, fostering curiosity and engagement while reducing the cognitive gap between mathematical abstraction and experimental interpretation. Although the case study focuses on ligand binding, the structure is general and can be adapted to other quantitative models in chemistry, biology, or physics. The implementation is lightweight, modular, and fully based on open-source software. A brief introduction to Flask-Babel is included to illustrate how the same dashboard can be internationalized for multilingual use. Beyond educational contexts, the dashboards described here can also be employed as flexible tools for data analysis, enabling instructors, researchers, and students to explore real datasets interactively with the same interface.

Key words Chemistry education, Interactive dashboards, Web applications, Plotly Dash, Python programming, Binding models, Data visualization, Active learning, Molecular modeling

1 Introduction

Mathematical models occupy a central role across the STEM disciplines as they provide a rigorous framework to describe, quantify, and predict the behavior of complex systems. Their power is enhanced by graphical representations, typically two- or three-dimensional plots, that make relationships between variables more accessible and allow patterns or regularities to emerge.

Robert E. Dempski (ed.), *Transforming Molecular Biology with Emerging Technologies*, Methods in Molecular Biology, vol. 3043, https://doi.org/10.1007/978-1-0716-5320-3_8,

Despite these advantages, models and graphs often remain challenging for learners. The teaching of molecular sciences, and chemistry in particular, poses specific challenges because many of the central concepts are abstract, multiparameter in nature [1], and not easily conveyed through static representations. Students often struggle to link the symbolic formalism of equations with the corresponding experimental behaviors that these equations describe. Difficulties may arise not only from an insufficient mathematical background [2] but also from the intrinsic abstraction of symbolic and graphical representations [3, 4]. Students can therefore focus on memorizing equations rather than grasping their conceptual meaning and struggle to connect equations, models, and plots into a coherent picture [3, 5, 6]. As a consequence, developing effective strategies to support the comprehension of mathematical models is a crucial objective in modern STEM education [7].

In recent years, the integration of digital technologies into science education has accelerated, driven both by the rapid evolution of software tools and by the need to provide students with more engaging and flexible learning experiences [8–10]. Within this context, interactive platforms have emerged as particularly powerful resources to overcome the limitations of static textbook representations. Unlike fixed graphs or numerical examples, interactive plots allow learners to manipulate parameters and immediately observe the consequences, creating a dynamic environment that promotes active exploration and supports deeper conceptual understanding [11–13]. These features make abstract models more tangible, encourage inquiry-based learning, and enhance critical thinking.

Although numerous platforms are already available [14–16], there may still be situations where custom applications are preferable. Tailor-made dashboards can be aligned more closely with specific teaching objectives and curricular requirements, while ensuring a consistent interface and notation, thereby offering educators greater flexibility in shaping the learning experience.

Web-based dashboards offer a particularly effective digital solution for STEM education, as they combine graphs, tables, and controls within a single interface. By manipulating parameters and observing the outcomes in real time, students can actively explore models, fostering deeper conceptual understanding and problem-solving skills. Compared to traditional software, dashboards ensure platform independence, simplified distribution, and continuous updates so that learners always interact with the most recent version [17, 18].

Open-source programming environments have become particularly attractive in this context. Python, in particular, has established itself as a versatile language that is both accessible to beginners and powerful for advanced scientific computation. Its ecosystem of numerical and visualization libraries allows instructors and researchers to rapidly prototype applications that combine quantitative modeling with high-quality graphics [19, 20]. Within this ecosystem, the Plotly Dash framework [21, 22] provides a distinctive solution by combining a web-based interface, interactive visualization, and server-side Python logic within a single coherent environment. Dash applications can be executed locally or deployed on institutional servers and cloud platforms, thereby lowering the barrier to adoption across a wide range of educational settings. Recent experiences in chemistry education, among others, have shown that dashboards of this type can support the understanding of complex models and encourage learning through parameter exploration [11, 13, 23].

Beyond their pedagogical role, dashboards of this kind also have potential as practical tools for research. The same features that make them effective for teaching, parameter manipulation, real-time visualization, and the possibility of comparing multiple models can be applied to the analysis of experimental datasets [18].

This chapter builds on these developments and proposes a practical method for developing web applications to teach chemistry models, taking ligand binding equilibria as an illustrative case. Ligand binding models represent a particularly suitable example because they combine relatively simple mathematical formulations with significant conceptual depth. Single-site binding, multi-site models, and cooperative behavior described by the Hill equation are all tractable within a unified framework but differ in ways that highlight key aspects of molecular recognition and cooperativity.

From a methodological perspective, the construction of the dashboard follows best practices for modular design: a clear separation of the mathematical models, the user interface components, and the callbacks that link them. This modularity ensures that the same approach can be easily adapted to other quantitative models in STEM disciplines.

The chapter is organized as follows: The Subheading 2 details the required software and libraries, as well as the setup of the working environment, the Subheading 3 describes the progressive construction of the dashboard, starting from the definition of the binding models to the assembly of the user interface and the implementation of callbacks, and the Subheading 4 provides additional explanations, best practices, and possible extensions for different deployment scenarios. While the examples focus on ligand binding models, the structure and methodology are generalizable to a wide range of applications in molecular science education and research.

2 Materials

The materials required for this protocol concern the software tools needed to develop the interactive application, together with the options for its deployment and use. The development and execution environment is entirely software-based and relies on components that are designed to be as platform-independent as possible, allowing development on different operating systems (Linux, Windows, and macOS).

Deployment can be carried out either locally, for testing or personal use, or online, to provide access to the application from any device with an Internet connection. In typical scenarios, the end user will not be required to install any software: The application will be accessed and used exclusively through a modern web browser.

All the software tools employed in this protocol are released under permissive open-source licenses (e.g., Python Software License, MIT, BSD, and Apache 2.0). These licenses do not require subscriptions or commercial fees and allow free use, modification, and redistribution. Such an approach ensures compatibility with institutional policies, facilitates reproducibility, and supports long-term maintenance without dependency on proprietary platforms (*see* **Note 1**). The source code accompanying this chapter is released under the GNU Affero General Public License v3.0 (*see* **Note 2**).

2.1 Python Environment and Required Packages

The development of the application requires a set of Python libraries that provide the fundamental capabilities for implementing the models, performing numerical computations, and creating the interactive web interface. They have been selected for their stability, active maintenance, and broad compatibility across different platforms, ensuring that the application can be developed and executed in heterogeneous environments without significant changes to the codebase. Other commonly used scientific Python libraries may also be valuable in extended implementations (*see* **Note 3**).

1. **Python interpreter** (version $\geq$ 3.11). Python is the core programming language used for developing the application. Version 3.11 or higher is recommended. For convenience and cross-platform compatibility, it is advised to use a scientific distribution such as *Anaconda* or *Miniconda* for both installing Python and managing its packages (*see* **Note 5**). Working within a dedicated virtual environment is also recommended to isolate dependencies and ensure reproducibility (*see* **Note 6**).
2. **NumPy**. NumPy is a fundamental package for scientific computing in Python, providing efficient multidimensional array objects and a comprehensive collection of mathematical functions to operate on them [24]. It is essential for implementing

the numerical computations underlying the models used in the application, ensuring both performance and reliability. NumPy is widely supported across platforms and Python distributions.

3. **Plotly Dash**. Plotly Dash is a Python framework for building interactive, web-based analytical applications without requiring direct knowledge of HTML, CSS, or JavaScript [21]. It combines the data visualization capabilities of the *Plotly* library [22] with the interactivity and layout control provided by the Dash framework, enabling the rapid creation of customized dashboards. In the context of this protocol, Plotly Dash is required to implement effectively the user interface, dynamic controls, and responsive plots for the application. It is cross-platform and works in any environment where Python is available.
4. **Flask-Babel**. Flask-Babel provides internationalization (i18n) and localization (l10n) capabilities for Flask-based applications, including Dash. It enables automatic language selection based on the user's browser settings and supports multiple translations of text within the application.

2.2 Development Tools

The development process can be carried out with minimal software requirements; however, using dedicated development tools can greatly improve efficiency, code quality, and maintainability. The following tools are not strictly required to run the application but are strongly recommended to facilitate the implementation, debugging, and long-term management of the project.

1. **Code editor or IDE**. In principle, any text editor can be used to write Python code. Nevertheless, it is strongly recommended to use a dedicated tool that provides syntax highlighting, code completion, integrated execution, and debugging capabilities. *JupyterLab* is particularly suitable for this protocol, as it allows interactive coding and rapid prototyping within notebooks.
2. **Version control system** (optional but strongly recommended). A version control system allows tracking changes in the source code, reverting to previous versions, and collaborating effectively with other developers. *Git* is the most widely used solution, integrates seamlessly with platforms such as GitHub, GitLab, and Bitbucket, and is integrated in many IDEs.
3. **Code quality and style checker**. Tools such as *pylint* or *flake8* help maintain code readability and compliance with the PEP 8 Python style guide, while also detecting potential errors before execution. Their integration into an IDE can provide real-time feedback during development.

2.3 Deployment

Once the application has been developed and tested locally, it can be deployed so that end users can access it. The deployment method should be chosen based on the intended audience, usage frequency, and available infrastructure. In all cases, the goal is to ensure that the application is accessible with minimal requirements on the client side with only a modern web browser without requiring the installation of additional software (*see* **Note 10**).

1. **Local deployment**. It is suitable for development, testing, demonstrations, and small seminars. The Dash server is started from the command line and accessed at http://localhost:8050. Availability is limited to the machine (or LAN) where the process is running and depends on that host remaining powered on and connected.
2. **Self-hosting on institutional servers**. An intermediate option is to deploy the application on a departmental or institutional server. This ensures wider availability to a defined community of users, while keeping control of the infrastructure within the institution. Configuration, maintenance, and security are managed locally by system administrators, making this solution suitable when institutional policies require that data and applications remain on premises.
3. **Deployment on Platform-as-a-Service (PaaS) solutions**. A Platform-as-a-Service (PaaS) is a cloud-based hosting service that provides a complete environment for running applications without requiring the user to manage the underlying infrastructure and ensure that all users access the most up-to-date version of the application (*see* **Note 11**).
4. **Containerized deployment**. In addition to traditional hosting methods, the application can be deployed in a containerized environment using platforms such as *Docker*. Containerization encapsulates the application, its dependencies, and the runtime environment into a single portable unit, ensuring reproducibility and simplifying distribution across different systems (*see* **Note 12**).

3 Methods

While the concrete code necessarily varies with the model under study and with specific educational or analytical goals, a general, reusable workflow can still be described and then adapted case by case. In some domains (e.g., stiff/nonlinear ODEs, stochastic simulations, PDEs, and streaming/large datasets), however, more substantial tailoring will be required (*see* **Note 3**). Accordingly, the method below is presented as a portable scaffold, not a one-size-fits-all recipe. In particular, each step will be illustrated with a binding equilibrium (ligand–receptor) model as an example of implementation.

Additional Dash-based dashboards developed in the same spirit are available (*see* **Note 4**).

3.1 Pedagogical Analysis and Definition of Learning Objectives

Careful planning of the pedagogical approach is essential to ensure that the development of the application is not limited to its technical aspects. A clear definition of learning objectives provides the framework for aligning the interactive features of the dashboard with meaningful educational outcomes.

1. Clearly identify the mathematical model(s) that describe the phenomenon of interest. In the case of ligand binding, the models considered are (a) single-site binding, described by a hyperbolic saturation curve, (b) binding to multiple equivalent independent sites, (c) binding to multiple nonequivalent independent sites, both modeled as the sum of independent binding equations, and (d) cooperative binding, described by the Hill equation, with the Hill coefficient n_H quantifying cooperativity. For each model, list the core equations, define the input variables (ligand concentration, binding constant(s), number of sites, and Hill coefficient), and specify the expected outputs (fractional occupancy and average number of occupied sites) (*see* **Note 13**).
2. Identify the typical conceptual challenges the model may present to students. For ligand–macromolecule binding, these include (a) understanding the quantitative relationship between [L] and θ or $\bar{\nu}$, (b) distinguishing between multiple independent sites and cooperative binding, (c) interpreting how K_a and n_H influence curve shape, and (d) linking mathematical parameters to their biological meaning (*see* **Note 14**).
3. A definition of clear, measurable learning objectives is crucial. Link each objective to specific application functionalities. After using the application, students should be able to (a) describe how θ changes with [L] in different binding scenarios, (b) compare binding curves for single-site, multi-site, and cooperative binding, both graphically and quantitatively, (c) interpret the effect of changing K_a, number of sites, and n_H on the curve shape, and (d) recognize differences between equivalent and nonequivalent sites from curve analysis (*see* **Note 15**).
4. When the application itself is also used as a learning tool for computational skills, students should be able to (a) translate a given mathematical model into working code, (b) separate the computational implementation of the model from the interface logic, (c) implement basic data analysis on the model outputs

(e.g., determine [L] at half-saturation), (d) document their code effectively to ensure clarity and reusability, and (e) understand how the computational structure supports the educational objectives.

3.2 Application Design and Implementation Planning

1. Translate the learning objectives into concrete application capabilities, including (a) model selection and parameter input, (b) real-time plot generation and comparison, (c) calculation and display of relevant quantitative indicators, and (d) exporting data and figures.
2. Design the general layout of the user interface (UI). For the binding example (*see* Fig. 1), plan the interface to include (a) a control panel for model selection, parameter setting, and [L] range definition, (b) a plot area capable of overlaying multiple curves and showing θ and $\bar{\nu}$ simultaneously, (c) interactive legends and tooltips to highlight curve features, and (d) a concise model description with clear mathematical notation (*see* **Note 18**).
3. Specify how the computational core will be separated from the interface logic, allowing reuse of model functions in different contexts. Plan for efficient calculation even with multiple curves and large datasets (e.g., using `NumPy`).
4. Ensure compatibility across devices and browsers (*see* **Note 19**); plan for multilingual support if relevant (e.g., dynamic translation of labels, descriptions, and tooltips).

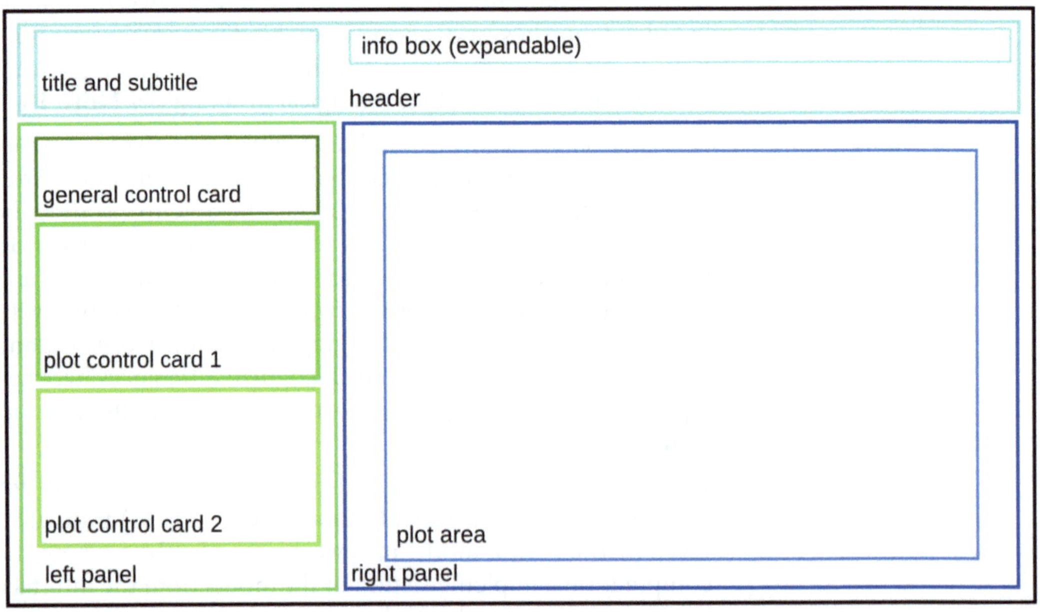

Fig. 1 Dashboard general layout

3.3 Setting Up the Development Environment

1. From the Anaconda prompt, the following command creates a new environment named `binding-dashboard` with Python 3.11 installed (*see* **Notes 6** and 7):

```
conda create -n binding-dashboard python=3.11 -y
```

The environment can be activated with:

```
conda activate binding-dashboard
```

Once activated, install the required packages (Dash, Plotly, NumPy, and Flask-Babel) with:

```
conda install dash plotly numpy flask-babel -y
```

This ensures that all dependencies are isolated within the environment and that the application can be easily reproduced on different systems by recreating the same environment (*see* **Note 21**).

2. **Initialize the project structure**. After activating the virtual environment, the project should be organized into a structured set of directories and files. For the `binding-dashboard` project, a minimal setup can be created as follows:

```
binding-dashboard/
|-- binding_dashboard.py
|-- models/
|   |-- binding_model.py
|-- translations/
```

The `binding_dashboard.py` file serves as the entrypoint of the dashboard. The `models/` directory contains the implementation of the scientific model (e.g., the binding equations) (*see* **Note 22**). The `translations/` directory is prepared for internationalization and will later be populated automatically when new languages are initialized.

3.4 Developing the Computational Model

1. A general Python function is implemented to compute the Hill binding isotherm (*see* **Note 23**). The function accepts ligand concentrations, K_a, and the Hill coefficient, n_H, and returns either the fractional occupancy θ or the average number of bound sites $\bar{\nu}$. For simplicity in this and all subsequent listings, type hinting is omitted (*see* **Note 24**), the docstring is

extremely simplified (*see* **Note 25**), and input validation code is not included (*see* **Note 26**).

Listing 1: binding_model.py (hill_binding)

```python
import numpy as np

def hill_binding(L, Ka, n_H=1.0, output="theta", sites=1):
    """Hill binding: return 'theta' or 'nu'."""
    binding_power = (Ka * L) ** n_H


    theta = binding_power / (1.0 + binding_power)
    return theta if output == "theta" else sites * theta
```

The input parameter `L` can be either a NumPy array of ligand concentrations or a scalar value; the function works in both cases without modification (*see* **Note 27**).

2. A new wrapper function is added to handle the case of a single ligand binding site. In this situation, the Hill equation reduces to the simplest form of a binding isotherm (*see* **Note 13**). The wrapper fixes $n_H = 1$ and $sites = 1$ and calls the general `hill_binding` function.

Listing 2: binding_model.py (single_site_binding)

```python
def single_site_binding(L, Ka, output="theta"):
    """Single-site binding (n_H=1, sites=1)."""
    return hill_binding(L, Ka, n_H=1.0, output=output, sites=1)
```

3. A wrapper function is now added to describe the case of *n* equivalent and independent binding sites. This is a direct extension of the single-site case, since the overall degree of binding can be obtained by multiplying the fractional occupancy of one site by the total number of sites (*see* **Note 13**).

Listing 3: binding_model.py (n_equivalent_sites_binding)

```python
def n_equivalent_sites_binding(L, Ka, n, output="theta"):
  """n equivalent, independent sites via hill_binding (n_H=1, sites=n)."""
  return hill_binding(L, Ka, n_H=1.0, output=output, sites=n)
```

4. Finally, a wrapper function is added to describe the case of heterogeneous independent site-classes. Each class i is characterized by its own association constant $K_{a,i}$ and multiplicity m_i. The overall degree of binding is obtained by summing the class contributions $\bar{\nu}_i$, and the global fractional occupancy follows as $\theta = \bar{\nu} / \sum_i m_i$ (*see* **Note 13**). Optionally, the function can also return the individual curves for each class (as a list of arrays), either for $\bar{\nu}_i$ or for $\theta_i = \bar{\nu}_i / m_i$.

Listing 4: binding_model.py (heterogeneous_independent_sites_binding)

```python
def heterogeneous_independent_sites_binding(L, Kas, multiplicities, output="theta",
    return_per_class=False):
  # Per-class nu_i via the n-equivalent-sites wrapper
  nu_i_list = []
  for Ka_i, m_i in zip(Kas, multiplicities):
    nu_i = n_equivalent_sites_binding(L, Ka=Ka_i, n=m_i, output="nu")
    nu_i_list.append(np.asarray(nu_i))

  # Total nu
  nu_total = np.sum(nu_i_list, axis=0)

  if output == "nu":
    return (nu_total, nu_i_list) if return_per_class else nu_total

  # Compute theta only if requested
  total_sites = int(np.sum(multiplicities))
  theta_total = nu_total / total_sites

  if return_per_class:
    # theta_i = nu_i / m_i (returned as a list, one array per class)
    theta_i_list = [nu_i / m_i for nu_i, m_i in zip(nu_i_list, multiplicities)]
    return theta_total, theta_i_list

  return theta_total
```

3.5 Designing the User Interface

The general layout of the application is organized into three main sections, as schematically illustrated in Fig. 1. In the upper part of the page a header is presented, where the title of the dashboard together with additional information about the model is reported. The central area of the interface is divided into two panels. On the left side a control panel is provided, in which the parameters of the selected models can be modified and the visualization options are managed. On the right side the plot area is shown, where the graphs corresponding to the chosen models are dynamically generated and updated. In this way a clear separation between the input elements and the graphical outputs is ensured, facilitating both the interaction and the interpretation of the model behavior (*see* **Note 17**).

1. The first step is to create the file `binding_dashboard.py`, which will contain the user interface of the application. The initial section of this file imports all the modules and functions required for computation (*see* **Note 28**), visualization (*see* **Note: 29**), the dashboard framework (*see* **Note 30**), internationalization support (*see* **Note 31**), and code organization (*see* **Note 32**).

Listing 5: binding_dashboard.py (imports)

```python
# dashboard components
import dash_bootstrap_components as dbc
import plotly.colors as pcolors
import plotly.graph_objs as go
from dash import callback, callback_context, Dash, dcc, html, Input, Output, State,e MATCH
    , ALL
from dash.dash_table import DataTable
from dash.exceptions import PreventUpdate
# needed for internationalization
from flask import request
from flask_babel import Babel, gettext


# others
import numpy as np
import re
from typing import Any, List, Dict, Union
# functions imported from ~/models/binding_model.py
from models.binding_model import hill_binding, single_site_binding,
    n_equivalent_sites_binding, heterogeneous_independent_sites_binding
```

2. The next step is to define a convenient translator function to handle internationalization of the dashboard text (*see* **Note 33**).

Listing 6: binding_dashboard.py (translator function)

```
# define translator function
_ = gettext
```

3. The general dashboard information is then defined as translatable strings: a title, a subtitle, and a concise informational text that will be rendered as Markdown with MathJax enabled, so that mathematical notation can be displayed (*see* **Note 35**).

Listing 7: binding_dashboard.py (dashboard information)

```
# Dashboard information (translatable)
title = _("Ligand binding models")
subtitle = _("Explore single-site, multi-site, and cooperative binding")

# Informational text (Markdown + MathJax)
info_text = _(
"""
A minimal reminder of the model equation:

$$
\\theta = \\frac{K_a [L]}{1 + K_a [L]}
$$
"""
)
```

4. The header is created in two steps: First, the individual components (title and subtitle, the informational Markdown block, and the "more info" button) are instantiated; then, these components are assembled into two columns and wrapped in a single row (*see* **Note 34**).

Listing 8: binding_dashboard.py (header)

```
################
# setup layout #
################

# --- define the header ---
title_html = html.H1(title, id="title")
subtitle_html = html.P(subtitle, id="subtitle")
info_button = dbc.Button(n_clicks=0, children=_("more info"), color="info", id="info-
    button")
# info will be dynamically updated with info_text on clicking on info_button
info = dcc.Markdown(" ", mathjax=True, id='info')
# put button and text area togheter
title_col = dbc.Col(dbc.Container([title_html, subtitle_html]), width='auto')
info_col = dbc.Col(dbc.Container([info, info_button]), width='auto')
header = dbc.Row([title_col, info_col])
```

5. The right-hand plot panel is defined by placing a `dcc.Graph` inside a lightweight container (`html.Div`) that provides inner padding via Bootstrap utilities and then wrapping it in a `dbc.Col` to occupy the available width.

Listing 9: binding_dashboard.py (right panel)

```
# --- define the right panel ---
plot = dcc.Graph(id="plot", style={"width": "100%", "height": "80vh"})
right_panel = dbc.Col(
  html.Div(
    plot,
    className="p-3" # add padding
  ),
  width=True,
  id="right_panel"
)
```

6. The left panel of the dashboard is defined to contain all general controls, as well as a placeholder for the per-plot control cards. The first element is a button that allows the user to add a new plot.

Listing 10 : binding_dashboard.py (left panel, add button)

```
# --- define the left panel ---
# --- general controls ---
add_button = dbc.Button(
  _("Add plot"),
  id="add-button",
  n_clicks=0,
  color="primary",
  className="mb-3"
)
```

7. A container is then created to specify the range of ligand concentrations $[L]$ to be explored. Two numerical inputs are provided: one for the minimum and one for the maximum value of $[L]$.

Listing 11 : binding_dashboard.py (left panel, L range)

```
L_inputs = dbc.Container([
  dbc.Label("[L] range"),
  dbc.Row([
    dbc.Col(
      dbc.Input(
        id="L-min",
        type="number",
        min=0,
        step=0.1,
        value=0.0,
        placeholder="Min [L]"
      ),
      width=6
    ),
    dbc.Col(
      dbc.Input(
        id="L-max",
        type="number",
        min=0,
        step=0.1,
        value=10,
        placeholder="Max [L]"
      ),
      width=6
    )
  ], className="g-2")
], className="mb-3")
```

8. The next control allows the user to choose which quantity will be plotted. A pair of radio buttons is used to select between θ (fractional saturation) and $\bar{\nu}$ (degree of binding). LaTeX syntax is supported through MathJax for rendering the labels.

Listing 12: binding_dashboard.py (left panel, quantity selector)

```
quantity_selector = dbc.Container([
  dbc.Label(_("Quantity to plot")),
    dbc.RadioItems(
      id="quantity",
      options=[
        {"label": dcc.Markdown(r"$\theta$ (fractional saturation)", mathjax=True), "value":
             "theta"},
        {"label": dcc.Markdown(r"$\bar\nu$ (degree of binding)", mathjax=True), "value": "
            nu"},
      ],
      value="theta",
      inline=True,
    )
], className="mb-3")
```

9. All general controls (button, range selector, and quantity selector) are grouped into a container, which is later placed on the left side of the dashboard layout.

Listing 13: binding_dashboard.py (left panel, general controls)

```
general_controls = dbc.Container(
  [add_button, L_inputs, quantity_selector],
  id="general-controls",
  fluid=True,
  className="mb-3"
)
```

10. A placeholder container for the per-plot controls is defined. This will dynamically host the control cards corresponding to the individual plots added by the user.

Listing 14: binding_dashboard.py (left panel, per-plot controls)

```
# --- per-plot controls container ---
controls_container = dbc.Container(
  id="controls-container",
  fluid=True
)
```

11. Finally, the left panel is assembled by combining the general controls and the per-plot container into a single column.

Listing 15: binding_dashboard.py (left panel assembly)

```
# --- left panel assembly ---
left_panel = dbc.Col(
  dbc.Container([general_controls, controls_container]),
  width="auto",
  id="left_panel"
)
```

12. The overall dashboard layout is now defined by combining the previously created `header`, `left_panel`, and `right_panel` within a single container. The layout is structured into two main columns: The left column contains the controls, while the right column contains the graphical output.

Listing 16: binding_dashboard.py (layout)

```
# --- Layout ---
layout = dbc.Container([
    header,
    html.Hr(),
    dbc.Row([
        dbc.Col(left_panel, xl=4, align='right'),
        dbc.Col(right_panel, xl=8, align='left')
        ])
    ],
    fluid=True,
    id='layout'
)
```

13. The creation of each control card is delegated to a dedicated *factory function*. Before defining the function, dictionaries are provided to describe the schema of the parameter tables, their default values, the available options in the model dropdown, and a helper set to distinguish single-row models from multi-row models.

Listing 17: binding_dashboard.py (schemas, defaults, and options)

```
# ---- Schemas and defaults ---
MODEL_COLUMNS = {
    "single_site": [{"id": "Ka", "name": "Ka", "type": "numeric"}],
    "n_equivalent_sites": [
        {"id": "Ka", "name": "Ka", "type": "numeric"},
        {"id": "n", "name": "n", "type": "numeric"},
    ],
    "cooperative_hill": [
        {"id": "Ka", "name": "Ka", "type": "numeric"},
        {"id": "n", "name": "n", "type": "numeric"},
        {"id": "n_H", "name": "n_H", "type": "numeric"},
    ],
    # Heterogeneous: each row is a site-class with its Ka and multiplicity m (labeled as n in table)
    "heterogeneous_sites": [
        {"id": "Ka", "name": "Ka", "type": "numeric"},
        {"id": "n", "name": "m (multiplicity)", "type": "numeric"},
    ],
}
DEFAULT_ROWS = {
    "single_site": [{"Ka": 1.0}],
    "n_equivalent_sites": [{"Ka": 1.0, "n": 2}],
    "heterogeneous_sites": [{"Ka": 1.0, "n": 1}, {"Ka": 0.1, "n": 1}],
    "cooperative_hill": [{"Ka": 1.0, "n": 2, "n_H": 2}],
}
MODEL_OPTIONS = [
    {"label": "Single-site", "value": "single_site"},
    {"label": "n equivalent sites", "value": "n_equivalent_sites"},
    {"label": "Heterogeneous sites", "value": "heterogeneous_sites"},
    {"label": "Cooperative (Hill)", "value": "cooperative_hill"},
]
SINGLE_ROW_MODELS = {"single_site", "n_equivalent_sites", "cooperative_hill"}
```

14. The factory function receives a unique integer `uid`, which is embedded in the IDs of all components belonging to the card.

This will ensure that each card is uniquely identifiable when callbacks are triggered.

Listing 18: binding_dashboard.py (card factory header)

```
# ---------- Card factory ----------
def controls_card_factory(uid: int):
    uid_label = dbc.Label(f'Trace {uid}')
    #model selection dropdown
    dropdown = dcc.Dropdown(
        id={"type": "model-dropdown", "uid": uid},
        options=MODEL_OPTIONS,
        value="single_site",
        clearable=False,
        style={"minWidth": "220px"}
    )
```

15. The header of the card contains three elements arranged in a row: a label displaying the trace number, the dropdown menu to select the model, and a delete button. The button allows users to remove the corresponding card and its associated plot.

Listing 19: binding_dashboard.py (card header row)

```
    # button to remove the current card
    clear_button = dbc.Button(_('Delete'), id={'type': 'clear-button', 'uid':uid}, color="danger", size="sm", className="ms-2")
    header_row = dbc.Row([dbc.Col(uid_label, width="auto"), dbc.Col(dropdown, width="auto"), dbc.Col(clear_button, width="auto")], className="g-2")
```

16. Below the header, a parameter table is defined. It is a `DataTable` initialized with the schema and default values for the single-site model (*see* Listing 8.17). The table is editable by the user and scrollable if needed.

Listing 20: binding_dashboard.py (parameter table)

```
# table with model parameters
table = DataTable(
    id={"type": "param-table", "uid": uid},
    columns=MODEL_COLUMNS["single_site"],
    data=DEFAULT_ROWS["single_site"],
    editable=True,
    row_deletable=False,
    cell_selectable=True,
    style_table={"overflowX": "auto"},
    style_cell={"padding": "0.25rem"},
)
```

17. For heterogeneous binding models, extra controls are needed. These are grouped into a row that is hidden by default and only shown when relevant. They consist of a button to add new site-classes and a toggle switch to plot contributions separately.

Listing 21: binding_dashboard.py (heterogeneous controls)

```
# extra control for the heterogeneous sites model
hetero_controls = dbc.Row(
    [ # buto to add a new row (e.g. a new class of sites)


        dbc.Col(dbc.Button("Add row", id={"type": "add-row", "uid": uid}, color="
            secondary", size="sm"), width="auto"),
        dbc.Col(
            dbc.Checklist( #True to plot single contributions
                options=[{"label": "Split contributions", "value": "split"}],
                value=[],
                switch=True,
                id={"type": "split-switch", "uid": uid},
            ),
            width="auto",
        ),
    ],
    className="g-2 mt-2",
    id={"type": "hetero-controls", "uid": uid},
    style={"display": "none"}, # display only when needed
)
```

18. Each card also includes a `dcc.Store`, which maintains its current state. The store keeps track of the selected model, the table data, and whether contributions should be split. This information is later used by callbacks to compute and draw the corresponding traces.

Listing 22: binding_dashboard.py (per-card state store)

```
# storage for model state
store = dcc.Store(id={"type": "plot-store", "uid": uid},
                  data={"model": "single_site", "params": DEFAULT_ROWS["single_site"], "
                      split": False})
```

19. Finally, the card body is assembled with the header row, the parameter table, the optional heterogeneous controls, and the hidden store. The complete card is returned and will be appended to the layout (*see* Fig. 2).

Listing 23: binding_dashboard.py (card assembly)

```
body = dbc.CardBody([header_row, html.Div(table, className="mt-2"), hetero_controls,
    store])
return dbc.Card(body, className="mb-3", id={"type": 'controls-card', "uid": uid})
```

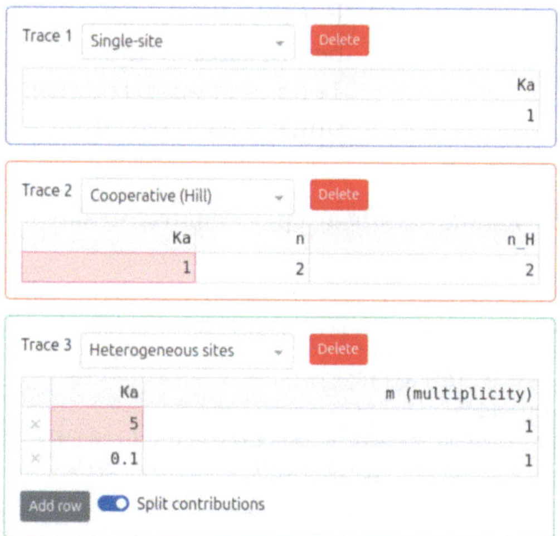

Fig. 2 Control cards for different binding models

20. The function `compute_traces` orchestrates the calculation of curves and their representation as Plotly `go.Scatter` traces. It receives the selected model, the parameter list extracted from the card, the ligand concentration array, the desired output (`"theta"` or `"nu"`), a label for the trace, a color, and an optional flag `split`.

Listing 24: binding_dashboard.py (function header and setup)

```
def compute_traces(model, params, L, quantity, name, color, split=False):
    """ compute traces for each binding model """
    line=dict(color=color)
    traces = []
```

21. When the model is `"single_site"`, the constant `Ka` is extracted from the first row of the parameter table. If present, the function `single_site_binding` computes the requested output, and a single line trace is appended (*see* Listing 8.2).

Listing 25: binding_dashboard.py (single-site case)

```
    if model == "single_site":
        Ka = params[0].get("Ka")
        if Ka:
            y = single_site_binding(L=L, Ka=Ka, output=quantity)
            traces.append(go.Scatter(x=L, y=y, mode="lines", name=name, line=line))
```

22. For the case `"n_equivalent_sites"`, both `Ka` and the number of sites `n` must be provided. The wrapper `n_equivalent_sites_binding` evaluates the function and returns a single trace (*see* Listing 8.3).

Listing 26: binding_dashboard.py (*n* equivalent sites)

```
    elif model == "n_equivalent_sites":
        Ka = params[0].get("Ka")
        n = params[0].get("n")
        if Ka and n:
            y = n_equivalent_sites_binding(L=L, Ka=Ka, n=n, output=quantity)
            traces.append(go.Scatter(x=L, y=y, mode="lines", name=name, line=line))
```

23. In the case of heterogeneous independent sites, each row of the parameter table represents a site-class, with its own `Ka` and multiplicity `n` (interpreted as *m*). The function builds lists of valid constants and multiplicities and then calls `heterogeneous_independent_sites_binding` (*see* Listing 8.4).

Listing 27: binding_dashboard.py (heterogeneous sites setup)

```
elif model == "heterogeneous_sites":
    # Sum of independent site-classes; each row has Ka and multiplicity "n" (m)
    Ka_list = []
    m_list = []
    for row in params:
        Ka = row.get("Ka")
        m = row.get("n")
        if Ka and m:
            Ka_list.append(Ka)
            m_list.append(m)
```

24. If valid lists are obtained, the binding routine is invoked. If `split=False`, only the total curve is plotted. If `split=True`, both the total and the per-class curves are returned. The per-class curves are distinguished using dash styles while maintaining the same base color.

Listing 28: binding_dashboard.py (heterogeneous sites traces)

```
    if Ka_list and m_list:
        y = heterogeneous_independent_sites_binding(L=L, Kas=Ka_list, multiplicities=m_list, output=quantity, return_per_class=split)
        if split:
            DASH_STYLES = ["dash", "dot", "dashdot", "longdash", "longdashdot"]
            y_total = y[0]
            y_partial = y[1]
            traces.append(go.Scatter(x=L, y=y_total, mode="lines", name=name, line=line))
            for i, yp in enumerate(y_partial):
                dash_style = DASH_STYLES[i % len(DASH_STYLES)]
                line["dash"]=dash_style
                traces.append(go.Scatter(x=L, y=yp, mode="lines", name=f'{name} - class {i+1}', line=line))
        else:
            traces.append(go.Scatter(x=L, y=y, mode="lines", name=name, line=line))
```

25. Finally, in the "cooperative_hill" case, the parameters Ka, number of sites n, and Hill coefficient n_H are extracted from the first row. If valid, the function hill_binding is called, and the resulting curve is plotted (*see* Listing 8.1).

Listing 29: binding_dashboard.py (cooperative Hill)

```
elif model == "cooperative_hill":
    Ka = float(params[0].get("Ka", 0))
    n = int(params[0].get("n", 0))
    n_H = float(params[0].get("n_H", 0))
    if Ka and n and n_H:
        y = hill_binding(L=L, Ka=Ka, n_H=n_H, output=quantity, sites=n)
        traces.append(go.Scatter(x=L, y=y, mode="lines", name=name, line=line))
return traces
```

3.6 Linking Model and Interface

A central mechanism that connects the user interface with the underlying computations in Dash is the *callback*. Callbacks are Python functions that are automatically triggered whenever specific inputs in the web interface change. They provide the link between user actions (e.g., selecting a model, editing parameters, or clicking a button) and the update of visual elements such as plots or tables (*see* **Note 39**). In this way, the interface remains fully reactive: Any change in the controls is immediately reflected in the dashboard outputs (*see* **Note 40**).

In the present dashboard, callbacks are employed to add and remove control cards, to update parameter tables when the model selection changes, to synchronize the internal state of each card with its table content, and finally to redraw the figure in the right panel whenever necessary.

1. The first callback manages the information box in the header (*see* Listing 8.7). It listens to clicks on the info-button and alternates between showing and hiding the explanatory text about the binding models. The outputs are the button label itself and the content of the information area.

Listing 30: binding_dashboard.py (show_info header)

```
@callback(Output("info-button", 'children'),
          Output("info", 'children'),
          Input("info-button", 'n_clicks')
    )
def show_info(n_clicks):
    '''show a short information about the model '''
```

2. The function body uses a simple parity check on the number of clicks. If the button has been pressed an odd number of times, the information text is displayed and the label changes to "less info." If it has been pressed again (even number), the text is hidden, and the label returns to "more info."

Listing 31: binding_dashboard.py (show_info logic)

```
    if n_clicks%2: # button pressed for an uneven number of times
        text = _(info_text)
        button_text = _('less info') # change the label
    else: # clicked again after showing, means hide the info
        text = ' '
        button_text = _('more info')
    return button_text, text
```

3. The second callback dynamically manages the set of control cards. It listens to clicks on the global `add-button` as well as on any of the per-card delete buttons (*see* Listings 8.8 and 8.19). Its output is the list of children inside the `controls-container`, which is updated whenever a card is added or removed (*see* **Note 41**).

Listing 32: binding_dashboard.py (update_controls_container header)

```
@callback(Output('controls-container', 'children'),
          [Input('add-button', 'n_clicks'),
           Input({'type':'clear-button', 'uid': ALL}, 'n_clicks')],
          State('controls-container', 'children'),
          prevent_initial_call=True,
         )
def update_controls_container(add_n_clicks, clear_n_clicks, controls_container_list):
    '''update container adding or removing controls as required'''
```

4. The function body distinguishes whether the trigger was the global add button or one of the per-card delete buttons. If the trigger corresponds to the add button, a new card is created

through the factory and appended. If instead a delete button was pressed, the corresponding card is removed from the container.

Listing 33: binding_dashboard.py (update_controls_container logic)

```
    if not controls_container_list:
        controls_container_list = []
    ctx = dash.callback_context # this is needed to know which button has been pushed
    trigger = ctx.triggered[0]
    if 'add-button' in trigger['prop_id']: # add a new panel
        uid = add_n_clicks # unique id
        new_card = controls_card_factory(uid=uid)
        return controls_container_list+[new_card]
    else: # some delete button has been pushed
        match = re.search(r'"uid":(\d+)',trigger['prop_id'])
        if match:
            uid = int(match.group(1)) # uid of the card
            return [c for c in controls_container_list if c['props']['id']['uid'] != uid]
```

5. The third callback reacts whenever the model selected in a card changes. Its purpose is to update the parameter table with the correct schema and default values, to toggle the visibility of the heterogeneous model controls, and to synchronize the per-card store. The callback listens to the `value` of each model selection dropdown and returns five outputs: the table columns, the table data, whether rows can be deleted, the style of the heterogeneous controls, and the store content.

Listing 34: binding_dashboard.py (on_model_change header)

```
@callback(
    Output({"type": "param-table", "uid": MATCH}, "columns"),
    Output({"type": "param-table", "uid": MATCH}, "data"),
    Output({"type": "param-table", "uid": MATCH}, "row_deletable"),
    Output({"type": "hetero-controls", "uid": MATCH}, "style"),
    Output({"type": "plot-store", "uid": MATCH}, "data"),
    Input({"type": "model-dropdown", "uid": MATCH}, "value"),
    State({"type": 'controls-card', "uid": MATCH}, "id"),
    prevent_initial_call=True,
)
def on_model_change(model_value, card_id):
```

6. The function body extracts the `uid` of the card and applies the logic: It selects the correct column schema and default rows from the dictionaries, determines whether the model is heterogeneous, and updates the corresponding style (visible or hidden). Row deletion is allowed only in the heterogeneous case. The store is then reset with the default parameters and the current model type.

Listing 35: binding_dashboard.py (on_model_change logic)

```
    uid = card_id['uid']
    new_columns = MODEL_COLUMNS[model_value]
    default_rows = DEFAULT_ROWS[model_value]
    is_hetero = (model_value == "heterogeneous_sites")
    hetero_style = ({} if is_hetero else {"display": "none"})


    row_deletable = is_hetero
    split_value = []
    store_data = {"model": model_value, "params": default_rows, "split": bool(split_value)}
    return new_columns, default_rows, row_deletable, hetero_style, store_data
```

7. The next callback enables row insertion in the parameter table, but only for the heterogeneous sites model (*see* Listing 8.21). It listens to clicks on the per-card "`add-row`" button and targets the "`data`" property of the corresponding table. The model type and current table content are read via `State`.

Listing 36: binding_dashboard.py (add_row header)

```
@callback(
    Output({"type": "param-table", "uid": MATCH}, "data", allow_duplicate=True),
    Input({"type": "add-row", "uid": MATCH}, "n_clicks"),
    State({"type": "model-dropdown", "uid": MATCH}, "value"),
    State({"type": "param-table", "uid": MATCH}, "data"),
    prevent_initial_call=True,
)
def add_row(n_clicks, model_value, data):
```

8. The function first guards against spurious executions: If no click has occurred, or if the active model is not "`heterogeneous_sites`", the update is prevented. Otherwise, a new blank row is created from the model's column schema and

appended to the existing table data. This keeps the table schema consistent and allows users to incrementally define additional site-classes.

Listing 37: binding_dashboard.py (add_row logic)

```
    if not n_clicks or model_value != "heterogeneous_sites":
        raise dash.exceptions.PreventUpdate
    base = {c["id"]: None for c in MODEL_COLUMNS[model_value]}
    return (data or []) + [base]
```

9. The following callback ensures that the internal store of each control card remains synchronized with the corresponding parameter table. Whenever the user edits table values or toggles the split switch, the callback updates the store so that the downstream plotting logic always receives consistent information.

Listing 38: binding_dashboard.py (on_table_edit header)

```
@callback(
    Output({"type": "plot-store", "uid": MATCH}, "data", allow_duplicate=True),
    Input({"type": "param-table", "uid": MATCH}, "data"),
    Input({"type": "split-switch", "uid": MATCH}, "value"),
    State({"type": "model-dropdown", "uid": MATCH}, "value"),
    prevent_initial_call=True,
)
def on_table_edit(table_data, split_value, model_value):
```

10. The function body simply assembles a dictionary containing the current model type, the updated parameter set, and the split flag. This dictionary is written into the store, making it available for the figure redraw callback. By centralizing the state in the store, the application avoids inconsistencies between table values, switches, and plotted curves.

Listing 39: binding_dashboard.py (on_table_edit logic)

```
    return {"model": model_value, "params": table_data, "split": bool(split_value)}
```

11. The figure is redrawn whenever any per-card store changes or when global controls such as [L] bounds or the quantity selector are modified. This callback returns two outputs: the Plotly figure for the right panel and the (possibly updated) styles for all control cards (used here to color their borders consistently with the corresponding traces). Inputs include all per-card stores, the [L] range, and the quantity; the IDs and current styles are passed as state.

Listing 40: binding_dashboard.py (redraw_figure header)

```
@callback(
    Output("plot", "figure"),
    Output({'type':'controls-card', 'uid': ALL}, 'style'),
    Input({"type": "plot-store", "uid": ALL}, "data"),


    Input("L-min", "value"),
    Input("L-max", "value"),
    Input("quantity", "value"),
    State({"type": "plot-store", "uid": ALL}, "id"),
    State({'type':'controls-card', 'uid': ALL}, 'style'),
    prevent_initial_call=False,
)

def redraw_figure(all_stores, L_min, L_max, quantity, all_ids, styles):
```

12. At the beginning of the function, a qualitative color palette is selected, containers for updated styles and the figure are created, and the ligand grid [L] is prepared. A small guard ensures that an empty `L_max` defaults to `0.1`.

Listing 41: binding_dashboard.py (redraw_figure setup)

```
    # colors for plotting different curves
    colors = pcolors.qualitative.Plotly
    new_styles = []
    fig = go.Figure()
    # TODO Keep stable color/order using uid
    if not L_max:
        L_max = 0.1
    L = np.linspace(L_min, L_max, 200)
```

13. The core loop zips together each store, its associated ID, and style. For every active card, a color is chosen based on its `uid`; the card border color is synchronized with the trace color; and

the appropriate model, parameters, and split flag are read from the store. The helper `compute_traces` (*see* Listing 8.24) is then called to obtain one or more traces, which are added to the figure. Cards with missing state are skipped safely.

Listing 42: binding_dashboard.py (redraw_figure loop)

```
for store, _id, style in zip(all_stores, all_ids, styles):
    if not style:
        style = {}
    if not store:
        continue
    i = _id['uid']-1
    uid = _id['uid']
    color = colors[i%len(colors)]
    style['border-color'] = color
    new_styles.append(style)
    model = store.get("model")
    params = store.get("params")
    split = bool(store.get("split", False))
    traces = compute_traces(model=model, params=params, L=L, quantity=quantity, name=f'
        Trace {uid}', color=color, split=split)
    # Suffix legend with [uid] to identify each card


    for tr in traces:
        fig.add_trace(tr)
```

14. Finally, the figure layout is adjusted (margins, axis titles, and legend placement), and both the figure and the updated card styles are returned. The legend is horizontal and placed above the plot to maximize plotting area, while axes are labeled with [L] and the selected quantity.

Listing 43: binding_dashboard.py (redraw_figure finalize)

```
fig.update_layout(
    margin=dict(l=20, r=10, t=10, b=20),
    xaxis_title="[L] (a.u.)",
    yaxis_title=quantity,
    legend=dict(orientation="h", yanchor="bottom", y=1.02, x=0),
)
return fig, new_styles
```

3.7 Final Application Setup and Local Execution

1. The execution of the dashboard begins in the entrypoint of the module. The conditional guard ensures that initialization occurs only when the file is executed as a script (*see* **Note 42**). Inside this block, the Dash app is created, and the layout is attached.

Listing 44: binding_dashboard.py (app entrypoint)

```
if __name__ == '__main__':
    ####################
    # Initilialize app #
    ####################
    # create the app
    app = Dash(__name__, external_stylesheets=[dbc.themes.BOOTSTRAP])
    app.layout = layout
```

2. Internationalization is implemented through Flask-Babel, which integrates seamlessly with the underlying Flask server of the Dash app (*see* **Note 43**). The locale selector ensures that the language is chosen according to the user's browser (*see* **Note 44**). The runtime environment then loads the appropriate message catalogs and applies translations transparently to the dashboard components (*see* **Note 45**).

Listing 45: binding_dashboard.py (Babel setup)

```
def get_locale():
    return request.accept_languages.best_match(LANGUAGES.keys())

# initialize Flask-babel
babel = Babel(app.server) # app.server is the Flask app inside the dash app.
with app.server.app_context():
    LANGUAGES = {l.language: l.get_language_name() for l in babel.list_translations()}
    babel.init_app(app.server, locale_selector=get_locale)
```

3. To run the dashboard locally, the built-in Dash development server is invoked inside the entrypoint. Configuration parameters such as `debug`, `host`, and `port` are specified (*see* **Note 46**). Once this line is executed, the dashboard can be accessed in a browser at http://localhost:8050.

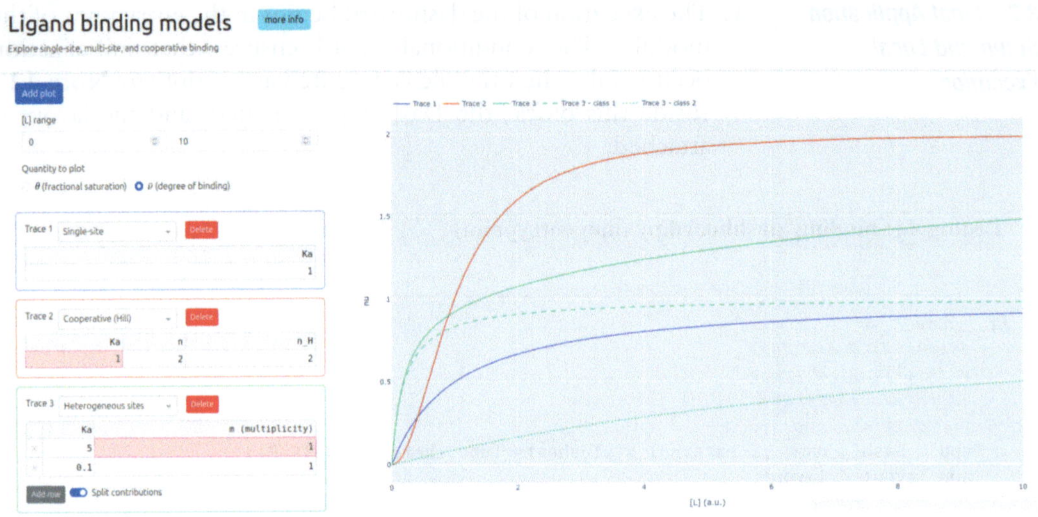

Fig. 3 Dashboard running in the browser. The left panel contains the general controls and model-specific cards, while the right panel displays the resulting binding curves for different models

Listing 46: binding_dashboard.py (local run)

```
app.run(debug=True)
```

4. The entire module can be executed from the command line:

```
python binding_dashboard.py
```

This command starts the development server and makes the dashboard available for interactive use, as shown in Fig. 3.

4 Notes

1. **License implications**. The permissive nature of the licenses also allows the development of applications that are not released as open-source, including proprietary or commercial software, provided that copyright notices are preserved. Conversely, if the applications are distributed under a permissive license, their potential for dissemination, collaborative development, peer review, and continuous improvement is greatly enhanced.

2. **Code license and availability.** Unless otherwise specified, the source code provided with this chapter is released under the GNU Affero General Public License v3.0 (AGPL-3.0). This license ensures that users who access the application over a network are granted the same freedom to obtain, inspect, and modify the source code. The license applies to the original code authored for this chapter and does not cover third-party libraries or the textual content of the chapter itself.
3. **Optional packages.** Other widely used scientific Python libraries, such as `SciPy` (advanced numerical methods) or `Pandas` (tabular data management), are not strictly required for the present application but may be useful in more complex implementations or extended projects.
4. **Additional dashboards available.** Additional Dash-based dashboards are available in a public repository at https://github.com/renato-lombardo/physchem-dashboards and can be explored as live demos at https://physchem-dashboards.vercel.app.
5. **Python installation methods**. There are several possible ways to install Python and the required packages, depending on the user's preferences, operating system, and existing setup:
   - *Anaconda/Miniconda* (recommended): scientific Python distributions that include or can install all the necessary packages; example: `conda install numpy pandas plotly dash`
   - *pip* (Python Package Index): the default Python package manager. It requires a working Python installation; example: `pip install numpy pandas plotly dash`.
   - *Operating system package managers*: convenient for quick installations but may not always provide the most recent versions; examples: `sudo apt install python3 python3-pip` (Debian/Ubuntu); `sudo dnf install python3` (Fedora); and `brew install python` (macOS with Homebrew)
6. **Virtual environments**. Using virtual environments is highly recommended to isolate project dependencies, avoid version conflicts, and ensure reproducibility. They allow multiple Python projects to coexist on the same system without interfering with each other. With *Anaconda* or *Miniconda*, a new environment can be created and activated as follows:

```
conda create -n myenv
conda activate myenv
conda install numpy plotly dash
```

With the built-in venv module in Python:

```
python -m venv myenv
source myenv/bin/activate     # Linux/macOS
myenv\Scripts\activate       # Windows
pip install numpy pandas plotly dash
```

Deactivating the environment restores the global Python context.

7. **Choosing Python version**. The explicit choice of a specific release, such as Python 3.11, ensures reproducibility of the computational environment and avoids incompatibilities that may arise with different versions of the interpreter, especially when working with scientific libraries.
8. **Plotly Dash packages**. A basic Dash application requires the installation of the core Dash framework (dash), which automatically includes the essential components for building layouts and adding interactivity:
   - dash-core-components provides a wide range of interactive controls (sliders, dropdowns, input fields, graphs, etc.).
   - dash-html-components allows the construction of page layouts using HTML tags as Python classes.
   - dash-table enables the display and editing of tabular data within the dashboard.

   The *Plotly* library (plotly) itself must also be installed, as it provides the underlying graphing and visualization engine. For extended styling and layout options, the package dash-bootstrap-components is recommended, allowing the use of Bootstrap themes and responsive design features. All these packages are cross-platform and can be installed together using:

   ```
   conda install dash plotly dash-bootstrap-compo-
   nents
   ```

   or with pip:

   ```
   pip install dash plotly dash-bootstrap-compo-
   nents
   ```

9. **Choice of editor or IDE** JupyterLab is recommended for exploratory work and rapid prototyping, especially when testing model functions independently of the full dashboard. Nonetheless, some Dash features (e.g., live reload) are more convenient when the code is run from a standard Python interpreter rather than a notebook. JupyterLab can be

enhanced through a large ecosystem of extensions that expand its capabilities far beyond notebook editing. Among the most relevant for this protocol are:

- *Version control integration*: Extensions such as `jupyter-lab-git` allow direct interaction with Git repositories, including committing, branching, and pushing changes from within the JupyterLab interface.
- *Code quality and linting*: Plugins can integrate tools like `pylint` or `flake8` to provide inline feedback on code style and potential errors.
- *Dash and interactive app development*: The `jupyter-dash` extension enables running and previewing Plotly Dash applications directly in a notebook environment.
- *Additional productivity tools*: File browsers, variable inspectors, integrated terminals, and real-time collaboration add to the flexibility of the environment.

JupyterLab can be installed in the current *conda* environment using:

```
conda install jupyterlab
```

For users who prefer a more traditional IDE, several cross-platform and free alternatives are available that can integrate Git, linting, testing frameworks, and other development tools. Visual Studio Code is an excellent multipurpose choice, offering integrated Git support and debugger. The choice of environment depends on personal workflow preferences, but in all cases, integration of version control, code quality tools, and testing frameworks is strongly recommended.

10. **Deployment.** Each method is typically associated with one or more providers, and every provider requires its own configuration steps, runtime specifications, and pricing model. Because these aspects are highly provider- and method-specific, they are not covered in this chapter. Readers should consult the official documentation of the chosen provider and, where appropriate, institutional IT support for detailed deployment instructions.
11. **PaaS convenience.** For the purposes of this protocol, PaaS solutions are particularly convenient because they allow the deployment of a Dash application with minimal setup, making them accessible to users without advanced system administration skills. Examples of widely used PaaS providers suitable for Python applications include Heroku, Vercel, Google App Engine, Microsoft Azure App Service, and AWS Elastic Beanstalk. These platforms typically support deployment directly from a source code repository (e.g., GitHub), automatically handle scaling and updates, and provide a public URL for

access via a standard web browser. They are especially well suited for educational and research contexts, as they remove the need for complex server configuration. Each provider has its own deployment procedure, pricing model, and resource limitations.

12. **PaaS versus Docker deployment**. PaaS may impose limitations in terms of available resources, costs for intensive workloads, and restrictions on supported technologies. Using *Docker* requires a slightly more complex initial configuration, but it offers the advantage of running the exact same container in multiple environments: locally, on a private or institutional server, or on an online hosting service. This ensures full reproducibility and consistency across platforms. When using Docker, it is recommended to include `gunicorn` as the production web server, replacing Dash's built-in development server to improve efficiency and reliability.

13. **Binding equations.** In ligand–macromolecule binding models, the symbol θ denotes the *fractional occupancy*, i.e., the fraction of binding sites that are occupied. Its value ranges from 0 (no site occupied) to 1 (all sites occupied). The symbol $\bar{\nu}$ denotes the *average number of occupied sites per macromolecule*, with values from 0 to n, where n is the total number of sites.

    For the case of a ligand binding to a single site, θ follows the hyperbolic equation

$$\theta = \frac{K_a[\mathrm{L}]}{1 + K_a[\mathrm{L}]},$$

    where K_a is the association constant and [L] the free ligand concentration.

    For n independent and equivalent sites, the same functional form applies to each site, and the average number of occupied sites per macromolecule is

$$\bar{\nu} = n\,\theta = \frac{n\,K_a[\mathrm{L}]}{1 + K_a[\mathrm{L}]}.$$

    If the n independent sites are not equivalent, each site i is characterized by its own constant $K_{a,i}$ and multiplicity m_i, giving

$$\bar{\nu} = \sum_i m_i \, \frac{K_{a,i}[\mathrm{L}]}{1 + K_{a,i}[\mathrm{L}]}.$$

    Cooperative binding can be described by the Hill equation, introducing the Hill coefficient n_H:

$$\theta = \frac{K_a[\mathrm{L}]^{n_H}}{1 + K_a[\mathrm{L}]^{n_H}}, \quad \bar{\nu} = n\,\theta.$$

Positive cooperativity corresponds to $n_H > 1$ and negative cooperativity to $n_H < 1$, and $n_H = 1$ reduces to the hyperbolic case.

14. **Binding models—illustrative examples.** A single-site binding model produces a hyperbolic θ versus [L] curve. For example, with $K_a = 10^6\ \mathrm{M}^{-1}$, $\theta = 0.5$ occurs at $[\mathrm{L}] = 1/K_a$.

For n independent and equivalent sites, the average number of bound ligands is

$$\nu = n\,\theta, \quad 0 \leq \nu \leq n,$$

and the curve retains the same hyperbolic shape, simply scaled vertically. If the sites are independent but not equivalent (e.g., $K_{a,1} = 10^6$ mol dm^{-3}, $K_{a,2} = 10^4$ mol dm^{-3}), $\bar{\nu}$ versus [L] shows two phases: a rapid initial rise as high-affinity sites fill, followed by a slower approach to saturation as low-affinity sites bind.

In cooperative binding, the Hill equation

$$\theta = \frac{K_a[\mathrm{L}]^{n_H}}{1 + K_a[\mathrm{L}]^{n_H}}$$

with $n_H > 1$ produces a sigmoidal curve, reflecting increased sensitivity in the midrange of [L]. A classic example is oxygen–hemoglobin binding, where a higher n_H enhances oxygen delivery efficiency by steepening the curve near physiological oxygen partial pressures. Such cases illustrate the need to connect abstract mathematical parameters (K_a, n_H) with their functional biological meaning.

15. **Assessing student achievement of binding-related learning objectives.** The following are examples of tasks and questions that can be used to verify whether students have achieved the intended learning outcomes after interacting with the application:

    1. *Single-site binding curve analysis.* Provide the plot of θ versus [L] for $K_a = 5 \times 10^5$ moldm^{-3}, and ask students to (i) identify the ligand concentration corresponding to half-saturation and (ii) explain how this relates to the value of K_a.

    2. *Comparison between equivalent and nonequivalent sites.* Show two curves for a macromolecule with $n = 2$ equivalent sites ($K_a = 10^6$ mol dm^{-3}) and with $n = 2$ nonequivalent sites ($K_{a,1} = 10^6$ mol dm^{-3} and $K_{a,2} = 10^4$ mol dm^{-3}). Ask students to describe the differences in the shapes and slopes of the curves and to relate them to site affinity.

    3. *Effect of cooperativity.* Present binding curves generated with the Hill equation for $n_H = 1$, $n_H = 2$, and $n_H = 0.7$ (the same K_a value). Students should (i) identify which

curve corresponds to positive, negative, and no cooperativity and (ii) explain how n_H affects the steepness and midpoint of the curve.

4. *Quantitative comparison*. Give numerical data from two different binding scenarios, and ask students to calculate θ or $\bar{\nu}$ at specified [L] values, verifying whether their results match the curves shown.
5. *Parameter manipulation*. Ask students to predict qualitatively how a twofold change in K_a would shift the binding curve, and then test their prediction by using the application and explaining any discrepancies.

These examples assess not only factual knowledge but also the ability to interpret, compare, and predict binding behavior in diverse scenarios, thus directly linking application use to the stated learning objectives.

16. **Using the application to foster computational thinking.** Examples of low-barrier activities include (i) outlining the inputs/outputs of the function that computes $\theta([L])$ and explaining where units enter, (ii) identifying which quantities are derived (e.g., $[L]_{1/2}$) and how they are computed, (iii) describing, in plain language, how a plot is produced from a vector of [L] values without referring to interface details, and (iv) explaining how the same computational core could be reused in a different context (e.g., a command-line script), reinforcing separation of concerns. These tasks support transparency, reusability, and gradual transfer from conceptual understanding to executable models.
17. **Possible improvements.** Thanks to the flexibility of Plotly Dash, the user interface can be easily enriched beyond the basic elements described above. Examples include the integration of interactive tables with options for data export (e.g., CSV or Excel), the embedding of supplementary explanatory resources such as images, hyperlinks, or videos, and the addition of dynamic components that adapt to the learner's input. This extensibility makes it possible to gradually increase the pedagogical effectiveness of the dashboard while keeping the initial design simple and manageable.
18. **Optional online assessment and integration with learning platforms.** It is possible to imagine an extended version that includes tools for assessing student understanding directly within the interactive environment. Such assessments could consist of short quizzes, guided exercises, or prediction tasks where students are asked to anticipate the effect of changing certain model parameters before observing the results. When appropriately designed, these activities could help verify

whether the intended learning objectives have been achieved, promote reflection, and encourage active engagement with the model.

For courses already using a Learning Management System (LMS) such as Moodle, the application could be linked so that results from these activities are automatically recorded in the platform, enabling teachers to monitor progress and provide timely feedback. This integration could also facilitate a consistent learning experience by keeping all resources (lesson notes, dashboards, and assessments) within the same digital environment. However, the development of such integrated assessment features is beyond the scope of this chapter, which focuses exclusively on the design and deployment of the interactive application itself.

19. **Browser compatibility.** Dash applications run reliably on modern "evergreen" browsers: Google Chrome, Mozilla Firefox, Microsoft Edge, and Apple Safari (macOS/iOS). Legacy browsers (e.g., Internet Explorer) are unsupported. A minimal pre-deployment check on (i) a desktop/laptop and (ii) a phone/tablet is recommended. Typical differences: On touch devices tooltips require a tap (not hover) and zoom/pan use pinch or two-finger gestures; file downloads may open in a new tab.
20. **Separation of computational core and interface.** Keeping the model implementation independent of the GUI clarifies *what* the science requires (equations, inputs/outputs, and units) and *how* it is operationalized in code (functions, parameters, and data flow). From a pedagogical perspective, the interface layer (layout, styling, callback wiring) is often of limited instructional value and may be treated as a fixed template/"black box"; the emphasis is placed on the scientific core. A practical pattern is to place pure functions and derived metrics in a `model.py` module and let the GUI call them via thin adapters. This enables (a) transparency, core functions can be read, tested, and reused in a notebook or a small CLI script without any interface code, and (b) assessment, computational learning goals (correct formulas, units, and limiting behaviors) can be evaluated independently of usability issues. Advanced courses may optionally open the GUI code, but for most learners a stable scaffold suffices to focus on modeling and analysis.
21. **Reproducibility of the environment.** To ensure that the same environment can be recreated on different machines, it is advisable to export the full specification of the Anaconda environment to a YAML file. This can be done with:

```
conda env export --no-builds > environment.yml
```

The resulting `environment.yml` file can be shared with collaborators or students, who may then recreate an identical environment using:

```
conda env create -f environment.yml
```

22. **Models directory**. In this specific example, all the models used to describe the binding will be included in a single file `binding.py`. Nevertheless, in a generic case, the models could be more complex and could be convenient to use different files.
23. **Naming conventions.** In all code listings, Python naming follows the PEP 8 guidelines: Function names are written in `snake_case` (e.g., `hill_binding`, `single_site_binding`), and variables are in lowercase with underscores if needed (e.g., `Ka`, `n_H`).
24. **Type hinting.** In Python, type hints allow one to annotate the expected data types of function arguments and return values. They improve readability, facilitate static analysis by editors and linters, and are generally considered a good programming practice. For example, the function defined in Listing 8.1 could be written with type hints as follows:

```python
def hill_binding(
  L: float | np.ndarray,
  Ka: float,
  n_H: float = 1.0,
  output: Literal["theta", "nu"] = "theta",
  sites: int = 1
) -> np.ndarray:
```

25. **Docstrings.** A docstring is a string literal placed as the first statement of a module, class, or function; it is stored in the object's `__doc__` attribute and exposed via `help()`. Docstrings serve as user-facing documentation and are consumed by IDE tooltips, documentation generators (e.g., Sphinx), and static analyzers. They are essential for maintainability and, in didactic contexts, for making code self-explanatory.

    In practice, multiline docstrings should be used to state the purpose of the object and to document *parameters* (names, types, and units), *return values*, *raised exceptions*, side effects, and short usage examples. Conventions from PEP 257 and a consistent format (e.g., NumPy or Google style) are recommended. In the listings of this chapter, docstrings are reduced to a single concise line for readability; in actual projects, extended docstrings with sections such as *Parameters*, *Returns*, and *Raises* are advisable.

26. **Input validation.** The listings omit explicit checks on the validity of input values (e.g., ensuring that $K_a > 0$, $n_H > 0$, and $[L] \geq 0$). Such checks are important in real applications to avoid silent errors and unexpected behavior. For instance, the function in Listing 8.1 could include simple validation lines such as:

```
if Ka <= 0:
  raise ValueError("Ka must be positive")
if n_H <= 0:
  raise ValueError("n_H must be positive")
if not isinstance(sites, int) or sites < 1:
  raise ValueError("sites must be an integer >= 1")
```

27. **Duck typing.** In Python, functions are often written according to the principle of "duck typing": Rather than enforcing a specific data type, the function assumes that the object passed as an argument implements the necessary operations. In Listing 8.1, the parameter `L` can be a scalar or a NumPy array because both support multiplication and exponentiation; the function will work as long as the input behaves "like a number" or "like an array." This makes the code more flexible but requires careful documentation to avoid confusion.

28. **Model import.** The functions `hill_binding`, `single_site_binding`, etc. are imported from the local `binding_model.py` module, where the mathematical definitions of the binding models are implemented.

29. **Plotly.** `plotly.graph_objs` and `plotly.colors` are used to generate publication-quality graphs with customizable styles and color palettes.

30. **Dash framework.** `dash`, `dash_bootstrap_components`, and `dash.dash_table` provide the interactive framework, prebuilt UI components, and tabular widgets, respectively. This allows a professional layout without writing low-level HTML or CSS.

31. **Flask.** `flask_babel` and `gettext` add multilingual support (internationalization), automatically adapting the dashboard content to the user's browser language.

32. **Typing.** `typing` provides type hints such as `Any`, `List`, and `Dict`, improving code readability and enabling static analysis.

33. **Gettext function.** The `gettext` function from the `flask_babel` package is assigned to the underscore symbol (_), so that every string to be translated can be simply wrapped as `_(text)`. This renaming is done for convenience and is the common practice when using Flask-Babel, as it allows the code to remain concise and easily readable.

34. **Rows and columns.** In Dash, the layout is typically organized using a grid system where `Row` defines a horizontal container and `Col` specifies how elements are arranged within that row. This design pattern, inherited from Bootstrap, allows developers to place multiple components side by side (in columns) and to control their relative widths and alignment. By nesting columns within a row, the interface can adapt more easily to different screen sizes, ensuring a responsive layout suitable for desktops, tablets, and mobile devices.
35. **Markdown and mathJax support.** Markdown is a lightweight markup language that allows rich text (headings, emphasis, lists, and links) using plaintext syntax. In Dash, mathematical expressions written in LaTeX syntax can be rendered via MathJax by enabling `mathjax=True` in `dcc.Markdown`. The example in Listing 8.7 deliberately uses a reduced informational text for simplicity; in practice, the content can be substantially richer (including multiple display equations, inline math, lists, and links).
36. **Right panel styling.** Wrapping `dcc.Graph` in a lightweight container (e.g., `html.Div`) allows controlling inner padding with Bootstrap utility classes (`p-2`, `p-3`, `p-4`), without affecting the grid gutters applied to `dbc.Col`. The graph uses style options, so it fills the column and scales with the viewport; alternative heights (e.g., `70vh`) can be chosen as needed.
37. **Role of the factory.** Each call to `controls_card_factory` produces a new card with a unique `uid` and all the needed components. This modular structure makes it possible to add or remove cards dynamically.
38. **The meaning of `uid`.** The IDs of all elements are dictionaries with two keys: `"type"` (defining the semantic role, e.g., `"param-table"` or `"plot-store"`) and `"uid"` (the unique integer assigned to the card). This design enables Dash pattern-matching. In practice, this means that multiple plots can coexist independently: Each has its own parameters, controls, and stored state, but the global logic can still access them collectively if needed.
39. **Callback mechanism.** Dash implements callbacks using the Python decorator syntax. A decorator `@callback` is placed directly above a Python function to define the reactive link between user actions and interface updates. Within this decorator, three categories of arguments are specified:
    - *Inputs* represent component properties that, when changed, trigger the callback execution (e.g., a button click or a modified table cell).

- *Outputs* indicate the component properties that will be updated by the callback (such as the children of a container, the content of a table, or the figure of a plot).
- *States* provide additional values that are passed to the function but do not themselves trigger execution.

This design allows the function to access contextual information (e.g., the current slider values or the active contents of a container) while keeping the callback execution controlled by explicit input changes.

40. **Pattern-matching IDs.** Dash supports *pattern-matching IDs*, which make it possible to dynamically address multiple instances of similar components (such as the control cards). Component IDs can be defined as dictionaries with keys like `"type"` and `"uid"`. In this context, `MATCH` is used to target only the component with the same `uid`, and `ALL` is used to apply a callback simultaneously to all components of a given type.
41. **Prevent_initial_call.** In Dash, a callback is normally executed once when the application is first loaded, even before the user interacts with any control. Setting the argument `prevent_initial_call=True` in the decorator disables this automatic first execution. The callback will only run after an explicit change in one of its inputs has occurred. This is needed to avoid errors when the initial state of the components is not yet meaningful.
42. **Entrypoint guard.** The conditional `if __name__ == __main__` ensures that the startup code is executed only when the module is run directly and not when it is imported elsewhere (e.g., another python script).
43. **Exposing the server.** Since Dash is built on `Flask`, the attribute `app.server` exposes the underlying Flask application that is needed to initialize Flask-Babel. All calls to `_()` are then translated at runtime, so that user-visible text automatically appears in the correct language (if available).
44. **Locale selector.** The function `get_locale()` reads the `Accept-Language` header from the client's browser and matches it with the available translations, activating the closest supported language.
45. **Translation catalogs.** Flask-Babel extends the Flask framework to support internationalization (i18n). All strings enclosed in the function call `_()` (an alias for `gettext`) are considered translatable. These marked strings are extracted into a template file, where they can be translated into one or more languages. Translations are stored as editable `.po` files and, once compiled, as binary `.mo` catalogs located under

translations/<lang>/LC_MESSAGES/messages.mo. At runtime, Flask-Babel loads the appropriate catalog according to the active locale (selected by the get_locale() function), so that every string wrapped in _() automatically appears in the correct language.

To create and maintain the catalogs, the typical workflow is:

- *Extraction:* Run pybabel extract to collect all strings into a template file (.pot).
- *Initialization:* Create a new catalog with pybabel init, which generates a .po file for the chosen locale.
- *Editing:* Translate the strings inside the .po file.
- *Compilation:* Convert the .po into a binary .mo with pybabel compile, so Flask-Babel can load it.
- *Update:* When code changes, rerun extraction and update existing catalogs with pybabel update.

If only a few Python files are present, it is sufficient to list them explicitly on the command line. For example:

```
pybabel extract -o messages.pot binding_dash-
board.py binding_model.py
pybabel init -i messages.pot -d translations -l
it
pybabel compile -d translations
```

This command extracts all strings marked with _() from the two Python files binding_dashboard.py and binding_model.py, initializes the Italian catalog under translations/it/LC_MESSAGES/, and compiles it for runtime use.

For brevity, many interface strings that could be wrapped in _() are left as plain literals in the listings; in a production application, UI labels, tooltips, error messages, and plot annotations should be consistently marked to be included in the translation catalogs.

46. **Running the app.** By default, running the application with app.run(debug=True) starts the server on localhost (127.0.0.1) and port 8050. In this configuration, the dashboard can be reached only from the same machine at the address http://127.0.0.1:8050 (or equivalently http://localhost:8050). If desired, the listening port can be changed by specifying the port parameter, e.g., app.run(debug=True, port=8080), which makes the application available at http://127.0.0.1:8080. To allow access from other computers on the same local network, the host parameter can be set to "0.0.0.0". In that case, the dashboard will be reachable

at http://%3CIP_of_the_host%3E:%3Cport%3EÊ, where <IP_of_the_host> is the machine's local network address (e.g., 192.168.1.5). This configuration should be used with caution, as it makes the application visible to external connections depending on firewall and network settings.

Acknowledgements

The author thanks the open-source community for the software, documentation, and public discussion forums that made this work possible. In particular, the sustained efforts of maintainers and contributors to Python, NumPy, Plotly Dash, Dash Bootstrap Components, and Flask-Babel, including testing, issue triaging, and timely releases under permissive licenses, enabled rapid prototyping, reproducible workflows, and accessible dissemination.
The author is also grateful to his students for constructive feedback during pilot classroom activities.

Competing Interest Statement The author has no conflicts of interest to declare that are relevant to the content of this chapter.

References

1. Johnstone AH (1991) Why is science difficult to learn? Things are seldom what they seem. J Comput Assist Learn 7(2):75–83. https://doi.org/10.1111/j.1365-2729.1991.tb00230.x
2. Becker N, Towns M (2012) Students' understanding of mathematical expressions in physical chemistry contexts: an analysis using Sherin's symbolic forms. Chem Educ Res Pract 13(3):209–220. https://doi.org/10.1039/C2RP00003B
3. Parobek AP, Chaffin PM, Towns MH (2021) Location-thinking, value-thinking, and graphical forms: combining analytical frameworks to analyze inferences made by students when interpreting the points and trends on a reaction coordinate diagram. Chem Educ Res Pract 22(3):697–714. https://doi.org/10.1039/D1RP00037C
4. Planinic M, Ivanjek L, Susac A, Milin-Sipus Z (2013) Comparison of university students' understanding of graphs in different contexts. Phys Rev Spec Top Phys Educ Res 9(2):020103. https://doi.org/10.1103/PhysRevSTPER.9.020103
5. Rodriguez JMG, Stricker AR, Becker NM (2020) Students' interpretation and use of graphical representations: insights afforded by modeling the varied population schema as a coordination class. Chem Educ Res Pract 21(2):536–560. https://doi.org/10.1039/C9RP00249A
6. Uzun MS, Sezen N, Bulbul A (2012) Investigating student's abilities related to graphing skill. Procedia Soc Behav Sci 46:2942–2946. https://doi.org/10.1016/j.sbspro.2012.05.594
7. Mainali B (2021) Representation in teaching and learning mathematics. Int J Educ Math Sci Technol 9(1):1–21
8. Haleem A, Javaid M, Qadri MA, Suman R (2022) Understanding the role of digital technologies in education: a review. Sustainable Oper Comput 3:275–285. https://doi.org/10.1016/j.susoc.2022.05.004
9. Adamoli M, Rondonotti M, Fedeli L, Scarinci A, Raviolo P (2024) The impact of the digital environment on faculty development: a literature review. In: Casalino G, Di Fuccio R, Fulantelli G, Raviolo P, Rivoltella PC, Taibi D, Toto GA (eds) Higher education learning methodologies and technologies online. Springer Nature Switzerland, Cham, pp 746–759. https://doi.org/10.1007/978-3-031-67351-1_49
10. Timotheou S, Miliou O, Dimitriadis Y, Sobrino SV, Giannoutsou N, Cachia R,

Monés AM, Ioannou A (2023) Impacts of digital technologies on education and factors influencing schools' digital capacity and transformation: a literature review. Educ Inf Technol 28(6):6695–6726. https://doi.org/10.1007/s10639-022-11431-8
11. Mahjour B, McGrath A, Outlaw A, Zhao R, Zhang C, Cernak T (2023) Interactive Python notebook modules for chemoinformatics in medicinal chemistry. J Chem Educ 100(12): 4895–4902. https://doi.org/10.1021/acs.jchemed.3c00357
12. Sermania B, VanFossen N, Steen R, Stewart M, Raston PL (2024) FTIR-SIS: a Fourier transform infrared scientific instrument simulator. J Chem Educ 101(4):1711–1717. https://doi.org/10.1021/acs.jchemed.3c01161
13. Lombardo R (2024) Python and Plotly Dash, a quick and convenient way to develop web apps for teaching physical chemistry models. J Chem Educ. https://doi.org/10.1021/acs.jchemed.3c01167
14. University of Colorado Boulder (2023) PhET: free online physics, chemistry, biology, earth science and math simulations. https://phet.colorado.edu/
15. Perri M (2015) Home | ChemCompute: free computational chemistry for undergraduates. https://chemcompute.org/
16. Khan Academy (2023) Khan Academy—Free Online Courses, lessons & practice. https://www.khanacademy.org/
17. Liermann V, Li S (2021) Dynamic dashboards. In: Liermann V, Stegmann C (eds) The digital journey of banking and insurance, Volume II: digitalization and machine learning. Springer International Publishing, Cham, pp 155–180. https://doi.org/10.1007/978-3-030-78829-2_9
18. Almasi S, Bahaadinbeigy K, Ahmadi H, Sohrabei S, Rabiei R (2023) Usability evaluation of dashboards: a systematic literature review of tools. Biomed Res Int 2023(1): 9990933. https://doi.org/10.1155/2023/9990933
19. Sundnes J (2020) Introduction to scientific programming with python, 1st edn. Springer, Berlin
20. Vaughan L (2023) Python tools for scientists: an introduction to using Anaconda, JupyterLab, and Python's Scientific Libraries. No Starch Press, San Francisco
21. Hossain S (2019) Visualization of bioinformatics data with dash bio. In: Proceedings of the 18th Python in science conference, pp 126–133. https://doi.org/10.25080/Majora-7ddc1dd1-012
22. Plotly Technologies (2015) Collaborative data science. https://plot.ly
23. Bravenec AD, Ward KD (2023) Interactive Python notebooks for physical chemistry. J Chem Educ 100(2):933–940. https://doi.org/10.1021/acs.jchemed.2c00665
24. Harris CR, Millman KJ, van der Walt SJ, Gommers R, Virtanen P, Cournapeau D, Wieser E, Taylor J, Berg S, Smith NJ, Kern R, Picus M, Hoyer S, van Kerkwijk MH, Brett M, Haldane A, del Río JF, Wiebe M, Peterson P, Gérard-Marchant P, Sheppard K, Reddy T, Weckesser W, Abbasi H, Gohlke C, Oliphant TE (2020) Array programming with NumPy. Nature 585(7825):357–362. https://doi.org/10.1038/s41586-020-2649-2

Chapter 9

Fluorescence Anisotropy Applied to Measure Aptamer-Protein Interactions

Maira Rivera and Maureen McKeague

Abstract

This chapter describes a fluorescence anisotropy-based method for quantifying the binding affinity between a nucleic acid aptamer and its protein target, using the thrombin-binding aptamer (TBA)-thrombin interaction as an example. Fluorescence anisotropy is a rapid and sensitive technique that monitors changes in a molecule's rotational speed upon binding. By fluorescently labeling an aptamer and titrating it with increasing concentrations of its target protein, users can generate a binding isotherm and determine the dissociation constant (K_D) with accuracy. The protocol includes detailed instructions for reagent preparation, plate setup, data acquisition, and analysis, with emphasis on minimizing experimental variability. Although demonstrated for the thrombin aptamer system, the approach is broadly applicable to other protein-nucleic acid interactions.

Key words Fluorescence anisotropy, Aptamers, Molecular interactions, Fluorescence polarization, Protein-nucleic acid, Binding affinity, G-quadruplex, Abasic site, Nucleic acid modification, Dissociation constant

1 Introduction

Almost every biological process within a cell relies on interactions with a binding partner, ranging from gene expression to cell-to-cell communication. Understanding these interactions at a biophysical level is key to advancing diagnostics and therapeutics. Fluorescence anisotropy is a sensitive, solution-based technique that quantifies binding by detecting changes in fluorophore rotation upon complex formation (Fig. 1a) [1]. In a fluorescence anisotropy experiment, polarized light is directed to the sample and excites the fluorescent molecules aligned in the plane. If the molecule is small, it will diffuse rapidly in solution, and therefore, the detected fluorescence emission in the parallel and perpendicular axes will be similar, resulting in depolarized emission and minimal anisotropy. When the fluorescent molecule binds to a larger molecule, forming a complex, its rotational diffusion slows down, producing stronger

Robert E. Dempski (ed.), *Transforming Molecular Biology with Emerging Technologies*, Methods in Molecular Biology, vol. 3043, https://doi.org/10.1007/978-1-0716-5320-3_9,

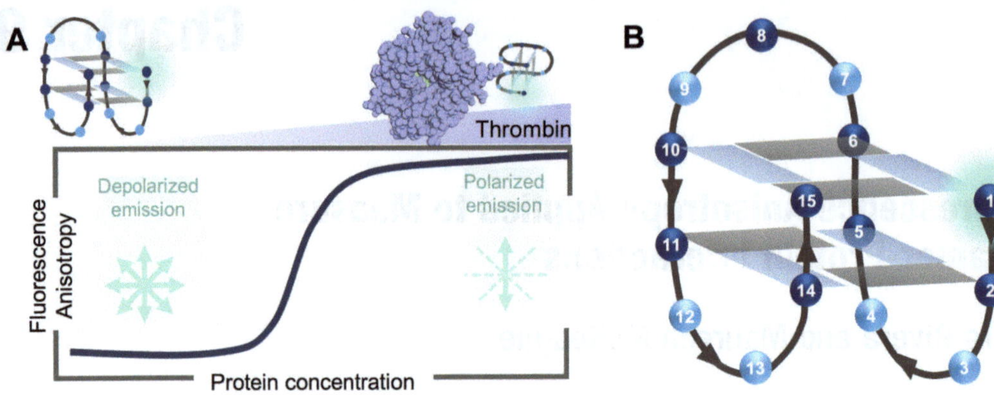

Fig. 1 Fluorescence anisotropy assay to measure binding and the thrombin-binding aptamer. (**a**) Principle of the fluorescence anisotropy assay. Upon excitation with polarized light, the free fluorescent aptamer emits a largely depolarized signal due to rapid molecular rotation. Binding to the larger thrombin protein slows rotation, increasing emission polarization. A binding isotherm generated from anisotropy values across a thrombin titration yields the dissociation constant (K_D). (**b**) Schematic of the 15-nucleotide thrombin-binding aptamer in its antiparallel G-quadruplex topology. Guanines (dark blue) and thymines (light blue) are indicated with a fluorescein label at G1 serving as the fluorescent probe

emission along the parallel axis and thus resulting in polarized emission and increased anisotropy. Mathematically, fluorescence anisotropy (r) is calculated from the ratio of parallel and perpendicular emission intensities and reflects the extent to which fluorophore rotation occurs.

The ratiometric design of fluorescence anisotropy makes the technique inherently robust, minimizing sensitivity to experimental fluctuations and instrument-specific biases [2]. Unlike surface plasmon resonance (SPR), which depends on immobilization, fluorescence anisotropy eliminates this source of artifact. In addition, compared to other common methods such as microscale thermophoresis (MST), isothermal titration calorimetry (ITC), and electrophoretic mobility shift assays (EMSA), the fluorescence anisotropy method requires less sample, accommodates complex buffers, allows rapid small-volume measurements, and scales readily to high-throughput formats [2, 3]. These advantages make it especially well-suited for studying nucleic acid recognition elements, where subtle structural changes and solution conditions can strongly influence binding behavior.

Aptamers are short, single-stranded nucleic acids experimentally generated through an in vitro selection process known as SELEX (systematic evolution of ligands by exponential enrichment) [4–7]. Although SELEX is the predominant method to identify new aptamers, alternative approaches have been developed. Computational strategies, for example, can be used to design more structurally complex nucleic acid pools [8] or combined with SELEX to facilitate high-throughput aptamer identification

[9]. In contrast, de novo aptamer design using purely computational approaches is an emerging field [10]. Experimental non-SELEX methods have also been explored, including nonequilibrium capillary electrophoresis of equilibrium mixtures (NECEEM), which avoids bias generated in PCR amplification [11], as well as competitive selection strategies such as CompELS [12] and SELCOS [13].

Following aptamer identification, it is still essential not only to characterize the binding performance of the aptamers to their intended targets but also to determine which nucleotides are critical for recognition and specificity. In this context, measuring aptamer affinities is a critical step, as it determines their utility as molecular tools, diagnostics, and therapeutic agents (e.g., biosensors, anticoagulants, targeted drug delivery systems) [3]. The thrombin-binding aptamer (TBA) is a classic example, forming an antiparallel G-quadruplex that binds exosite I of thrombin and inhibits clotting [14–17] (Fig. 1b). While fluorescence anisotropy is widely used for proteins, peptides, and small molecules, aptamers require dedicated optimization because they are particularly sensitive to ionic conditions, sequence-dependent folding, and surface adsorption, which can distort binding measurements.

Previous protocols have demonstrated the versatility and robustness of fluorescence anisotropy-based methods for studying protein-ligand interactions [1, 18]. Here, we present a detailed protocol to study the interaction of fluorescently labeled aptamers with a protein target, using the thrombin-TBA system as a representative model. A significant challenge in these assays is ensuring the accuracy of the measurements, as experimental errors, including nonspecific binding of aptamers to microplate surfaces, can interfere with the results. We provide step-by-step instructions for sample preparation, plate setup, data acquisition, and data analysis to accurately determine the dissociation constant (K_D). In addition, we demonstrate how site-specific aptamer modifications—specifically the introduction of abasic internal spacers (dSpacer) at positions 12, 13, and 14—can be used to probe structural requirements and conformational flexibility of the aptamer-target binding interface, and how their impact on binding affinity is quantitatively assessed using the fluorescence anisotropy assay.

2 Materials

1. All solutions must be prepared in molecular biology grade water, such as highly purified deionized water (with a resistivity of 18.2 MΩ·cm at 25 °C), such as Milli-Q water.
2. All reagents must be of molecular biology grade (*see* **Note 1**).

2.1 Instrumentation

1. *Multimode Microplate Reader*: This protocol has been tested on a Tecan Spark, a BioTek Synergy H1, and a Cytation 5. Synergy H1 and Cytation 5 are compatible with the FP filter cube, part number 8040561.

2.2 Preparation of Stock Solutions

1. *5 M NaCl*: Weigh 146 g of NaCl and dissolve in ~450 mL of ultrapure water in a glass beaker with agitation. Transfer to a graduated cylinder, adjust volume to 500 mL, mix gently, and filter through a 0.22 μm PES membrane using a vacuum filtration unit.
2. *1 M $CaCl_2$*: Weigh 14.7 g of $CaCl_2 \cdot 2H_2O$ (MW 147.02 g/mol), dissolve in ~90 mL ultrapure water with agitation. Adjust to 100 mL total volume, mix, and filter as above.
3. *1 M Tris-HCl pH 7.4*: Weigh 24.2 g of Tris base (MW 121.14 g/mol), dissolve in ~150 mL of ultrapure water with agitation. Adjust pH to 7.4 at room temperature using HCl. Allow pH to stabilize for 2 min before adjusting final volume to 200 mL and filtering.
4. Commercial solutions (Invitrogen):
   (a) 1 M $MgCl_2$.
   (b) 2 M KCl.

2.3 Thrombin--Binding Buffer (TBB)

1. TBB composition: 140 mM NaCl, 20 mM Tris-HCl, pH 7.4, 20 mM KCl, 1 mM $CaCl_2$, 1 mM $MgCl_2$, and 0.05% (w/v) Triton X-100 (*see* **Note 2**).
2. To prepare 40 mL of a 5X TBB solution, in a 50 mL Falcon tube, add the following:
   (a) 5.6 mL of 5 M NaCl.
   (b) 4 mL of 1 M Tris-HCl pH 7.4.
   (c) 2 mL of 1 M KCl.
   (d) 200 μL of 1 M $CaCl_2$.
   (e) 200 μL of 1 M $MgCl_2$.
3. Add 28 mL of ultrapure water to the tube, and mix by inverting.
4. Using an analytical balance, weigh 100 mg of Triton X-100 directly into the cap of the Falcon tube (*see* **Note 3**).
5. Gently and using a P-1000 micropipette, mix the detergent from the cap with the buffer inside the tube. Pipette the buffer up into the cap to dissolve the Triton X-100 and then transfer it back into the tube. Repeat until all the detergent is dissolved (*see* **Note 4**).
6. Once no visible detergent remains in the cap, close the tube and mix again by inverting it 5 times.

7. To prepare 25 mL of a 1X TBB solution, in a 50 mL Falcon tube, add 20 mL of ultrapure water and 5 mL of 5X TBB (*see* **Note 4**).

2.4 Thrombin

1. Thrombin from human plasma.
2. Reconstitute the protein in the corresponding amount of TBB to reach a final concentration of 400 nM. Usually, for a 100 U vial, it will be about 2.6 mL. Store reconstituted thrombin at 4 °C for 5 to 7 days (*see* **Notes 5** and **6**).

2.5 Fluorescently Labeled DNA Aptamers

1. Fluorescently labeled aptamers can be synthesized and ordered from any nucleic acid manufacturer, e.g., Integrated DNA Technologies (IDT Coralville, IA, USA) (*see* **Notes 7** and **8**).
2. Resuspend in nuclease-free water to 100 μM. Store at −20 °C for up to 2 years (*see* **Note 9**).

2.6 Plates

1. Low-binding 96-well conical black plate (*see* **Note 10**).

3 Methods

3.1 Thrombin and Aptamers Preparation

1. Prepare a twofold serial dilution starting from the 400 nM thrombin stock solution in TBB to obtain 11 concentrations (400 nM to 390 pM).
   (a) For this assay, we recommend at least one technical replicate (duplicates). This allows three different aptamers to be tested on the same plate (Fig. 2a). Here, we compared TBA, a negative control aptamer, and three test sequences containing abasic sites at positions 12, 13, and 14 (Table 1).
   (b) Testing five aptamers in duplicate requires two plates. To minimize pipetting error, prepare the following volumes:

Table 1
Sequences of the aptamers studied. All aptamers are 5′-labeled with 6-FAM (called /56-FAM/ from IDT)

Name	Sequence (5′-3′) and IDT codes
Thrombin- Binding Aptamer	/56-FAM/GGTTGGTGTGGTTGG
Negative control	/56-FAM/GGTTGGTGTTGTTTG
Test Sequence A1 (prev A2)	/56-FAM/GGTTGGTGTGG/idSp/TGG
Test Sequence A2 (prev A1)	/56-FAM/GGTTGGTGTGGT/idSp/GG
Test Sequence A3	/56-FAM/GGTTGGTGTGGTT/idSp/G

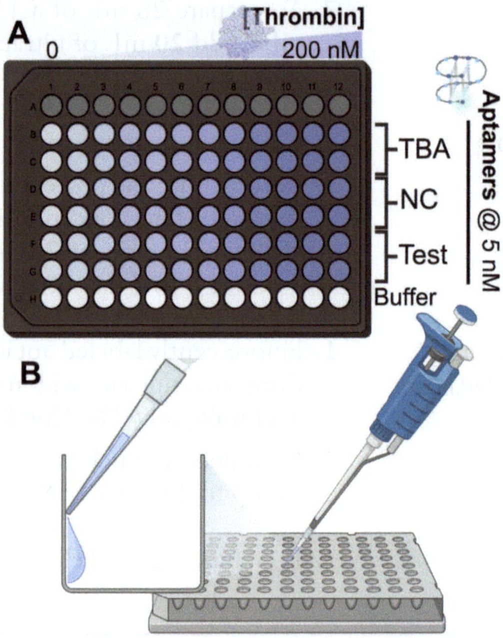

Fig. 2 96-well plate setup and bubble prevention during sample loading. (**a**) Example layout showing thrombin concentration gradients (390 pM to 200 nM), buffer and aptamer controls (column 1), buffer-only control (row H), and three aptamer samples (TBA, NC, and Test) at 5 nM final concentration, arranged in duplicate. (**b**) Recommended pipetting technique: hold pipette at a 45° angle, touching the well wall to minimize the introduction of air bubbles

(i) *Thrombin dilutions*: Prepare 11 tubes each containing 400 μL TBB. Add 400 μL of the previous concentration to each tube to generate the dilution series, starting from the 400 nM reconstituted protein.

(ii) *10 μM aptamer stocks*: For each aptamer, prepare a 10 μM stock by adding 90 μL of TBB to an Eppendorf tube, followed by 10 μL of the 100 μM aptamer solution.

(iii) *10 nM aptamer working solution*: Dilute the 10 μM stock 1:1000 in TBB to prepare 800 μL to 1 mL total volume. A larger volume is recommended when using multichannel pipettes to reduce pipetting error.

3.2 96-Well Plate Setup

1. Refer to Fig. 2a for a representative plate map. Fill the plate as follows:

   (a) Add 30 μL of TBB to all wells in column 1. Avoid introducing bubbles (*see* **Note 11** and Fig. 2b).

(b) Add 60 μL of TBB to all wells in row G (*see* **Note 12**).

(c) Starting from column 2, add 30 μL of each thrombin dilution (beginning with the lowest concentration) to the appropriate wells.

(d) Add 30 μL of the 10 nM aptamer solution to every well that contains either thrombin or TBB (*see* **Note 13**).

2. After filling the plate, gently mix by swirling by hand. Cover the plate with aluminum foil to protect from light, then measure fluorescence (*see* **Notes 14** and **15**).

3.3 Fluorescence Measurements

1. Configure the plate reader to record fluorescence intensity from both the parallel ($I_{\parallel}$) and perpendicular ($I_{\perp}$) emission axes.
2. For 5′ 6-FAM-labeled aptamers, use an excitation wavelength of 492 nm and an emission wavelength of 520 nm (as specified by IDT).
3. If using a fluorescence plate reader requiring a fluorescence polarization filter cube, select the correct cube on the software.
4. Export the raw fluorescence intensities obtained for each axis to a spreadsheet for further analysis.

3.4 Determination of the Fluorescence Anisotropy and Dissociation Constants

1. Calculate fluorescence anisotropy (r) for each well using:

$$r = \frac{I_{\parallel} - I_{\perp}}{I_{\parallel} + 2 \cdot I_{\perp}}$$

where $I_{\parallel}$ and $I_{\perp}$ are the measured intensities in the parallel and perpendicular channels, respectively.

2. Calculate the change in anisotropy:

$$\Delta r = r - r_0$$

where r_0 is the anisotropy value of the aptamer in the absence of thrombin (aptamer + buffer control).

3. Plot Δr values against thrombin concentration using graphing software (e.g., GraphPad Prism). Note: mixing thrombin and aptamer solutions on the plate results in a 1:2 dilution of the thrombin. Therefore, the highest final thrombin concentration is 200 nM with a serial dilution down to 195 pM.
4. Fit the binding curve to determine the dissociation constant (K_D) for each aptamer (Fig. 3b) using (*see* **Note 16**):

$$\Delta r = \Delta r_{\max} \frac{[\text{target}]}{K_D + [\text{target}]}$$

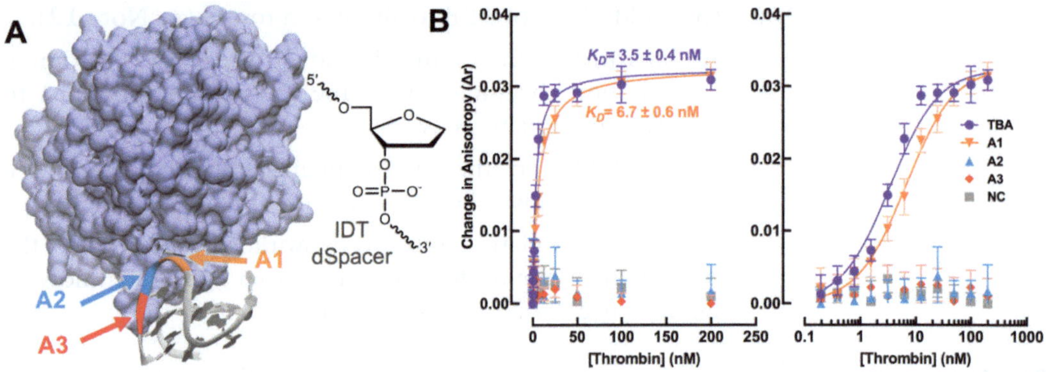

Fig. 3 Binding affinity of thrombin-binding aptamer variants determined using fluorescence anisotropy. (**a**) Structure of thrombin (light blue) in complex with the thrombin-binding aptamer (TBA, gray) (PDBid: 1HAO), highlighting the three nucleotide positions modified with an internal 1′2′-dideoxyribose modification (dSpacer). The chemical structure of the dSpacer is shown. These modifications generated variants A1 (orange), A2 (blue), and A3 (red). (**b**) Fluorescence anisotropy binding curves for the aptamer variants, shown with linear (left) and logarithmic (right) thrombin concentration scales. Data were obtained by plotting fluorescence anisotropy (*r*) against increasing concentrations of thrombin. Dissociation constants (K_D) were determined by fitting the data to a one-site specific binding model. The wild-type aptamer (TBA) and variant A1 showed similar binding affinities, whereas the non-binding control (NC) and variants A2 and A3 showed no significant binding to thrombin. Error bars represent the standard deviation of technical triplicates. Data shown are representative of a single experiment; the protocol can be applied to other aptamer-target pairs to determine binding affinity

4 Notes

1. If working with RNA aptamers, use nuclease-free water or DEPC-treated water to prevent degradation.
2. In our experience, including Triton X-100 is essential for accurate fluorescence anisotropy measurements.
3. To add Triton X-100 as v/v, 100 μL can be dispensed using a P-200 micropipette. However, due to the high viscosity of the detergent, extra care is required. We recommend cutting the pipette tip with a previously cleaned pair of scissors and carefully aspirating the detergent from the surface of the solution, while avoiding contact with the outer edge of the tip.
4. Be gentle to avoid foaming, as Triton X-100 is a detergent. The order of addition is important; adding water first helps avoid foaming when you add the 5X TBB.
5. In our experiments, we used thrombin from human plasma (Sigma, Cat #T6884). Each batch comes with its own certificate of analysis (CoA), which reports the specific activity in units per milligram of protein. For example, for a vial containing 100 U of thrombin with a specific activity of 2500 U/mg, the protein mass is calculated to be 40 μg (considering a

molecular weight of 37.4 kDa). This is equivalent to 1.07 nmol of protein. Therefore, to achieve a final concentration of 400 nM, the vial must be resuspended in 2.67 mL of TBB.

6. Measuring protein concentration by absorbance at 280 nm is not possible when the protein is in TBB because Triton X-100 also absorbs at this wavelength. Instead, the protein can be resuspended in a lower volume of TBB without the detergent. After the concentration has been measured and the protein has been diluted to the desired final concentration, Triton X-100 can be added by supplementing with a 10% (w/v) stock solution.

7. When selecting fluorophores and labeling sites for other aptamer systems, consider both the photophysical properties of the dye and the aptamer's folding and binding interface. While the 5′ end is often chosen for convenience, either terminus can be used provided that labeling does not interfere with secondary or tertiary structure. The optimal site should be determined empirically for each aptamer-target pair. Common dyes such as FAM, TMR, and Texas Red differ in quantum yield, fluorescence lifetime, size, and charge, which can influence both baseline anisotropy and the relative anisotropy change upon binding [19, 20]. In this protocol, we use FAM because it is widely detectable on standard fluorescence instruments and has demonstrated reliable performance in our lab for both RNA-DNA and DNA-protein interaction studies.

8. Consider including a negative control aptamer sequence. Sometimes, negative control sequences are reported in papers with published aptamers. If unknown, a good starting point for a negative control is a scrambled sequence of the same length and relative A/C/G/T.

9. When reconstituting DNA, account for the final amount supplied (in nmol). For example, for 10 nmol DNA, add 100 μL nuclease-free water for a 100 μM stock. Confirm concentration by spectrophotometry, taking into account the molar extinction coefficient and any modifications.

10. We recommend using treated black 96-well plates as they provide more consistent results compared to untreated plates, in particular we have used plates from Thermo Fisher (Cat#249945). Additionally, both clear-bottom and solid black plates have been successfully used in our laboratory, with U-bottom and V-bottom formats yielding reliable results. White plates can be used in exceptional cases but may introduce a higher background.

11. Because the TBB buffer contains Triton X-100, bubbles can easily form and are difficult to remove. To prevent this, add the solution slowly to the well at a 45° angle, making sure the

pipette tip is touching the well wall. Avoid using the final push on the micropipette, as this often introduces bubbles. This step is critical because bubbles will interfere with an accurate fluorescence measurement.

12. This buffer row (e.g., H) is used for subtracting background fluorescence from the buffer. For TBB, this is negligible, but we recommend including it if a different buffer is used.
13. In this step, a multichannel micropipette can be used. Keep following the recommendation from **Note 11**.
14. The thrombin-binding aptamer reaches binding equilibrium in <20 min. For other aptamer-protein pairs with unknown kinetics, we recommend an initial measurement immediately after mixing, followed by measurements at 15, 30, 60, and 120 min at room temperature. For further validation, incubate overnight at 4 °C and measure again after returning to room temperature. The incubation time at which the dissociation constant (K_D) no longer changes is the minimum time needed for equilibrium.
15. The optimal incubation time also depends on protein stability, which can be affected by ionic strength, pH, and required additives.
16. During data analysis, watch for fitting artifacts. Biphasic binding curves can indicate multiple binding sites, conformational heterogeneity, or aggregation. Poor fits may also arise from incorrect baseline subtraction, inaccurate concentration values, or insufficient equilibration time.

5 Final Remarks

- This assay is well-suited for evaluating the effects of chemical modifications on aptamers, such as fluorophore labeling, abasic spacers, or backbone alterations. Its solution-based format and sensitivity to changes in molecular size and flexibility make it ideal for detecting how such modifications influence binding affinity. In fact, we have successfully used this method for undergrad laboratory activities studying the TBA-thrombin-binding [21].
- Any aptamer-protein binding pair can theoretically be analyzed with this method. However, it is important to note that not all proteins are available in lyophilized form. Therefore, we recommend checking the recommended or optimal protein's buffer composition and assessing its compatibility with the assay. Components that increase buffer viscosity (e.g., glycerol) can interfere with the measurements by reducing the overall

rotational speed, leading to artifactual results and inaccurate K_D values.

- If the protein to be tested is suspended in 50% glycerol (as is common for many commercially available enzymes), we recommend performing buffer exchange into an anisotropy-compatible buffer that preserves protein stability. This can be achieved by dialysis, desalting columns, or centrifugal filters. In the latter case, the molecular weight of the functional protein should be considered when selecting the appropriate cutoff filter size.
- If the protein is produced in the laboratory by heterologous expression, further validation of structural integrity is recommended. For example, SDS-PAGE can be used to confirm purity, circular dichroism to assess proper secondary structure, and size-exclusion chromatography to verify the correct aggregation state (in the case of globular proteins).

Acknowledgements

We are grateful to members of the McKeague lab for useful discussions, in particular thanks to Sietse Munting for his assistance in testing this protocol during the writing of this chapter. We also thank the National Science and Engineering Research Council of Canada (NSERC, Discovery Grant), FRQNT Center for Green Chemistry and Catalysis, the Canada Foundation for Innovation (CFI, JELF, and IF 2023), the Canada Research Chairs Program, and D2R McGill for funding.

Conflict of Interest This study was funded by Médicament Québec (RQM00158), the Natural Sciences and Engineering Research Council of Canada (RGPIN-2019-04949), and the Canada Research Chairs Program (CRC-2023-00110).

References

1. Rossi AM, Taylor CW (2011) Analysis of protein-ligand interactions by fluorescence polarization. Nat Protoc 6:365–387. https://doi.org/10.1038/nprot.2011.305
2. Xiao X, Zhen S (2022) Recent advances in fluorescence anisotropy/polarization signal amplification. RSC Adv 12:6364–6376. https://doi.org/10.1039/D2RA00058J
3. Bottari F, Daems E, De Vries AM, Van Wielendaele P, Trashin S, Blust R, Sobott F, Madder A, Martins JC, De Wael K (2020) Do aptamers always bind? The need for a multifaceted analytical approach when demonstrating binding affinity between aptamer and low molecular weight compounds. J Am Chem Soc 142:19622–19630. https://doi.org/10.1021/jacs.0c08691
4. Chan KY, Kinghorn AB, Hollenstein M, Tanner JA (2022) Chemical modifications for a next generation of nucleic acid aptamers. Chembiochem 23:e202200006. https://doi.org/10.1002/cbic.202200006
5. Liu S, Xu Y, Jiang X, Tan H, Ying B (2022) Translation of aptamers toward clinical

diagnosis and commercialization. Biosens Bioelectron 208:114168. https://doi.org/10.1016/j.bios.2022.114168

6. Lyu M, Chan CH, Chen Z, Liu Y, Yu Y (2025) Advantages, applications, and future directions of in vivo aptamer SELEX: a review. Mol Ther Nucleic Acids 36:102575. https://doi.org/10.1016/j.omtn.2025.102575
7. DeRosa M, Lin A, Mallikaratchy P, McConnell E, McKeague M, Patel R, Shigdar S (2023) In vitro selection of aptamers and their applications. Nat Rev Methods Primers 3:55. https://doi.org/10.1038/S43586-023-00247-6
8. Luo X, McKeague M, Pitre S, Dumontier M, Green J, Golshani A, Derosa MC, Dehne F (2010) Computational approaches toward the design of pools for the in vitro selection of complex aptamers. RNA 16:2252–2262. https://doi.org/10.1261/rna.2102210
9. Sun D, Sun M, Zhang J, Lin X, Zhang Y, Lin F, Zhang P, Yang C, Song J (2022) Computational tools for aptamer identification and optimization. TrAC Trends Anal Chem 157: 116767. https://doi.org/10.1016/j.trac.2022.116767
10. Fasogbon IV, Ondari EN, Tusubira D, Rangasamy L, Venkatesan J, Musyoka AM, Aja PM (2025) Recent focus in non-SELEX-computational approach for de novo aptamer design: a mini review. Anal Biochem 699: 115756. https://doi.org/10.1016/J.AB.2024.115756
11. Berezovski M, Musheev M, Drabovich A, Krylov SN (2006) Non-SELEX selection of aptamers. J Am Chem Soc 128:1410–1411. https://doi.org/10.1021/ja056943j
12. Tapp MJN, Slocik JM, Dennis PB, Naik RR, Milam VT (2018) Competition-enhanced ligand selection to identify DNA aptamers. ACS Comb Sci 20:585–593. https://doi.org/10.1021/acscombsci.8b00048
13. Kushwaha A, Takamura Y, Nishigaki K, Biyani M (2019) Competitive non-SELEX for the selective and rapid enrichment of DNA aptamers and its use in electrochemical aptasensor. Sci Rep 9:6642. https://doi.org/10.1038/s41598-019-43187-6
14. Harvison PJ (2007) Thrombin as a target. In: xPharm: the comprehensive pharmacology reference. Elsevier, pp 1–6
15. Pica A, Russo Krauss I, Merlino A, Nagatoishi S, Sugimoto N, Sica F (2013) Dissecting the contribution of thrombin exosite I in the recognition of thrombin binding aptamer. FEBS J 280:6581–6588. https://doi.org/10.1111/febs.12561
16. Russo Krauss I, Merlino A, Giancola C, Randazzo A, Mazzarella L, Sica F (2011) Thrombin–aptamer recognition: a revealed ambiguity. Nucleic Acids Res 39:7858–7867. https://doi.org/10.1093/nar/gkr522
17. Murray MT, Wetmore SD (2024) Unlocking precision in aptamer engineering: a case study of the thrombin binding aptamer illustrates why modification size, quantity, and position matter. Nucleic Acids Res 52:10823–10835. https://doi.org/10.1093/nar/gkae729
18. Moerke NJ (2009) Fluorescence polarization (FP) assays for monitoring peptide-protein or nucleic acid-protein binding. Curr Protoc Chem Biol 1:1–15. https://doi.org/10.1002/9780470559277.ch090102
19. Zhao Q, Tao J, Feng W, Uppal JS, Peng H, Le XC (2020) Aptamer binding assays and molecular interaction studies using fluorescence anisotropy - a review. Anal Chim Acta 1125: 267–278. https://doi.org/10.1016/j.aca.2020.05.061
20. Gokulrangan G, Unruh JR, Holub DF, Ingram B, Johnson CK, Wilson GS (2005) DNA aptamer-based bioanalysis of IgE by fluorescence anisotropy. Anal Chem 77:1963–1970. https://doi.org/10.1021/ac0483926
21. Forder M, Rivera M, Chen Y, Phénix J, Shomali T, Munter LM, McKeague M (2025) Exploring molecular binding: a fluorescence anisotropy lab for undergraduates. J Chem Educ 102:796–802. https://doi.org/10.1021/acs.jchemed.4c01023

Chapter 10

Gamification in Enzymology with CARBGAME (CARd and Board GAmes in Medical Education)

Krishna Mohan Surapaneni

Abstract

Gamification has emerged as a promising pedagogical strategy in medical education, fostering active engagement, collaborative learning, and improved knowledge retention. Clinical enzymology, a core yet conceptually challenging area of biochemistry for medical students and also students majoring in the life sciences and related fields often suffers from poor comprehension and recall when taught using traditional didactic methods. To address these challenges, *CARBGAME (CARd and Board GAmes in Medical Education)* was developed as a structured gamified approach designed to reinforce enzymology concepts through an interactive board game format. This chapter provides a reproducible methodology for implementing CARBGAME in undergraduate medical and life science/other related disciplines classrooms. The protocol outlines essential materials, step-by-step instructions for designing and conducting the game, strategies for formative assessment, and critical troubleshooting tips. When applied in classroom settings, CARBGAME has been shown to significantly improve both short-term and long-term learning outcomes, enhance student motivation, and create a collaborative environment conducive to deeper learning. The format described here enables educators across institutions to replicate and adapt CARBGAME to different biomedical topics, thereby integrating gamification as an innovative teaching strategy in medical curricula.

Key words Gamification, Biochemistry education, Clinical enzymology, Medical education, Active learning, Board game, Card game, teaching, Knowledge retention, Collaborative learning

1 Introduction

The teaching of biochemistry, particularly enzymology, presents a distinctive challenge in the medical and health professions education. Enzymology represents the bridge between basic biochemical concepts and their clinical application, yet many students experience difficulty in mastering this domain. The subject requires integration of theoretical foundations in enzyme kinetics, inhibition, isoenzyme variation, and regulatory mechanisms with the diagnostic and therapeutic implications of enzymes in human disease. Such integration often demands abstract reasoning, a strong background in chemistry, and the ability to connect molecular events with

Robert E. Dempski (ed.), *Transforming Molecular Biology with Emerging Technologies*, Methods in Molecular Biology, vol. 3043, https://doi.org/10.1007/978-1-0716-5320-3_10,

pathophysiological processes. For many medical students and also students majoring in the life sciences and related fields, these requirements contribute to misconceptions, fragmented understanding, and limited long-term retention of knowledge [1, 2].

Traditional didactic lectures remain the predominant mode of instruction in most medical curricula. While effective in delivering factual content, passive learning formats often fail to sustain attention, stimulate critical thinking, or foster deep conceptual understanding [3]. Furthermore, the cognitive load associated with learning large numbers of enzyme names, functions, and associated clinical conditions frequently leads to rote memorization, with limited ability to transfer knowledge into clinical reasoning contexts [4]. Consequently, there is a pressing need for innovative strategies that transform enzymology from a perceived difficult subject into an engaging, accessible, and memorable learning experience.

Gamification has emerged as a promising pedagogical strategy to address these challenges. By incorporating elements of play, competition, and interactivity, gamification enhances learner motivation, promotes collaboration, and fosters active participation [5, 6]. Educational research consistently demonstrates that game-based methods improve engagement, facilitate retention of knowledge, and support problem-solving in complex subjects [7]. Within the context of medical education, gamification provides an additional benefit of creating a non-threatening environment where learners can apply concepts, receive immediate feedback, and correct misconceptions with minimal penalty [8].

In recent years, multiple studies have reported the successful application of games in anatomy, physiology, pathology, and pharmacology education [9–11]. These innovations demonstrate that games can be integrated effectively into rigorous curricula without compromising academic standards [12]. However, the application of gamification in biochemistry, particularly in enzymology, remains relatively underexplored. The distinctive features of enzymology, like its reliance on abstract kinetic principles, clinical correlations, and diagnostic applications, make it an ideal candidate for gamified approaches that allow iterative reinforcement of knowledge and collaborative exploration of complex concepts.

To address this pedagogical gap, *CARBGAME (CARd and Board GAmes in Medical Education)* was conceptualized and developed as a structured gamification tool for teaching clinical enzymology. CARBGAME is designed around a 10 × 10 board with integrated question and instruction cards, complemented by dice-based progression and case-based problem-solving. The format combines chance, competition, and knowledge application, thereby simulating a dynamic learning environment. In controlled educational studies, CARBGAME has demonstrated significant improvement in pre- to post-intervention test scores, enhanced student engagement, and overwhelmingly positive perceptions from both learners and faculty [13, 14]. Beyond its immediate application to enzymology, the method serves as a reproducible template for gamification strategies that can be adapted across other domains of biochemistry and the biomedical sciences.

This chapter outlines a stepwise, reproducible protocol for implementing CARBGAME in undergraduate medical education, life sciences, and related disciplines. The format is presented according to the Methods in Molecular Biology series, ensuring sufficient detail for replication across diverse institutional contexts. By following the outlined materials, procedures, and notes, educators can integrate CARBGAME into their teaching to enhance comprehension, application, and retention of enzymology concepts, while simultaneously fostering an interactive and collaborative classroom environment.

2 Materials

The successful implementation of CARBGAME requires the preparation of standardized game components, student assessment tools, and logistical arrangements. The following subsections list the essential materials needed to conduct the protocol. All items should be prepared in advance to ensure smooth execution.

2.1 CARBGAME Board

1. A 10×10 square board with 100 numbered boxes (Fig. 1).
2. Seventy boxes should contain content-related questions on clinical enzymology.
3. Thirty boxes should include safe zones, bonus rewards, or instruction prompts.
4. Eleven of the 70 question boxes should be marked with a red dot, indicating that the question must be answered individually by the player who rolls the dice without team discussion.
5. The board may be printed on A3 or A2 laminated chart paper or even as larger table top flex material sheet for durability, or projected digitally for larger classrooms.

2.2 Question Cards

1. Questions can be printed on the board or A set of 70 cards corresponding to the 70 question boxes can be made.
2. Topics should/can include:
   (a) Enzyme structure and function.
   (b) Classification of enzymes.
   (c) Enzyme kinetics and Michaelis-Menten principles.
   (d) Enzyme inhibition.
   (e) Isoenzymes.

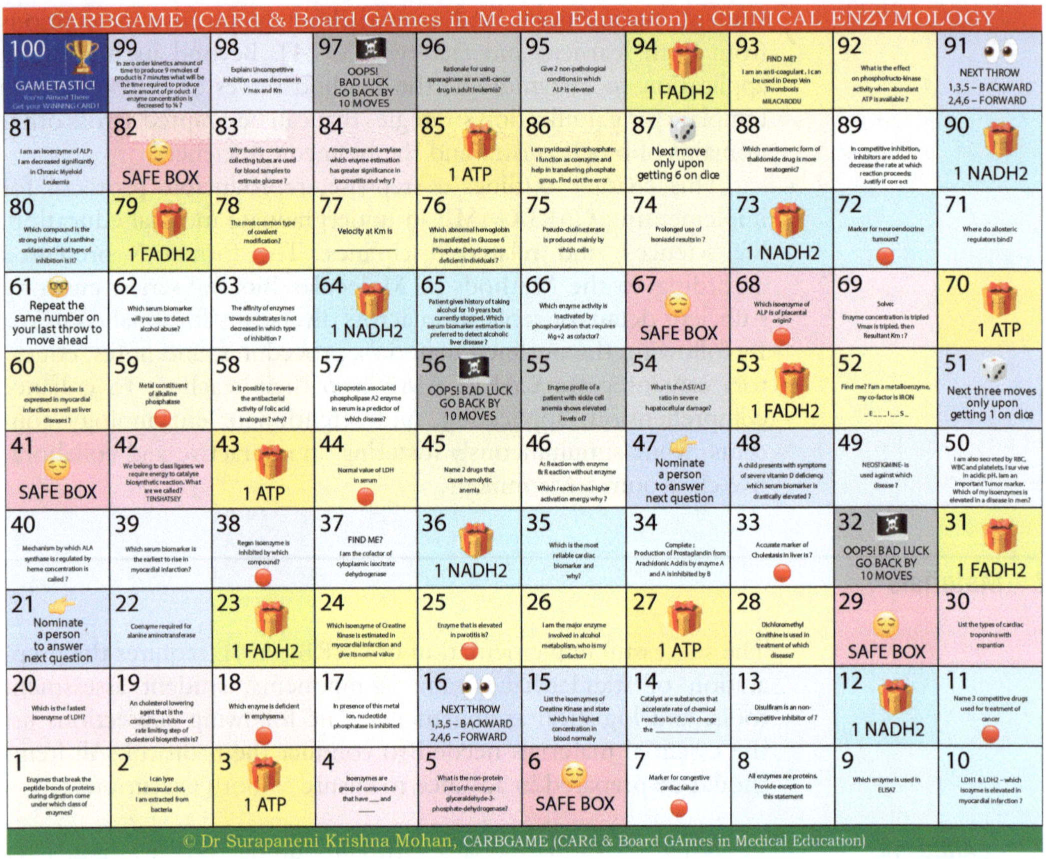

Fig. 1 CARBGAME Clinical Enzymology Game Board. Image from Surapaneni KM. "CARBGAME" (CARd & Board GAmes in Medical Education) as an innovative gamification tool for learning clinical enzymology in biochemistry for first-year medical students. Biochem Mol Biol Educ. 2024;52(6):666–675. doi:10.1002/bmb.21857 [14]

(f) Regulation of enzymatic activity.
(g) Enzyme disorders and deficiencies.
(h) Diagnostic enzyme assays.
(i) Enzyme markers in disease.
(j) Therapeutic enzymes.

3. Each card should contain a single multiple-choice or short-answer question, with the correct answer indicated on the reverse side.

2.3 Instruction, Reward Boxes and Skipping Star Cards

1. Instruction boxes ($n = 10$): direct players to move forward, backward, or skip turns as per predefined rules.
2. Reward boxes ($n = 15$): categorized into ATP, $NADH_2$, and $FADH_2$ cards, enabling players to move 1, 2, or 3 spaces ahead, respectively.

3. Skipping Star cards ($n = 3$ per team): allow students to bypass particularly difficult questions.

2.4 Dice and Game Tokens

1. One standard six-faced dice per group.
2. Distinctive tokens (coins, buttons, or markers) for each team to track progress on the board.

2.5 Student Assessment Tools

1. Pretest and posttest questionnaire: Ideally, 20 multiple-choice questions divided into knowledge-based and application-based on clinical enzymology.
2. Feedback questionnaire: a 32-item survey on a five-point Likert scale, structured into domains of design, environment, motivation, learning experience, feedback, and peer collaboration (Table 1).
3. Rating scale: a ten-point scale assessing perceptions of critical thinking, application, knowledge sharing, self-assessment, and motivation (Table 2).

2.6 Facilitator Requirements

1. Students can be divided into small groups of six. For interactive, small-group teaching (like PBL, simulation, or gamified assessment), 4–6 students per group is usually considered optimal.
2. One faculty member per group of 6 students to supervise gameplay, clarify rules, and validate answers. When faculty might not be available, a large number of groups, senior

Table 1
Example for student perception survey domains for evaluating CARBGAME (32 items, five-point Likert scale)

Domain	Example statements (strongly agree → strongly disagree)	No. of items
Design	"The questions in CARBGAME were appropriate to my level of knowledge." "The rules were clear and consistent"	7
Gaming environment	"The game created a fun and engaging environment." "The game aroused curiosity and exploration"	6
Motivation	"The game increased my motivation to learn enzymology." "The game made my learning experiential"	4
Learning experience	"CARBGAME helped me recall concepts studied in class." "The game enabled transfer of learning to real contexts"	5
Assessment and feedback	"The game provided immediate and effective feedback." "I could assess my own knowledge during play"	4
Collaboration and peer learning	"CARBGAME improved my interaction with peers." "The game facilitated teamwork and knowledge sharing"	6

Table 2
Example for confidence and self-efficacy questionnaire for clinical enzymology (administered pre- and post-CARBGAME)

Domain assessed	Example item	Rating scale
Recall of enzyme structure and classification	"I can recall and classify enzymes accurately"	1 (very low confidence) to 10 (very high confidence)
Application of enzyme kinetics	"I can apply Michaelis-Menten principles to solve problems"	1–10
Understanding enzyme inhibition	"I can differentiate between competitive and noncompetitive inhibition"	1–10
Interpretation of enzyme markers	"I can use enzyme markers such as CK-MB, LDH, or ALT to interpret clinical cases"	1–10
Integration with clinical contexts	"I feel confident connecting enzymology concepts to disease processes"	1–10
Communication of enzymology concepts	"I can explain enzymology concepts clearly to peers"	1–10

The authors have no conflicts of interest to declare that are relevant to the content of this chapter

students, or residents can be trained as facilitators to help guide small groups.

3. A central coordinator to oversee overall conduct and maintain consistency across groups.

2.7 Classroom Setup

1. A well-ventilated room with sufficient seating like round table set up for group work.
2. One CARBGAME set per group of 4–6 students.
3. Writing materials for recording answers, scoring, and observations.

2.8 Data Analysis Tools

1. Statistical software (e.g., SPSS, R, or equivalent) for paired t-test, ANOVA, and nonparametric analysis where appropriate.
2. Spreadsheet software (e.g., Microsoft Excel) for tabulating pre- and posttest scores, Likert-scale responses, and qualitative comments.

3 Methods

Carry out all procedures in a classroom or teaching laboratory setting, unless otherwise specified. Ensure that game materials are fully prepared prior to beginning the session.

3.1 Draft the Learning Objectives

Before commencing the CARBGAME session, draft the learning objectives, and it should be clearly communicated to students.

Example

By the end of the activity, learners should be able to:

1. Recall the fundamental principles of enzyme structure, classification, and function.
2. Explain key concepts of enzyme kinetics, including Km, Vmax, and Michaelis-Menten dynamics.
3. Differentiate between types of enzyme inhibition, like competitive, noncompetitive, uncompetitive, and allosteric.
4. Identify isoenzymes and their clinical diagnostic relevance.
5. Apply enzymology concepts to the interpretation of clinical enzyme assays.
6. Analyze case-based scenarios involving enzyme markers in disease diagnosis.
7. Evaluate the therapeutic uses of enzymes in clinical practice.
8. Demonstrate collaborative problem-solving and peer-to-peer learning in an interactive environment.
9. Reflect on personal confidence in enzymology and recognize areas for further improvement.

Clear articulation of these objectives ensures alignment between instructional goals, gameplay design, and assessment strategies.

3.2 Preparation of the CARBGAME Set

1. Prepare a 10 × 10 square game board with 100 numbered boxes (*see* **Note 1**). Of these, 70 boxes should be allocated for knowledge- or application-based questions in enzymology, 11 of which are designated with a red dot to indicate that no team discussion is permitted. The remaining 30 boxes should be distributed as follows: 5 safe boxes, 15 reward boxes (ATP, $NADH_2$, $FADH_2$), and 10 instruction boxes.
2. Print the game board on A3 or A2 size chart paper and laminate or print in the required size as flex sheet for durability. Alternatively, a projected digital version may be used if groups are large (*see* **Note 2**).
3. Print the questions on the board or Prepare 70 question cards corresponding to the 70 question boxes. Each card should include one question on the front and the correct answer on the reverse side for facilitator use. Ensure a balanced distribution of topics:

   (a) Structure and function of enzymes.
   (b) Enzyme classification.
   (c) Enzyme kinetics (Km, Vmax, Michaelis-Menten equation).
   (d) Enzyme inhibition (competitive, noncompetitive, uncompetitive, allosteric).
   (e) Isoenzymes and diagnostic utility.
   (f) Regulation of enzymatic activity.
   (g) Enzyme disorders.
   (h) Diagnostic assays (AST, ALT, ALP, CK, LDH, amylase, lipase).
   (i) Enzyme markers in clinical disease.
   (j) Therapeutic enzymes.
4. This CARBGAME is also adaptable to other topics and could be explored for electronic implementation or using other online gamification platforms (*see* **Notes 15, 16**).
5. Prepare reward boxes consisting of 5 ATP (advance one move), 5 $FADH_2$ (advance two moves), and 5 $NADH_2$ (advance three moves).
6. Prepare ten instruction boxes with short directives such as "Skip one turn," "Go back three spaces," or "Exchange place with another team."
7. Provide each team with three Skipping Star cards, which allow bypassing of a difficult question without penalty.
8. Provide distinct tokens (e.g., buttons, coins, or small markers) for each team and one standard six-faced dice per group.
9. Print scoring sheets for recording dice rolls, correct/incorrect responses, and final outcomes (*see* **Note 3**).

3.3 Pregame Assessment

1. Distribute a 20-question pretest prior to beginning the game (*see* **Notes 4, 17**). The test should contain:
   (a) Ten knowledge-based questions (recall of definitions, classifications, or enzyme properties).
   (b) Ten application-based questions (clinical case vignettes requiring interpretation of enzyme levels).
2. Allot 20 min for completion. Collect responses and record scores anonymously using coded identifiers.
3. Administer a confidence/self-efficacy questionnaire using a ten-point scale for each of the following domains:
   (a) Confidence in recalling enzyme structures, classifications, and functions.

(b) Confidence in applying enzyme kinetics to problem-solving.
(c) Confidence in explaining enzyme inhibition.
(d) Confidence in interpreting enzyme markers in disease states.
(e) Confidence in transferring enzymology knowledge into clinical contexts.

4. Collect baseline confidence ratings and retain for comparison with postgame data (*see* **Note 5**).

3.4 Group Allocation and Orientation

1. Divide participants into groups of four to six students each. Group size should not exceed seven, as larger groups may reduce active participation (*see* **Note 2**).
2. Assign one faculty facilitator per group. The facilitator's role is to monitor adherence to rules, validate answers using the card key, and provide feedback when necessary (*see* **Notes 6, 13, 14**).
3. Provide each group with a CARBGAME set: one laminated board, one dice, one set of question/reward/instruction cards, three Skipping Star cards, one scoring sheet, and a set of tokens.
4. Allocate 5 min at the beginning of the session to explain rules, demonstrate dice rolling, and clarify the scoring system.

3.5 Rules and Conduct of the Game

1. To begin, players must roll a one on the dice to start movement.
2. Players roll the dice sequentially in a clockwise order.
3. On landing on a question box, the facilitator draws the corresponding card and reads the question aloud. The group provides an answer, except in the case of red-dot boxes, where the individual player must answer without discussion (*see* **Note 7**).
4. If the response is correct, the team remains in place and resumes play at the next turn.
5. If the response is incorrect, the team moves backward by five spaces.
6. On landing on a safe box, the player remains in position with no penalty or reward.
7. On landing on a reward box, the facilitator issues the corresponding reward card. The player advances the number of steps indicated by the card.
8. On landing on an instruction box, the facilitator issues the instruction card, and the player follows the directive (e.g., "Move back three steps" or "Skip next turn").

9. Teams may use a Skipping Star card to bypass a difficult question. Once used, the card is surrendered to the facilitator (*see* **Note 8**).
10. Gameplay continues until one team reaches the 100th box.
11. The team that first reaches the 100th box must solve a case-based clinical question provided by the facilitator. Example:
    "A 50-year-old male presents with chest pain. Serum enzyme levels show markedly elevated CK-MB and LDH-1. What is the most likely diagnosis, and what is the basis of using these enzymes as markers?" (*see* **Note 12**).
12. The team that answers correctly is declared the winner.
13. The estimated time for gameplay is 45–50 min, including discussion and rule enforcement (*see* **Note 11**).

3.6 Postgame Assessment

1. Readminister the same 20-question test as the post-test. Allocate 20 min.
2. Collect responses anonymously using the same coded identifiers as the pretest.
3. Readminister the confidence/self-efficacy questionnaire. Compare pre- and postgame ratings to evaluate the change in perceived competence.
4. Distribute a 32-item (or any other convenient validated item number) structured perception questionnaire rated on a five-point Likert scale, grouped under the following six domains: (*see* **Note 18**).
    (a) Design: clarity of rules, appropriateness of difficulty, relevance of questions.
    (b) Gaming environment: engagement, enjoyment, sense of curiosity.
    (c) Motivation: whether the activity increased interest in enzymology.
    (d) Learning experience: ability to recall, reinforce, and apply concepts.
    (e) Assessment and feedback: opportunity for self- and peer-assessment, immediate correction.
    (f) Collaboration: teamwork, communication, peer learning.
5. Supplement with a ten-point global rating scale assessing (*see* **Note 19**):
    (a) Development of critical thinking in enzymology.
    (b) Application of enzymology concepts to clinical contexts.
    (c) Extent of peer knowledge sharing.

(d) Self-assessment of learning gaps.
(e) Interest generated in enzymology.

6. Optionally, conduct small-group interviews (10–15 min) to obtain qualitative impressions from both students and facilitators (*see* **Note 9**).

3.7 Data Analysis

1. Enter all pretest and posttest scores into a spreadsheet. Calculate mean ± SD.
2. Use a paired t-test to compare pre- and posttest scores for statistical significance.
3. Analyze confidence scores with a paired t-test or Wilcoxon signed-rank test if data are non-parametric.
4. Analyze perception survey responses descriptively. Report frequencies and percentages for each item.
5. Aggregate open-ended responses and interview transcripts into thematic categories such as "motivation," "engagement," and "clinical application."
6. Compare the outcomes with historical control groups taught by traditional methods, if available, to demonstrate relative effectiveness (*see* **Note 10**).
7. For suggestions regarding Cultural adaptation:, Faculty workload considerations, Game fairness, Evaluation beyond immediate outcomes, Limitations and Ethical considerations (*see* **Notes 20–25**).

4 Notes

1. Game board preparation (*see* Subheading 3.2, **step 1**): The CARBGAME board should be visually appealing and easy to interpret at a glance. Using laminated or digital projection formats prevents wear and allows repeated use. Boards printed in color enhance student engagement.
2. Group size (*see* Subheading 3.3, **step 1**): Optimal group size is six. Smaller groups may reduce collaborative discussion, whereas larger groups (>7) risk passive participation of weaker students.
3. Scoring sheets (*see* Subheading 3.2, **step 9**): Maintaining written records of responses, dice rolls, and penalty moves ensures transparency. These sheets are also useful for later reflection on student performance and errors.
4. Pretest construction (*see* Subheading 3.2, **step 1**): The 20-question pretest should cover all domains of enzymology

that will appear in the game. Questions must be of comparable difficulty to ensure the validity of pre-post comparisons.

5. Confidence/self-efficacy measures (*see* Subheading 3.2, **step 4**): Assessing confidence levels provides important information about affective learning outcomes, beyond cognitive gain. This may be particularly relevant in biochemistry, where students often perceive enzymology as intimidating.
6. Facilitator training (*see* Subheading 3.3, **step 2**): Facilitators should be oriented in advance regarding game rules, scoring procedures, and the nondirective role they must maintain. Facilitators should not supply answers but may guide reasoning if misconceptions are identified.
7. Red-dot boxes (*see* Subheading 3.4, **step 3**): These boxes encourage individual accountability and prevent "free riding" in team discussions. Faculty should strictly enforce the no-discussion rule to maintain fairness.
8. Use of Skipping Star cards (*see* Subheading 3.4, **step 9**): These cards prevent frustration when students repeatedly fail to answer a question. However, overuse should be discouraged; three cards per group is sufficient to maintain challenge.
9. Qualitative interviews (*see* Subheading 3.5, **step 6**): Short post-session interviews with students and facilitators provide insights into the acceptability and perceived value of the method. These comments can guide local adaptations of CARBGAME.
10. Comparison with control cohorts (*see* Subheading 3.6, **step 6**): Where possible, comparing CARBGAME outcomes with traditional lecture-based instruction strengthens the evidence for its effectiveness and can be used to justify curricular integration.
11. Time management (general) (*see* Subheading 3.5, **step 13**): Each game session should be strictly limited to 60 min. Longer sessions risk fatigue and reduced motivation, while shorter sessions may not allow sufficient exposure to questions.
12. Case-based final question (*see* Subheading 3.4, **step 11**): The final case should be carefully designed to integrate multiple enzymology concepts. This reinforces application and provides a clinically relevant "endgame" to the session.
13. Engagement strategies (*see* Subheading 3.4, **step 2**): Facilitators should encourage all team members to participate actively. Rotating the responsibility of answering questions helps prevent dominance by stronger students.
14. Handling misconceptions (*see* Subheading 3.4, **step 2**): Incorrect answers provide valuable opportunities for discussion.

Facilitators should briefly explain the rationale for the correct answer immediately, to prevent reinforcement of errors.

15. Adaptability to other topics (*see* Subheading 3.2, **step 4**): Although developed for enzymology, CARBGAME can be adapted for other biochemistry units (e.g., metabolism, molecular biology) by modifying the question cards while retaining the board and rules.
16. Online adaptation (*see* Subheading 3.2, **step 4**): During periods of remote teaching, CARBGAME can be digitized using online platforms (e.g., Kahoot, Google Slides, or custom-built apps). In such cases, facilitators should ensure that competitive dynamics and collaborative discussion are preserved.
17. Assessment reliability (*see* Subheading 3.3, **step 1**): To enhance the reliability of pre- and post-tests, questions should be pilot-tested on a small group of students not participating in the intervention. This ensures clarity of wording and appropriate difficulty level.
18. Feedback tool design (*see* Subheading 3.5, **step 4**): The 32-item perception questionnaire can be modified for local needs, but domains of design, environment, motivation, learning, assessment, and collaboration should be retained for comparability across institutions.
19. Confidence assessment interpretation (*see* Subheading 3.6, **step 5**): Gains in confidence do not always correlate directly with gains in test scores. Confidence measures should be interpreted as complementary indicators of affective learning outcomes.
20. Cultural adaptation (*see* Subheading 3.7, **step 7**): The motivational impact of games may vary across educational settings. Institutions should adapt the tone of questions and rewards to align with cultural expectations while preserving the essential gamified structure.
21. Faculty workload considerations (*see* Subheading 3.7, **step 7**): While CARBGAME requires one facilitator per group, recruitment of faculty from multiple departments can reduce individual burden. Interdepartmental collaboration also promotes faculty engagement with innovation.
22. Game fairness (*see* Subheading 3.7, **step 13**): Disputes may arise regarding the interpretation of questions or answers. In such cases, the facilitator's decision should be final, based on authoritative reference texts.
23. Evaluation beyond immediate outcomes (*see* Subheading 3.7, **step 7**): For sustained evidence, retention of knowledge should be reassessed during subsequent internal assessments or semester examinations.

24. Limitations (*see* Subheading 3.7, **step 7**): CARBGAME should be used as a supplement, not a replacement, for traditional teaching. While effective in improving engagement and application, it should be integrated with lectures, tutorials, and self-study for optimal results.
25. Ethical considerations (*see* Subheading 3.7, **step 7**): In research or curricular evaluation contexts, informed consent and anonymization of data are essential. Institutional ethics approval should be obtained if data are to be published.

References

1. Elhousni Z, Zerhane R, Janati-Idrissi R (2025) A look at biochemistry learning difficulties in the undergraduate level. Multidiscip Sci J 7(2): 2025049
2. D'Souza JMP, Raghavendra U, D'Souza DH, D'Souza ND (2013) Teaching learning of biochemistry in undergraduate medical curriculum: perceptions and opinions of medical students. Educ Med J 5(2):e45–e53
3. Challa KT, Sayed A, Acharya Y (2021) Modern techniques of teaching and learning in medical education: a descriptive literature review. MedEdPublish 10:18
4. Young JQ, Van Merrienboer J, Durning S, Ten Cate O (2014) Cognitive load theory: implications for medical education: AMEE Guide No. 86. Med Teach 36(5):371–384
5. Krishnamurthy K, Selvaraj N, Gupta P, Cyriac B, Dhurairaj P, Abdullah A, Krishnapillai A, Lugova H, Haque M, Xie S et al (2022) Benefits of gamification in medical education. Clin Anat (New York, N.Y.) 35(6): 795–807. https://doi.org/10.1002/ca.23916
6. Nevin CR, Westfall AO, Rodriguez JM, Dempsey DM, Cherrington A, Roy B, Patel M, Willig JH (2014) Gamification as a tool for enhancing graduate medical education. Postgrad Med J 90(1070):685–693. https://doi.org/10.1136/postgradmedj-2013-132486
7. Muthiyan G, Kasat P, Vij V, Solanki RS, C K, Sontakke B (2023) Effectiveness of an innovative card game as a supplement for teaching factual content to medical students: a mixed method study. Cureus 15(10):e47768. https://doi.org/10.7759/cureus.47768
8. Wang YF, Hsu YF, Fang KT, Kuo LT (2024) Gamification in medical education: identifying and prioritizing key elements through Delphi method. Med Educ Online 29(1):2302231. https://doi.org/10.1080/10872981.2024.2302231
9. Karimian Z, Momeni M, Zarifsanaiey N (2025) Enhancing radiographic interpretation: effects of gamification on medical students' knowledge, skills, and satisfaction - a quasi-experimental study. BMC Med Educ 25(1): 958. https://doi.org/10.1186/s12909-025-07523-x
10. Twist KE, Ragsdale JW (2022) Candy gland: a diabetes board game for medical students. MedEdPORTAL 18:11294. https://doi.org/10.15766/mep_2374-8265.11294
11. Bawazeer G, Sales I, Albogami H, Aldemerdash A, Mahmoud M, Aljohani MA, Alhammad A (2022) Crossword puzzle as a learning tool to enhance learning about anticoagulant therapeutics. BMC Med Educ 22(1): 267. https://doi.org/10.1186/s12909-022-03348-0
12. Khazaei MR, Hosseini MA, Ramezani G (2025) Gamification in medical sciences education: a scoping review. Recenti Prog Med 116(5):291–301. https://doi.org/10.1701/4495.44949
13. Surapaneni KM (2024) "CARBGAME" (CARd & Board GAmes in medical education): a gamification innovation to foster active learning in biochemistry for medical students. Adv Physiol Educ 48(1):97–101. https://doi.org/10.1152/advan.00214.2023
14. Surapaneni KM (2024) "CARBGAME" (CARd & Board GAmes in Medical Education) as an innovative gamification tool for learning clinical enzymology in biochemistry for first year medical students. Biochem Mol Biol Educ 52(6):666–675. https://doi.org/10.1002/bmb.21857

Chapter 11

Developing an Educational Game on Protein Degradation

Natalia A. Szulc and Wojciech Pokrzywa

Abstract

This chapter details the development methodology employed in the creation of DEGRADATOR, a 2D computer game designed to teach the intricacies of the ubiquitin-proteasome system (UPS), a crucial cellular pathway responsible for protein degradation. The UPS, with its multistep enzymatic processes and critical role in health and disease, presents both a challenge and an opportunity for engaging education. Rooted in constructivist game-based learning principles, DEGRADATOR translates abstract molecular mechanisms into an engaging and interactive learning experience. This chapter presents the pedagogical design framework, technical implementation, and iterative development process, incorporating beta testing and user feedback. We aim to provide a comprehensive account of the development process, offering insights and considerations for educators and developers interested in leveraging game-based methodologies to enhance science education and foster public understanding of complex, yet vital, scientific topics like protein degradation.

Key words DEGRADATOR, Game-based learning, Science education, Protein degradation, Ubiquitin-proteasome system

1 Introduction

The development of DEGRADATOR (https://degradator-game.com) [1] was driven by the need to effectively communicate the complex, yet fundamentally important, processes of cellular protein degradation, specifically the ubiquitin-proteasome system (UPS) (Fig. 1a). The UPS is not just a passive waste disposal system; it is a highly regulated and dynamic machinery responsible for selectively removing damaged, misfolded, or no longer needed proteins [2].

This process is essential for maintaining cellular homeostasis, regulating cell cycle progression, controlling immune responses, and influencing a vast array of other functions [3]. Dysregulation of the UPS is implicated in numerous diseases, including cancers, neurodegenerative disorders, and immune system dysfunctions [4–6]. Furthermore, understanding the UPS is crucial for grasping the

Robert E. Dempski (ed.), *Transforming Molecular Biology with Emerging Technologies*, Methods in Molecular Biology, vol. 3043, https://doi.org/10.1007/978-1-0716-5320-3_11,

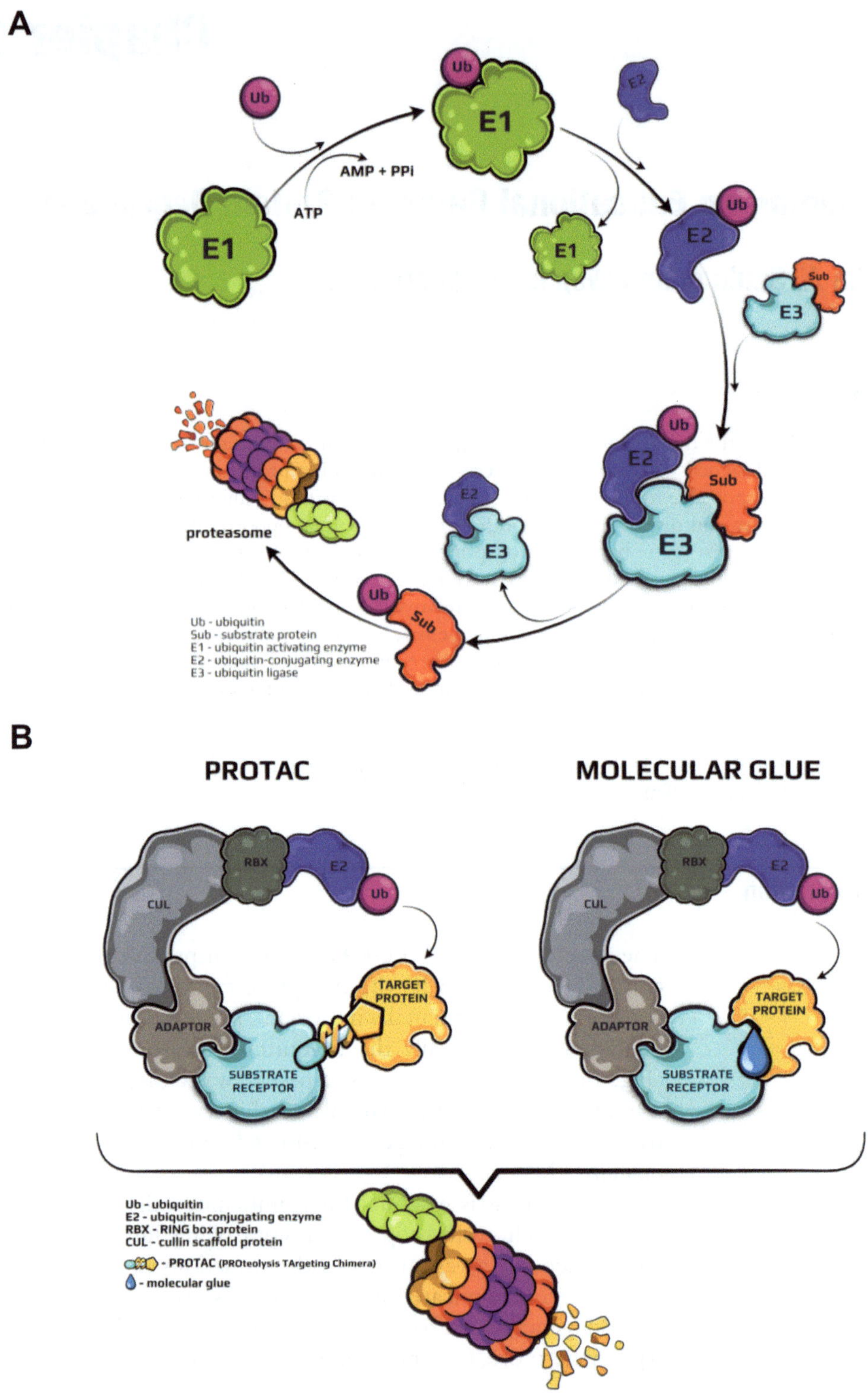

Fig. 1 A schematic representation of the UPS and targeted protein degradation strategies. (**a**) The UPS is a tightly regulated pathway that controls protein degradation in cells. It operates through a cascade of three enzymes: E1 ubiquitin-activating enzymes, E2 ubiquitin-conjugating enzymes, and E3 ubiquitin ligases. The process begins with E1 enzymes, which activate ubiquitin in an ATP-dependent manner. Activated ubiquitin is

mechanism of action of emerging therapeutic strategies like PROTACs (PROteolysis TArgeting Chimeras), which harness the UPS to selectively degrade disease-causing proteins (Fig. 1b) [7, 8]. This inherent complexity, combined with its critical importance in both health and disease, and the potential for interactive gameplay mechanics based on its enzymatic steps, made the UPS an ideal, yet challenging, subject for a game-based learning approach. DEGRADATOR was conceived and built upon constructivist learning principles [9], strategically designed to translate these abstract molecular mechanisms into a learning experience that is both entertaining and intrinsically motivating.

1.1 Constructivist Game-Based Learning Framework

DEGRADATOR pedagogical design is deeply rooted in constructivist learning theory, which emphasizes that learners actively *construct* their own understanding through experience and reflection, rather than passively receiving information. Game-based learning is a powerful tool for constructivist education because it inherently provides engaging, interactive experiences that promote active learning [10, 11]. DEGRADATOR leverages the motivational power of games to transform the complex molecular processes of the UPS into a series of interactive challenges, fostering deeper comprehension and improved knowledge retention. This approach aligns with research demonstrating game-based learning effectiveness, particularly in STEM, for developing procedural thinking and problem-solving skills [12, 13].

The core principles of constructivist pedagogy are directly reflected in DEGRADATOR design:

1. *Learning by Doing (Experiential Learning):* DEGRADATOR lets players *experience* the UPS, rather than just being told about it. Players control an E3 ligase, maneuvering it within a dynamic cellular environment to interact with substrates. These probabilistic interactions mirror real-world molecular collisions, fostering an intuitive understanding of enzyme-substrate interactions.

Fig. 1 (continued) then transferred to E2 enzymes, which act as carriers, passing it to E3 ligases that recognize specific target proteins and attach ubiquitin to them. This tagging process is usually repeated multiple times, forming a polyubiquitin chain, which serves as a signal for the 26S proteasome. The proteasome then recognizes the tagged protein and degrades it into small peptide fragments, ensuring proper protein turnover and maintaining cellular balance. This process can be reversed by deubiquitinating enzymes (DUBs), which remove ubiquitin from substrates, preventing their degradation [2]. (**b**) Molecular glues and PROTACs represent two distinct strategies for targeted protein degradation. Molecular glues are small molecules that enhance the interaction between an E3 ligase and a target protein, promoting its ubiquitination. PROTACs, in contrast, are bifunctional compounds with one end binding a specific E3 ligase and the other recognizing the target protein. Both approaches ultimately lead to the ubiquitination and degradation of the target protein, leveraging the cell's natural protein disposal system [7, 8]. (Figure adapted from [1])

2. *Developing Mental Models (Conceptual Understanding):* Game mechanics are designed to build accurate mental models. Level two introduces enzyme specificity, requiring players to distinguish between correct substrates and "decoy" proteins. Success requires applying, not just memorizing, the concept of specificity.
3. *Problem-Solving and Iteration (Adaptive Learning):* DEGRADATOR presents increasingly complex challenges, requiring players to refine strategies, manage time, prioritize targets, and adapt to changing conditions. This iterative process of trial, error, and adaptation is key to constructivist learning.
4. *Scaffolding Knowledge:* The game's structure inherently scaffolds learning, starting with fundamental concepts that progressively increase in complexity.
5. *Metacognition and Reflection:* Integrated quizzes serve a dual purpose: assessment and reflection. They prompt players to articulate their understanding, reinforcing learning and highlighting areas needing further exploration. The lesson plan complements this by encouraging student-driven exploration and verbalization, promoting metacognitive awareness and mirroring the scientific method's emphasis on observation and description.

1.2 Learning Objectives and Target Audience

DEGRADATOR was designed to address specific learning gaps in biology education, targeting students aged 12 and above, particularly high school biology students (aged 15–19) who are beginning to explore molecular biology. The game assumes a basic understanding of cells and proteins, but no prior knowledge of the UPS is required. The primary learning objectives are to help students:

1. *Understand the Core Function of the UPS:* Grasp the UPS as a *central* cellular mechanism for maintaining protein homeostasis through *selective and regulated* protein degradation. This includes understanding the roles of ubiquitin, the proteasome, and the key enzymes: E1 (ubiquitin-activating), E2 (ubiquitin-conjugating), and E3 (ubiquitin ligases).
2. *Master the Principle of Enzyme-Substrate Specificity:* Understand how E3 ubiquitin ligases selectively recognize and interact with their specific target proteins, ensuring that only the correct proteins are degraded.
3. *Grasp the Energy-Dependent Nature of Intracellular Protein Degradation:* Distinguish between nonselective extracellular digestion and the highly regulated, ATP-dependent process of intracellular protein degradation via the UPS.
4. *Understand Key Mechanisms of Enzyme Regulation:* Explore the regulation of enzyme activity through examples such as

the inhibition of deubiquitinating enzymes and the self-regulation of E3 ubiquitin ligases via auto-ubiquitination.

5. *Explore the Therapeutic Potential of Targeted Protein Degradation:* Learn about cutting-edge therapeutic approaches like PROTACs and molecular glues that leverage the UPS to treat diseases, particularly cancer, and address potential misconceptions about these new therapies.

DEGRADATOR prioritizes accessibility and engagement. It is freely available via web browsers, ensuring broad compatibility across devices (laptops, tablets, and smartphones), minimizing technological barriers. The game features intuitive controls, in-game tutorials, and a rewarding game-like structure (time-based challenges, star-based rewards, and custom soundtrack) to optimize usability and engagement for all users, regardless of prior gaming experience.

1.3 Translating Complex Science into Engaging Gameplay

Developing DEGRADATOR presented the challenge of transforming intricate molecular mechanisms into engaging and pedagogically sound game mechanics. This required carefully balancing scientific accuracy with effective game design, strategically abstracting and gamifying core biological processes. Key design decisions included:

1. *E3 Ubiquitin Ligase—the Player's Avatar:* Players embody the E3 ubiquitin ligase, the central regulator of protein degradation. The *first-person perspective* provides an embodied, experiential understanding of substrate specificity, and the E3's role in initiating degradation. The player directly *becomes* the enzyme, making choices and experiencing the consequences within the UPS. This goes beyond simple interaction; it fosters identification with the process.
2. *Ubiquitination—the Core Loop:* The primary gameplay loop centers on "quadruple ubiquitination," mirroring the polyubiquitin chain formation required for proteasomal degradation. Repetitive tagging reinforces the multistep process as the *signal* for protein breakdown. The game mechanic *directly represents* the biological process, making it intuitive and memorable.
3. *ATP—the Fuel for Degradation:* ATP is a collectible resource required to replenish ubiquitin on E2 enzymes. This directly links gameplay to cellular bioenergetics, emphasizing that intracellular protein degradation, unlike extracellular digestion, is an active, energy-dependent process. Players *feel* the cost of degradation, making the concept more concrete.
4. *PROTACs—Power-Ups with a Purpose:* PROTAC molecules act as collectible power-ups, introducing players to targeted protein degradation in an interactive way. Each PROTAC has

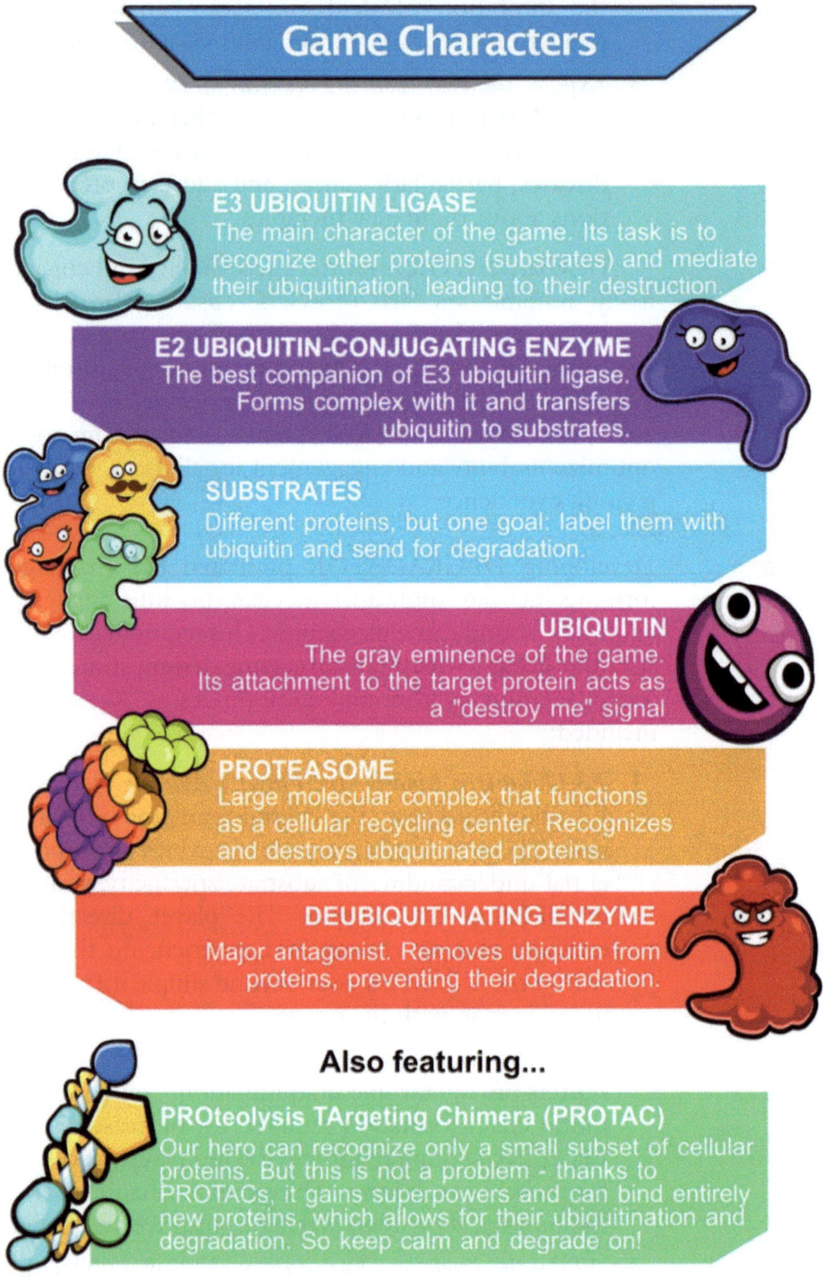

Fig. 2 Key characters in DEGRADATOR and their roles in gameplay. (Originally published in [1])

two ends—one that fits the E3 ligase like a key in a lock, matching its shape and color, and another that varies in color and shape to match the target protein (neo-substrate). Players use PROTACs to link the E3 ligase to proteins that would not normally be degraded, triggering their breakdown. This simplifies a real pharmacological process into an engaging gameplay mechanic, making complex science easier to grasp.

5. *DUBs—Antagonists with a Regulatory Role:* Deubiquitinating enzymes (DUBs) are cast as antagonists, introducing challenge while highlighting their crucial role in the UPS by removing ubiquitin from substrates, thereby rescuing them from degradation or editing ubiquitin chains to fine-tune signaling outcomes. Players must actively circumvent DUBs, experiencing the dynamic balance between ubiquitination and deubiquitination that maintains protein homeostasis. This "cat-and-mouse" gameplay reinforces the concept of opposing forces within the cellular environment.
6. *Quiz Levels—Strategic Reflection Points:* Dedicated quiz levels are strategically placed throughout the game, providing opportunities for knowledge consolidation and active recall. These quizzes prompt players to reflect on the gameplay and the underlying biological concepts, reinforcing learning and offering immediate formative assessment. The quizzes are not just about memorization; they ask players to apply their understanding, for example, by predicting the outcome of specific scenarios.

The main characters in DEGRADATOR are illustrated in Fig. 2.

1.4 Game Narrative and Theme

DEGRADATOR's narrative and visual theme are integral to its pedagogical strategy, designed to bridge the gap between abstract molecular processes and concrete, relatable experiences. The game casts players as "molecular cleanup crew" members within a futuristic, dark-blue cellular environment, achieving several key objectives:

1. *Visual Learning and Cognitive Load:* The visually engaging, futuristic aesthetic (avoiding dry textbook representations) leverages visual learning principles. The consistent color palette and clear representations of molecules (e.g., distinct shapes for E1, E2, E3, ubiquitin, proteasome, substrates, and decoys) reduce cognitive load, allowing students to focus on the processes rather than struggling to decipher complex visuals. This is crucial for understanding dynamic systems like the UPS, as research shows that well-designed visuals can improve comprehension and recall of complex scientific information.
2. *Analogical Reasoning and Conceptual Understanding:* The "molecular cleanup crew" metaphor facilitates analogical reasoning. By relating the unfamiliar (UPS function) to the familiar (cleaning), the game provides a cognitive scaffold, making the abstract concept of protein degradation more concrete and memorable. This is particularly effective for learners

with limited prior knowledge. For example, understanding that misfolded proteins are like "broken equipment" that needs to be removed helps students grasp why selective degradation is essential.

3. *Agency, Embodiment, and Deeper Processing:* The player's active role as an E3 ligase promotes agency and embodied cognition. Instead of passively observing, players become the enzyme, making decisions about which proteins to ubiquitinate. This active participation leads to deeper processing of information. Studies in cognitive science demonstrate that active learning, where students are directly involved in the learning process, significantly improves retention and understanding compared to passive learning methods like lectures. The act of maneuvering the E3, choosing targets, and experiencing the consequences (success or failure) creates a more meaningful and memorable learning experience.
4. *Motivation and Framing Effects:* The narrative subtly frames protein degradation as a vital cellular function, not just waste removal. Casting the player as an "essential worker" reinforces this importance. This framing can positively influence motivation and engagement. Students are more likely to invest effort in learning about a process they perceive as crucial. The game's challenges (time limits, DUB antagonists) are framed as obstacles to overcome, promoting a sense of accomplishment and reinforcing the importance of efficient protein degradation.
5. *PROTACs and Real-World Relevance:* The inclusion of PROTACs as power-ups directly connects the game to cutting-edge research and therapeutic applications. This not only increases engagement but also demonstrates the real-world relevance of understanding the UPS, potentially inspiring future scientific pursuits.

In summary, DEGRADATOR's narrative and visual design are not superficial additions; they are carefully chosen elements that leverage principles of cognitive science and educational psychology (visual learning, analogical reasoning, embodied cognition, and motivational framing) to create a more effective and engaging learning experience.

2 Materials

The technical development of DEGRADATOR was undertaken with a focus on creating a robust, cross-platform game accessible via standard web browsers, minimizing barriers to entry for educators and students. The choice of game engine, programming languages, and design considerations was all driven by the pedagogical goals and the need for broad accessibility.

DEGRADATOR was developed using Construct 3, a predominantly 2D game engine known for its visual scripting interface and HTML5 output capabilities. Construct 3 was selected for several key reasons:

1. *HTML5 and Cross-Platform Compatibility:* Construct 3's HTML5 output ensures that DEGRADATOR runs directly within web browsers without requiring any downloads, plugins, or specific operating systems. This *zero-installation* approach is *critical* in educational settings, guaranteeing compatibility across diverse devices (Windows, macOS, ChromeOS, iOS, and Android) and simplifying deployment in schools and homes with varying IT infrastructure and potential administrative restrictions.
2. *Visual Scripting for Rapid Prototyping and Iteration:* Construct 3's visual scripting system, based on an event-sheet model, allows for rapid prototyping and iteration of game mechanics and level designs. This visual approach facilitated efficient development cycles and allowed for quicker incorporation of feedback during the iterative design process.
3. *2D Game Focus:* Given the pedagogical goals of DEGRADATOR, a complex 3D environment was deemed unnecessary and potentially distracting. Construct 3's strengths in 2D game development aligned perfectly with the desired scope and visual style of the project, enabling a focus on clear visual representation of molecular components and gameplay mechanics without the added complexity of 3D asset creation and rendering.
4. *JavaScript Customization:* While Construct 3 visual scripting is powerful, the ability to integrate custom JavaScript code was essential for implementing specific game behaviors and interactions that went beyond the engine's built-in capabilities. For instance, custom JavaScript was used to manage the probabilistic interactions between enzymes and substrates, ensuring a realistic simulation of molecular collisions. This extensibility provided the flexibility to tailor the game precisely to the pedagogical requirements.

The development environment primarily utilized Construct 3's visual editor for level design, character animation, user interface creation, and event scripting. External image editing software was used for creating and refining game assets. Audio assets, including the custom soundtrack and sound effects, were created and integrated using audio editing software and Construct 3's built-in audio engine.

3 Methods

The methodological approach to DEGRADATOR's development was rooted in iterative design, combining pedagogical theory with user-centered testing to ensure both educational effectiveness and gameplay engagement. Each stage of the process—from defining learning objectives to implementing feedback from classroom use—was guided by evidence-based practices in science education and digital learning.

3.1 Level Design and Progression

DEGRADATOR's ten levels are structured to progressively introduce the UPS and targeted protein degradation, building complexity, and reinforcing concepts in a pedagogically sound manner. Each level builds upon prior knowledge, introducing new mechanics and challenges gradually. Table 1 summarizes the design and learning objectives of each level.

The level progression follows a deliberate scaffolding approach:

1. *Foundational Understanding (Levels 1–2):* These levels introduce the core components of the UPS (E3 ligase, E2 enzymes, ubiquitin, and proteasome) and the energy-dependent nature of the process. Simple mechanics and clear tutorials ensure accessibility for all players. The focus is on building a solid mental model of the basic pathway.
2. *Introducing Complexity and Therapeutics (Levels 4–6):* These levels build upon the foundation by introducing PROTACs (as power-ups) and DUBs (as antagonists). Level 4 allows players to experience the mechanism of PROTAC action, while Level 6 introduces the regulatory role of DUBs, requiring strategic adaptation.
3. *Advanced Regulation and Mechanisms (Levels 8–9):* These levels further expand on UPS regulation, introducing DUB inhibitors (as counter-antagonists to DUBs) and E3 ligase auto-ubiquitination (as a self-regulatory mechanism). This introduces more nuanced control mechanisms within the system.
4. *Strategic Reinforcement (Quiz Levels 3, 7, and 10):* Quiz levels are strategically interspersed throughout the game to reinforce learning and assess comprehension. These quizzes prompt active recall and application of knowledge gained in the preceding gameplay levels. Level 10 serves as a comprehensive assessment, covering all key concepts, including those detailed in the in-game *Great Encyclopedia of Protein Degradation.*

Each gameplay level (non-quiz) features a 3-minute time limit, adding a challenge and encouraging efficient gameplay. A star-based reward system (one to three stars based on completion

Table 1
Summary of game levels in DEGRADATOR

Level	Description	Introduced elements	Learning objective
1	Introduces the player as an E3 ubiquitin ligase tasked with ubiquitinating substrate proteins using E2 enzymes. Proteasome degradation occurs after four rounds of ubiquitination.	Substrate protein, E2 enzyme, ubiquitin, proteasome.	Understand the basic concept of protein degradation and the role of the UPS.
2	Introduces proteins that are not E3's substrates and the ATP power-up, which respawns ubiquitin on the nearest E2 enzyme.	Non-substrate protein, and ATP power-up.	Learn about substrate specificity and the role of ATP in molecular processes.
3	Features a quiz with five single-choice questions based on the Levels 1 and 2.	Quiz 1	Reinforce understanding of the UPS and substrate specificity through knowledge assessment.
4	Demonstrates PROTAC drug mechanism, enabling E3 to recognize non-native substrate proteins via a stick-like PROTAC power-up.	PROTAC power-up	Introduce the concept of targeted protein degradation and its potential in therapeutic applications.
5	Combines mechanics from previous levels, requiring ubiquitination of various proteins using correspondingly colored PROTACs.	Various colored substrates, distinct PROTACs.	Develop understanding of the versatility of E3 ligases and the adaptability of molecular systems in targeting proteins.
6	Introduces DUB that removes ubiquitin from proteins, impeding degradation.	DUB antagonist	Explore the role of DUBs in regulating protein degradation and maintaining cellular balance.
7	A quiz with five single-choice questions based on the Levels 4 to 6.	Quiz 2	Assess comprehension of targeted protein degradation and DUB functions.
8	Introduces a DUB inhibitor that irreversibly disables its activity. Multiple DUBs must be avoided while ubiquitinating substrates.	DUB inhibitor	Understand the concept of enzyme inhibition and its applications in regulating cellular processes.
9	Introduces auto-ubiquitination, where the E3 ligase itself can be ubiquitinated if idle with an E2 enzyme for too long.	Auto-ubiquitination mechanic.	Learn about the self-regulation of enzymes.
10	Final quiz with 12 single-choice questions, covering gameplay concepts and content from the *Great Encyclopedia of Protein Degradation*.	Final quiz	Consolidate knowledge of the UPS, targeted degradation, and regulatory mechanisms through a comprehensive assessment.

Adapted from [1]

time) motivates players to refine strategies and deepen their understanding to improve performance. Varying time thresholds for star ratings across levels encourage continuous improvement and reinforces the modeled biological processes. This combination of structured progression, time-based challenges, and performance rewards enhances replayability and reinforces learning in an engaging and intrinsically motivating way.

3.2 User Interface and User Experience (UI/UX)

Throughout the development of DEGRADATOR, significant attention was paid to user interface (UI) and user experience (UX) design to ensure the game was intuitive, accessible, and enjoyable for the target audience of students and educators. Key UI/UX considerations included:

1. *Intuitive Controls (Cross-Device Compatibility):* Simple, responsive mouse, or touch controls maximize accessibility across devices (laptops, tablets, and smartphones) and user preferences. This allows players to focus on gameplay and learning, rather than struggling with complex controls.
2. *Clear Visual Hierarchy and Information Presentation (Minimizing Cognitive Load):* The UI employs a clear visual hierarchy, prioritizing key information (level objective, timer, star rating, and tutorials). Concise, graphical presentation minimizes cognitive load, facilitating rapid comprehension of gameplay elements and objectives. Visual cues, such as color-coding of different enzyme types, further enhance understanding.
3. *Animated Tutorials and Just-in-Time Guidance (Scaffolding Learning):* Each level starts with a brief tutorial introducing any new mechanics, providing just-in-time guidance to ensure a seamless learning experience without interrupting gameplay flow. Additionally, post-level summaries reinforce key concepts by summarizing the objectives and functions of newly introduced mechanics. Based on beta tester feedback, the Level 1 tutorial was redesigned, replacing static screens with animations to enhance clarity and engagement, allowing for a smoother entry into the game.
4. *Multilingual Support (Inclusivity):* DEGRADATOR supports both English and Polish, easily switchable from the game menu. This broadens accessibility and caters to diverse learners.
5. *In-Game Encyclopedia (Supporting Deeper Learning): The Great Encyclopedia of Protein Degradation*, accessible from the menu, provides supplementary learning materials directly within the game. This resource offers simplified explanations of key concepts, allowing players to explore the biological background at their own pace and according to their individual

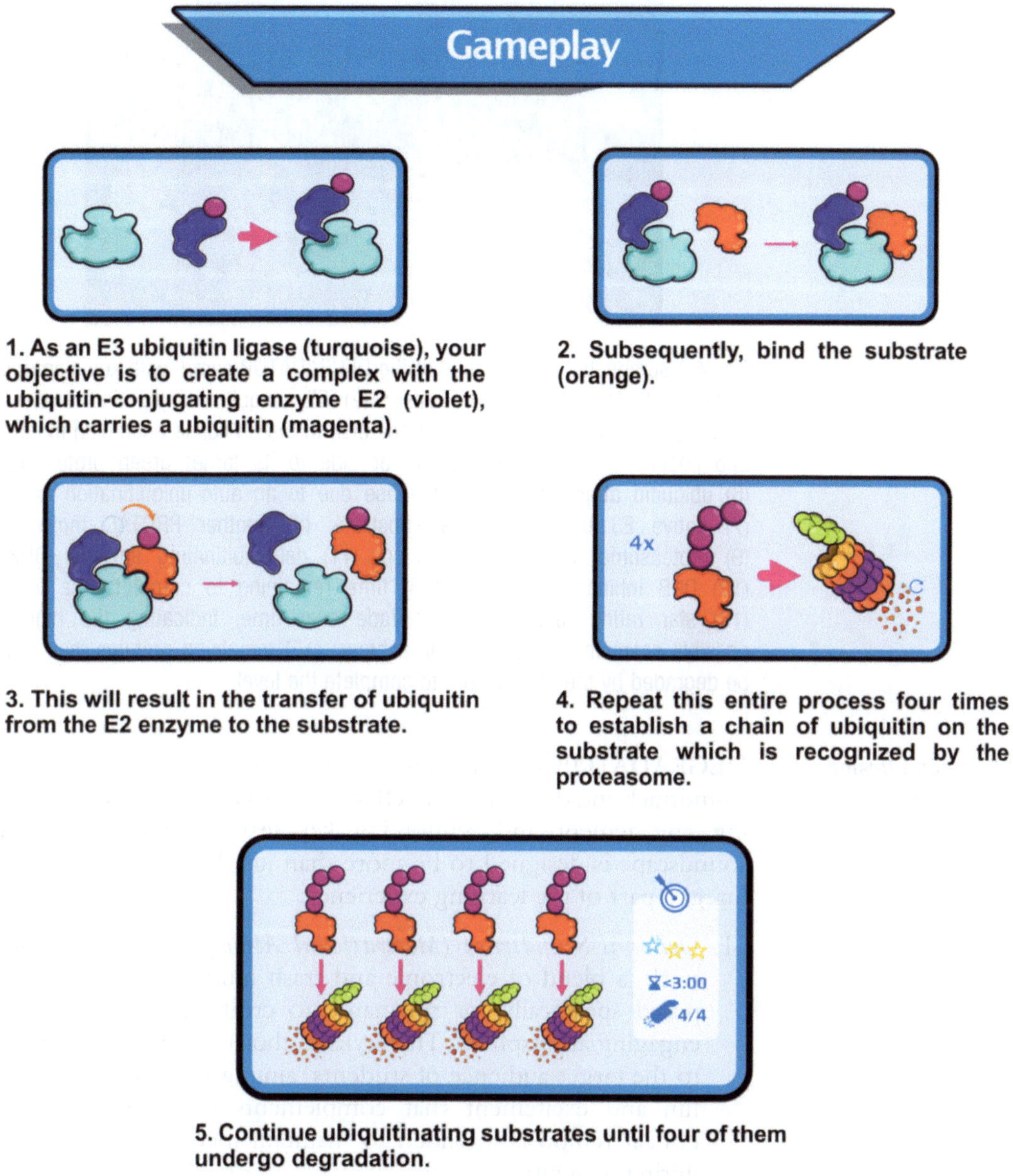

Fig. 3 A schematic representation of the core gameplay mechanics in DEGRADATOR. (Originally published in [1])

needs. This caters to different learning styles and levels of prior knowledge.

Figure 3 presents a detailed schematic of the core gameplay mechanics used throughout all non-quiz levels, while Fig. 4 highlights the key gameplay elements in Level 9, the final playable stage of DEGRADATOR, where all mechanics introduced progressively from Level 1 are fully integrated.

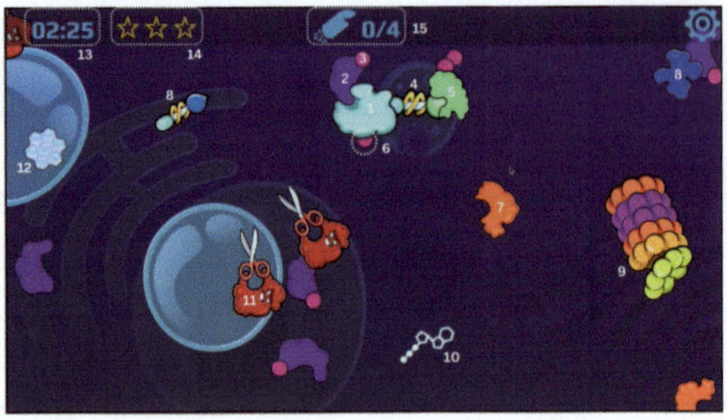

Fig. 4 A screenshot from Level 9 of the DEGRADATOR game, with key elements highlighted. (1) E3 ubiquitin ligase; (2) E2 ubiquitin-conjugating enzyme; (3) ubiquitin; (4) PROTAC bound on one side to the E3 ligase (matching in color and pocket shape) and on the other side to its target green protein (5); (6) ubiquitin attached to the E3 ligase due to an auto-ubiquitination event; (7) native E3 ubiquitin ligase substrate; (8) another PROTAC molecule; (9) proteasome; (10) ATP power-up; (11) deubiquitinating enzyme (DUB); (12) DUB inhibitor power-up; (13) time remaining to complete the level; (14) star rating feedback—stars fade over time, indicating the highest possible score the player can still achieve; (15) remaining proteins that must be degraded by the proteasome to complete the level

3.3 Sound Design and Soundtrack

DEGRADATOR's audio design, encompassing both a custom soundtrack and distinct sound effects, plays a crucial role in enhancing engagement and reinforcing key learning moments. The soundscape is designed to be more than just background; it is an integral part of the learning experience:

1. *Original Soundtrack (Motivational Atmosphere):* The soundtrack, a blend of electronic and trash music styles, was composed specifically for the game to create an energetic and engaging atmosphere. This stylistic choice was made to appeal to the target audience of students, aiming to create a sense of fun and excitement that complements the gameplay. The upbeat tempo is intended to maintain focus and motivation during gameplay.
2. *Distinct Sound Effects (Auditory Feedback and Reinforcement):* Specific sound effects are strategically used to highlight key in-game events, providing auditory feedback for player actions and reinforcing the significance of the underlying biological processes:
   (a) *Successful Ubiquitination:* A distinct, positive click-like sound effect plays each time a protein is successfully ubiquitinated, providing immediate positive reinforcement and highlighting the core game mechanic.
   (b) *Proteasomal Degradation:* A more substantial crunching-like sound effect accompanies the degradation of a protein

by the proteasome, emphasizing the culmination of the degradation pathway.

(c) *Power-Up Activation (PROTACs):* A unique sound effect signals the capture of a PROTAC power-up, drawing attention to this important therapeutic mechanism.

(d) *DUBs Inhibition:* A distinct freezing-like sound effect plays when a DUB is blocked, clearly signaling its inactivation.

These auditory cues were designed to create a multisensory learning experience, linking actions to sounds and reinforcing the connection between gameplay and the underlying biological concepts.

3.4 Iterative Development and Evaluation

The development of DEGRADATOR followed an iterative design process, emphasizing continuous testing and feedback incorporation throughout the game's creation (Fig. 5). This methodology was crucial for ensuring both the pedagogical effectiveness and the overall playability of the final product.

3.5 Prototyping and Early Stage Testing

The initial phases of development involved rapid prototyping of core game mechanics and level designs within the Construct 3 engine. Early prototypes focused on testing the feasibility and engagement of the ubiquitination mechanic, the movement and control of the E3 ligase character, and the basic level progression structure. These early prototypes were internally tested by the authors to identify and address fundamental usability issues, refine core gameplay loops, and validate the overall concept's potential for engaging and educating players about protein degradation.

3.6 Beta Testing and Feedback Integration

A formal beta testing phase was conducted with high school students from the TE Vizja High School with Bilingual Branches in Warsaw. These student beta testers represented the primary target audience for DEGRADATOR, providing valuable feedback from the intended user demographic. Beta testing focused on several key areas:

1. *Playability and Engagement:* Beta testers were asked to assess the overall enjoyment of the game, the intuitiveness of the controls, the clarity of the game objectives, and the level of challenge.
2. *Pedagogical Effectiveness:* Testers were asked to evaluate the game's effectiveness in explaining the concepts of protein degradation and the UPS. Feedback was sought on the clarity of in-game tutorials, the helpfulness of *the Great Encyclopedia of Protein Degradation*, and the perceived learning value of the gameplay experience.

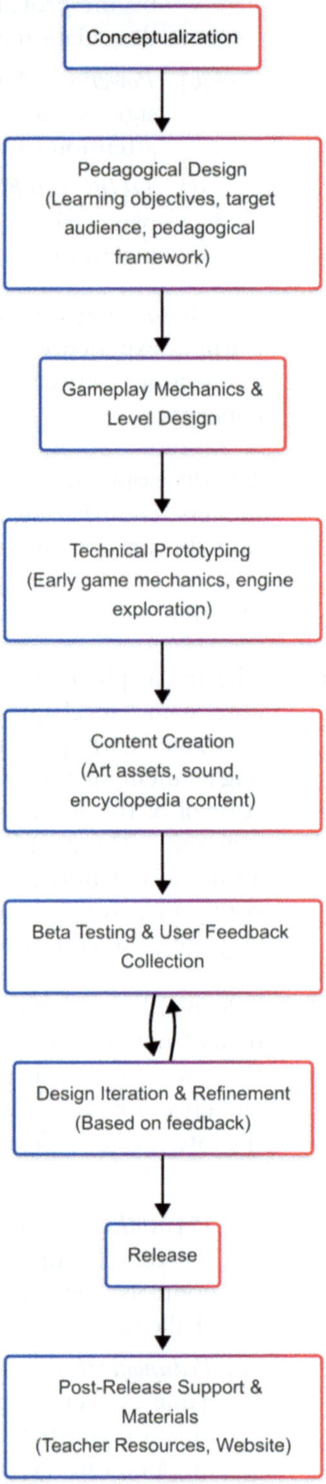

Fig. 5 Flowchart of DEGRADATOR's development process. (Created using Mermaid.js and visualized in mermaidchart.com)

3. *Usability and Technical Issues:* Beta testers were encouraged to identify any technical bugs, UI/UX issues, or areas where the game felt confusing or frustrating to play.

Feedback from beta testers was collected through structured surveys and direct observation of gameplay sessions. Analysis of beta testing feedback directly informed several key design iterations. As previously mentioned, feedback highlighting the initial tutorial for Level 1 as insufficiently clear led to a complete redesign incorporating animated visual guidance. Beta testers also provided valuable suggestions for improving level pacing, clarifying in-game instructions, and refining the user interface elements. This iterative process, driven by user feedback, was essential for polishing DEGRADATOR and ensuring its accessibility and effectiveness for the target student audience.

3.7 Classroom Case Study and Findings

To further evaluate the educational impact of DEGRADATOR in a classroom setting, a case study was conducted with 97 high school students (aged 15–19) at Władysław IV High School No. 8 in Warsaw. Students played DEGRADATOR individually during a single biology lesson, with the teacher providing minimal pregame instruction to assess the standalone impact of the game. Post-gameplay surveys were administered to collect quantitative and qualitative data on student engagement and perceived learning.

The results of the classroom study indicated positive reception and pedagogical impact of DEGRADATOR. Over 75% of students completed the game within 30 minutes (Fig. 6a), suggesting good accessibility and engagement. A significant majority (63%) rated the game highly (4 or 5 out of 5) for its ability to enhance their understanding of protein degradation (Fig. 6b). Furthermore, in-game quiz performance was encouraging, with 59% of students scoring 10 or more out of 12 points on the final quiz (Fig. 6c), indicating substantial knowledge retention after gameplay. Qualitative feedback from students highlighted the enjoyable and informative nature of the gameplay experience, with interactive mechanics effectively enhancing engagement with the subject matter. These findings provided empirical support for the pedagogical effectiveness of DEGRADATOR as a tool for introducing complex molecular biology concepts in an engaging and accessible manner.

3.8 Addressing Teacher Feedback and Developing Support Materials

Feedback from biology teachers involved in the case study revealed a need for additional support, particularly given their potential unfamiliarity with the UPS. To address this, a suite of supplementary educational materials was developed:

1. *Detailed Lesson Plans:* Comprehensive lesson plans tailored for both high school/secondary school students (aged 15+) and upper elementary grades (aged 12–15), providing structured frameworks for integrating DEGRADATOR into existing

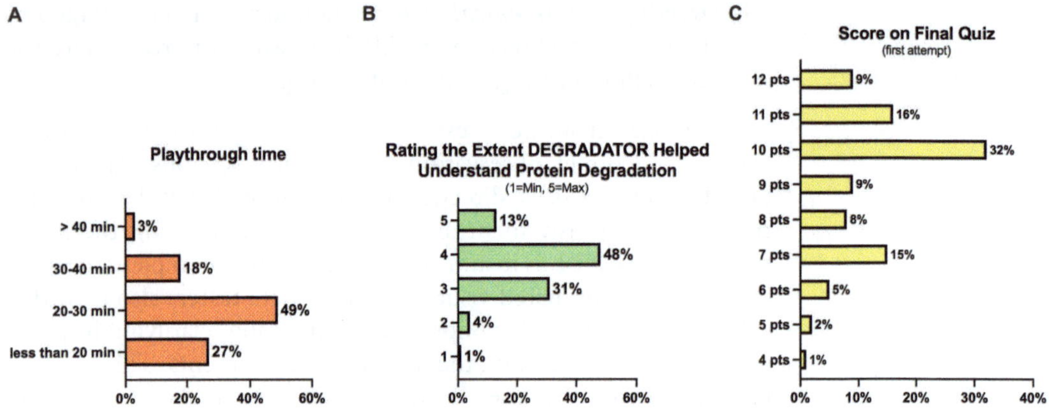

Fig. 6 Survey results from 97 high school students (aged 15–19) who individually played DEGRADATOR during biology class. (**a**) Bar graph depicting the distribution of playthrough times. (**b**) Bar graph showing students' ratings on how much DEGRADATOR improved their understanding of protein degradation. (**c**) Bar graph illustrating students' scores on the Final Quiz (first attempt only; max: 12 pts.; no student scored below 4 pts). (Adapted from [1])

biology curricula. These lesson plans offer step-by-step guidance, discussion prompts, and activity suggestions to maximize the educational impact of gameplay.

2. *The Great Encyclopedia of Protein Degradation (In-Game and Standalone):* The in-game encyclopedia, providing simplified descriptions of key UPS concepts, was also made available as a standalone, printable resource for teachers and students to access outside of the game environment.
3. *Educational Comic on DUBs:* A visually engaging educational comic was created to provide a deeper understanding of DUBs, portraying them not as antagonists but as essential regulators of cellular protein homeostasis. The comic highlights their crucial roles in maintaining balance within the UPS, ensuring they are not negatively perceived but rather recognized for their fundamental biological importance.
4. *Multimedia Presentations:* Accompanying PowerPoint presentations were developed for educators to use in conjunction with the lesson plans, providing visual aids and supplementary information to enhance classroom instruction.

These freely available support materials, accessible through the DEGRADATOR game website (https://degradator-game.com) and integrated into LabXchange platform, are intended to empower educators to confidently and effectively utilize DEGRADATOR as a valuable tool for teaching protein degradation and related molecular biology concepts.

4 Notes

The development of DEGRADATOR, a game-based learning tool for teaching about the UPS, provided crucial insights into designing effective educational resources for complex scientific topics. This process highlighted both successes and limitations, offering valuable lessons for future endeavors in science education and public engagement.

A core strength of the project was its foundation in constructivist game-based learning principles. This guided design decisions, ensuring game mechanics actively fostered learning about complex protein interactions and cellular processes. The iterative, user-centered design was equally vital. Beta testing with students and feedback from teachers directly shaped the game, revealing usability issues (addressed, for example, with animated tutorials) and informing level pacing and UI refinements. This iterative cycle, prioritizing user feedback, is fundamental to effective educational game development. Furthermore, accessibility was paramount. DEGRADATOR's browser-based HTML5 design and intuitive UI/UX maximized its reach, particularly in diverse educational settings. This mirrors the emphasis on accessibility and inclusivity found in work exploring augmented reality (AR) for biomolecule visualization, where student-generated AR tools were shown to enhance learning and enthusiasm [14]. Just as readily available smartphone technology democratized AR creation in that study, DEGRADATOR's web-based platform aims to democratize access to complex scientific concepts. Finally, comprehensive teacher support materials (lesson plans, a standalone encyclopedia, an educational comic, and presentations) were essential for empowering educators to integrate the game effectively.

However, limitations emerged. Resource constraints are a significant challenge; even a relatively simple 2D game requires substantial time and resources (DEGRADATOR's budget was approximately $28,000 USD). While the classroom case study provided valuable data, more rigorous, large-scale evaluations, including randomized controlled trials, would be needed to definitively quantify long-term learning outcomes. Maintaining sustained engagement over time and across diverse learners also remains an ongoing challenge.

Key lessons center on balancing scientific accuracy with engaging gameplay. Simplifications are necessary but must not distort the underlying science. DEGRADATOR's approach—gamifying PROTACs as power-ups and DUBs as antagonists, while accurately representing their core functions—exemplifies this principle. This challenge of transforming traditionally "dry" or abstract scientific content into an interactive and engaging experience is a common thread in innovative educational approaches.

For instance, [15] demonstrated that augmented reality could significantly increase enthusiasm and learning outcomes even for topics often perceived as unengaging, such as laboratory safety. Their work highlights the power of immersive technology to overcome motivational barriers, a principle that also underpins DEGRADATOR's game-based approach.

These experiences inform future directions. DEGRADATOR's foundation allows for expansion: new levels and mechanics could explore related processes (autophagy and lysosomal degradation) or disease contexts (neurodegeneration and cancer). Adapting the game for different age groups (younger learners, undergraduates, and graduates) could broaden its reach. Integration with learning management systems would facilitate adoption in formal settings, enabling progress tracking and curriculum integration. Multiplayer or collaborative features could enhance engagement and peer learning.

Beyond its educational applications, DEGRADATOR demonstrates game-based learning's potential to bridge the gap between complex science and public understanding. It makes molecular biology more accessible and enjoyable, leveraging the engaging nature of video games. In an era of rapid scientific advancement and the need for increased scientific literacy, DEGRADATOR exemplifies how games can promote informed discourse and combat misinformation. This extends the potential of interactive technologies in education to encompass not only traditionally visual subjects like protein structures [14], but also abstract and complex processes. The game's recognition at the 12th International Educational Games Competition validates its design and potential impact. Ultimately, DEGRADATOR represents a step toward a future where game-based learning plays a central role in fostering a more scientifically literate and engaged public, by adopting a holistic approach that merges a strong and accurate scientific base, with funny gameplay, accessibility for a wide audience, constant iterative design, and solid teacher support.

Acknowledgments

The development of this educational computer game was funded in Poland through the Polish State Budget under the Ministry of Education and Science program entitled *Social Responsibility of Science* (grant number SONP/SP/546507/2022). The preparation of this chapter was supported by the National Science Centre, Poland (SONATA BIS grant number 2021/42/E/NZ1/00190).

Conflicts of Interest The authors have no conflicts of interest to declare that are relevant to the content of this chapter.

References

1. Szulc NA, Olchowik A, Jaszczak P et al (2025) DEGRADATOR: a gaming expedition into targeted protein degradation therapies. J Chem Educ. https://doi.org/10.1021/acs.jchemed.4c00823
2. Glickman MH, Ciechanover A (2002) The ubiquitin-proteasome proteolytic pathway: destruction for the sake of construction. Physiol Rev 82:373–428. https://doi.org/10.1152/physrev.00027.2001
3. Ciechanover A, Schwartz AL (1998) The ubiquitin-proteasome pathway: the complexity and myriad functions of proteins death. Proc Natl Acad Sci USA 95:2727–2730
4. Hanna J, Guerra-Moreno A, Ang J, Micoogullari Y (2019) Protein degradation and the pathologic basis of disease. Am J Pathol 189:94–103. https://doi.org/10.1016/j.ajpath.2018.09.004
5. Osinalde N, Duarri A, Ramirez J et al (2019) Impaired proteostasis in rare neurological diseases. Semin Cell Dev Biol 93:164–177. https://doi.org/10.1016/j.semcdb.2018.10.007
6. Wang J, Maldonado MA (2006) The ubiquitin-proteasome system and its role in inflammatory and autoimmune diseases. Cell Mol Immunol 3:255–261
7. Zhao L, Zhao J, Zhong K et al (2022) Targeted protein degradation: mechanisms, strategies and application. Signal Transduct Target Ther 7:113. https://doi.org/10.1038/s41392-022-00966-4
8. Békés M, Langley DR, Crews CM (2022) PROTAC targeted protein degraders: the past is prologue. Nat Rev Drug Discov 21:181–200. https://doi.org/10.1038/s41573-021-00371-6
9. Cooperstein SE, Kocevar-Weidinger E (2004) Beyond active learning: a constructivist approach to learning. Ref Serv Rev 32:141–148. https://doi.org/10.1108/00907320410537658
10. Adipat S, Laksana K, Busayanon K et al (2021) Engaging students in the learning process with game-based learning: the fundamental concepts. Int J Technol Educ 4:542–552. https://doi.org/10.46328/ijte.169
11. Wang M, Zheng X (2021) Using game-based learning to support learning science: a study with middle school students. Asia-Pac Educ Res 30:167–176. https://doi.org/10.1007/s40299-020-00523-z
12. Cheung SY, Ng KY (2021) Application of the educational game to enhance student learning. Front Educ 6. https://doi.org/10.3389/feduc.2021.623793
13. Vlachopoulos D, Makri A (2017) The effect of games and simulations on higher education: a systematic literature review. Int J Educ Technol High Educ 14. https://doi.org/10.1186/s41239-017-0062-1
14. Argüello JM, Dempski RE (2020) Fast, simple, student generated augmented reality approach for protein visualization in the classroom and home study. J Chem Educ 97:2327–2331. https://doi.org/10.1021/acs.jchemed.0c00323
15. Zhu B, Feng M, Lowe H et al (2018) Increasing enthusiasm and enhancing learning for biochemistry-laboratory safety with an augmented-reality program. J Chem Educ 95:1747–1754. https://doi.org/10.1021/acs.jchemed.8b00116

Chapter 12

Automated Assessment of Argumentation Skills in Chemistry-Related Socioscientific Issues Using AI Chatbot

Sri Rahayu and Ananta Ardyansyah

Abstract

Socioscientific issues (SSI) require strong argumentation skills to support sound decision-making. Toulmin's Argument Pattern (TAP) is effective for assessing argument quality; however, manual evaluation is often time-consuming and prone to bias. Leveraging GPT offers a solution for developing automated assessments that are efficient, objective, and reliable. This article provides a guide for creating automated assessments of argumentation skills in chemistry-related SSI. This automated assessment was developed using Claude. The app produced by Claude to evaluate arguments is fully functional. This guide can be used with the free package provided.

Key words Argumentation, Artificial intelligence, Automated assessment, Chemistry, GPT, Socio-scientific issues, Test

1 Introduction

Socioscientific issues (SSI) are socially relevant problems that are connected to science [1]. SSI encourages moral engagement and ethical considerations, which often lead to debates and controversies in their resolution [2]. SSI issues are increasingly prevalent and easier to encounter today, such as palm oil plantations, the use of artificial sweeteners, and climate change. These cases require not only strong knowledge of chemistry but also a good understanding of other aspects [3, 4]. Addressing these issues requires a literate society capable of engaging in debates and public discussions about the best stance to take regarding existing SSI phenomena, where argumentation skills are essential. This is crucial to ensure that everyone can make the best decisions. Strong argumentation skills help individuals make decisions through a process that requires critical evaluation and consideration of multiple perspectives [5–7]. Given the importance of this skill, it is also essential to develop

Robert E. Dempski (ed.), *Transforming Molecular Biology with Emerging Technologies*, Methods in Molecular Biology, vol. 3043, https://doi.org/10.1007/978-1-0716-5320-3_12,

related assessments that can evaluate a person's argumentation ability.

The use of an appropriate framework in assessing argumentation is very important, especially regarding SSI, considering that argumentation related to SSI differs from that of ordinary scientific issues [8]. One argumentation framework considered to produce appropriate and well-structured arguments, and also the most established, is Toulmin's Argument Pattern (TAP) framework [9]. The Toulmin argument framework consists of claim (C), data (D), warrant (W), backing (B), qualifier (Q), and rebuttal (R). With its relatively complex structure, this framework is highly suitable for assessing multifaceted issues related to SSI. Compared to other, simpler argumentation frameworks [10], Toulmin's framework can capture the complexity of SSI situations. The Toulmin framework requires an understanding of counterclaims, which are important in an argument. However, this framework cannot be directly quantified. Therefore, an argument assessment rubric based on the Toulmin framework, as developed in previous research, is used [11]. This rubric describes the levels of students' argumentation skills, ranging from level 1 to level 5.

Although guidelines for assessing arguments already exist, argumentation assessment often takes a considerable amount of time and is prone to bias, as it is also influenced by an individual's subjectivity, making the evaluation of this skill challenging [12–14]. Assessments that require a long time often become an obstacle in open-ended questions, such as argumentation tasks. This becomes even more complex considering that assessing argumentation requires understanding each component from claim and data to rebuttal. With many argumentation questions and a high number of students, evaluating argumentation skills often presents a unique challenge. The complexity and scale of argumentation assessment can affect the accuracy of a person's evaluation. Moreover, assessor subjectivity plays a significant role, as an evaluator may hold personal views that lean toward one side, potentially leading to inaccurate assessments.

In addressing this issue, the use of artificial intelligence can help create valid and reliable evaluations. An AI chatbot is currently capable of understanding language well, which enables it to assess open-ended responses such as arguments. In addition, the latest GPT models are equipped with strong coding capabilities, allowing them to develop simple applications [15–18]. By leveraging this capability, anyone can design an AI-based automated assessment for open-ended questions such as argumentation. In this manuscript, it will be explained how to create an automated argumentation assessment for scientific chemistry-related SSI with the help of AI. Readers will later be able to develop an argumentation assessment using SSI passages that they can customize. In this instruction, the author uses an AI chatbot with the free package, so the entire readership should be able to follow these instructions as long as they have access to the AI chatbot.

2 Materials

In creating an AI-based argumentation assessment for chemistry-related SSI, three main components are needed: a designed SSI passage, a spreadsheet file, a text editor, and an AI chatbot.

2.1 Socioscientific Issues (SSI) Passage Used

Argumentation is often influenced by a person's interest, familiarity, and relevance to the issue [19, 20], so a well-designed SSI passage is needed to truly engage someone in argumentation. In general, SSI are complex and controversial, but specifically, common characteristics that can be found include being cultural, case-based, discursive, related to the nature of science, requiring personal cognition, and involving moral considerations [21]. These issues can be sourced from the Internet, but their content needs to be verified. The presented issue should provide data supporting both the pro and con sides so that both perspectives can argue equally. Naturally, the text must be ensured not to present a narrative biased toward one side (here, the teacher or assessor needs to ensure they remain as neutral as possible when forming the narrative). Then, ensure that the provided passage is not too long. In addition, practitioners need to ensure that students have prior knowledge related to the topic being discussed. Taking these considerations into account, creating a good SSI passage requires quite a lot of effort. However, considering that chemical knowledge is closely related to everyday life, it should not be too difficult to determine this. Some suitable issues can be found in previous studies [4, 22].

2.2 Spreadsheet File

A spreadsheet file is used to facilitate the input of all assessment components (SSI discourse, questions or instructions, and numbering). This spreadsheet file is very useful when the assessment contains many questions. However, it should be remembered that the basic idea of using this type of file is to make it easier for the AI to compile the automated assessment. Therefore, using other similar file formats, such as Doc or PDF formats that display tables, is also fine.

2.3 AI Chatbot to Build Apps

Most AI chatbots today can already automatically create applications that can run well. However, based on the author's experience, among several existing AI chatbots, Claude has the best ability in delivering apps and assessing essays. Claude can produce applications with a good appearance and full functionality. These applications are then fully supported by Claude's capabilities, enabling it to understand the inputted essay answers. This is important for providing accurate assessments. Therefore, in this manuscript, Claude is used (although similar results may also be possible with other AI chatbots).

3 Methods

3.1 Preparing the SSI Discourse

1. Determine an SSI that is likely relevant to the participants, considering the context, the issue's virality, and its controversy (*see* **Note 1**).
2. Ensure that students have prior knowledge that can support their understanding and engagement with the SSI.

   For example, issues related to the use of artificial sweeteners and sugar alcohols require students to understand carbon compounds, carbohydrates, and their structures.
3. Create an introductory paragraph that provides an initial understanding of the SSI, such as what the issue is and why it is given that name. This opening paragraph should be neutral and not show any pros or cons.

   Example: "*As the name suggests, sugar alcohol is a combination of sugar and alcohol molecules. However, this substance does not contain alcohol like ethanol. Sugar alcohols, such as erythritol and sorbitol, can be naturally found in fruits and vegetables. For example, erythritol is a sugar alcohol found in stevia leaves. Its structure consists of four carbon chains with four hydroxyl groups. This structure differs from common sugars, which are typically aldoses or ketoses.*"
4. Provide data that supports both the pro and con sides of the issue.

   Example: "*Sugar alcohols are widely used as sweeteners by the industry because they have a higher sweetness level than regular sugar (25–100% of sugar) but contain fewer calories. This is related to their structural differences from regular sugar, as most sugar alcohols are not well digested in the digestive tract. However, sugar alcohols have some drawbacks that need to be considered. The main issue with sugar alcohols is digestive problems, especially when consumed in excessive amounts. The digestive reactions caused by sugar alcohol consumption include bloating and diarrhea. In addition, there are other issues, such as an unpleasant aftertaste and a higher price compared to regular sugar. Some sugar alcohols, like xylitol and maltitol, may produce specific symptoms.*"
5. Ensure there are more than two supporting reasons for both pro and con sides to make sure students can produce multiple counterarguments.
6. Create an engaging title.

Number	Passage	Instruction
1	Sugar Alcohols: A Good Sugar Substitute? As the name suggests, sugar alcohols are a combination of sugar and alcohol molecules. However, they do not contain alcohol like ethanol. Sugar alcohols, such as erythritol and sorbitol, can be naturally found in fruits and vegetables. For example, erythritol is a sugar alcohol found in stevia leaves. Its structure consists of four carbon chains with four hydroxyl groups. This structure differs from common sugars, which are typically aldoses or ketoses.	If you are a sweet beverage seller, would you use sugar alcohols? Explain why you would or would not use them! Organize your argument carefully

Fig. 1 An example of a spreadsheet or table file containing all aspects of each question item

Example: "*Sugar Alcohols: A Good Sugar Substitute?*"

7. Provide questions and instructions related to the SSI that can encourage them to present arguments, such as "give your argument" or "explain why you chose that" after asking them to choose or state an opinion (*see* **Note 2**).

   Example: "*If you are a sweet beverage seller, would you use sugar alcohols? Explain why you would or would not use them! Organize your argument carefully.*"

8. After creating several discourses, make a table in a spreadsheet containing three columns: number, discourse, and instruction, like in Fig. 1.
9. Place all test items to be assessed in the spreadsheet. Then save the spreadsheet file (*see* **Note 3**).

3.2 Creating the Application through an AI Chatbot

1. Log in to the AI chatbot account you are using, such as Claude.
2. Then enter a prompt to create a React application.

   Prompt:

> *Create a React application to assess the quality of students' arguments based on the Erduran Rubric. Use the SSI data from the spreadsheet.*
>
> *Application specifications:*
>
> - *Clean and modern interface.*
> - *Each SSI includes a reading passage, a question, and a textarea for answers.*
> - *Use the Claude API for automated scoring based on the Erduran Rubric (Level 1–5).*
> - *Display assessment results with detailed feedback.*
> - *Include result statistics (average score, highest level, distribution).*
> - *Responsive design.*

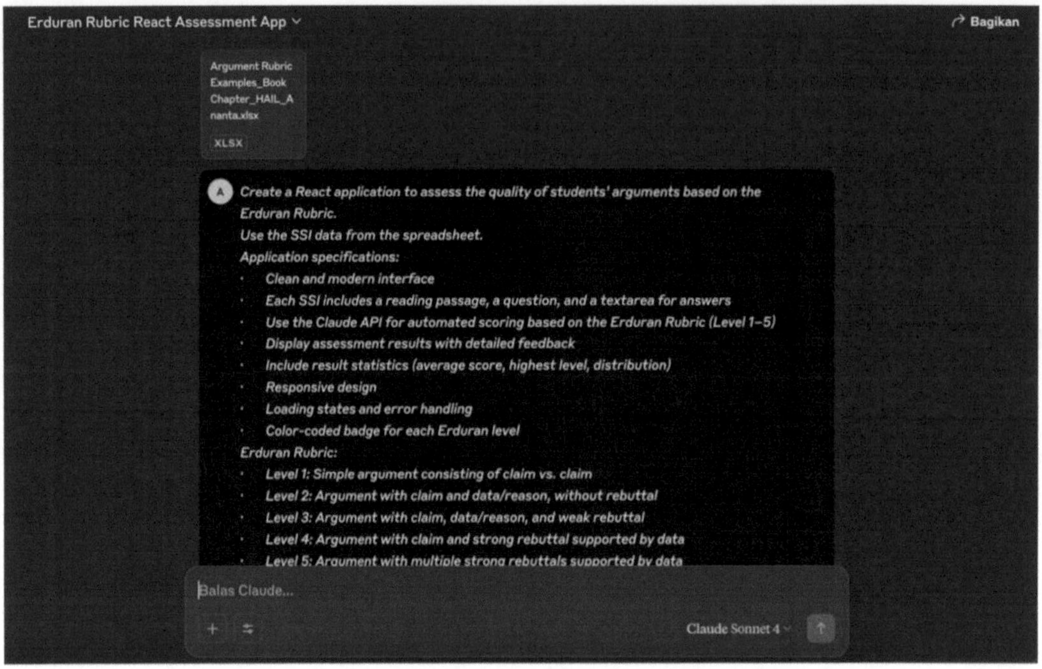

Fig. 2 The view displayed when the prompt and spreadsheet file have been entered

> - *Loading states and error handling.*
> - *Color-coded badge for each Erduran level.*
>
> *Erduran Rubric:*
>
> - *Level 1: Simple argument consisting of claim* vs. *claim.*
> - *Level 2: Argument with claim and data/reason, without rebuttal.*
> - *Level 3: Argument with claim, data/reason, and weak rebuttal.*
> - *Level 4: Argument with claim and strong rebuttal supported by data.*
> - *Level 5: Argument with multiple strong rebuttals supported by data.*
>
> *Create the application in a ready-to-use React artifact.*

3. Add the previously prepared spreadsheet file as Fig. 2 shows, then let the AI work.
4. Wait until the AI chatbot creates the application.

5. When the artifact has been created, there is usually a "publish" option.
6. Copy the provided link and ensure the application functions properly.
7. In the created artifact, you can view the code of the application that has been generated. Copy the code, then save it using a text editor application you have, such as Sublime Text. This is useful for saving or transferring the application, or if the user wants to edit it externally.

3.3 Implementing the Automated Assessment

1. In implementing the automated assessment you have created, you can create a short link or convert it into a barcode that students can scan.
2. Then, ensure that students have Internet access and a Claude account. Once confirmed, you can share the link for them to complete the task.
3. Provide guidance and supervision during the assessment. Supervision is carried out by setting rules at the beginning that students must work independently and may not use any sources other than those provided in the test text. Ensure that students only open the designated application during the exam.
4. After students have completed the task, you can ask them to take a screenshot of their results and not refresh the web page (*see* **Note 4**).
5. You can then record the students' results. Teachers can record student test results in a spreadsheet file containing student identities and their outcomes. The assessment results will display a student summary showing the average, highest score, and distribution of argumentation levels, as shown in Fig. 3. The previously taken screenshots can serve as evidence of the test implementation.

4 Notes

Several points to note are:

1. Highly relevant and personal discourses are usually closely related to health topics (although other topics may also be relevant). Health issues can be adopted from highly credible websites, such as Healthline. Practitioners can also adopt SSI discourses from existing news reports. However, the most important thing is to double-check the data and facts provided.

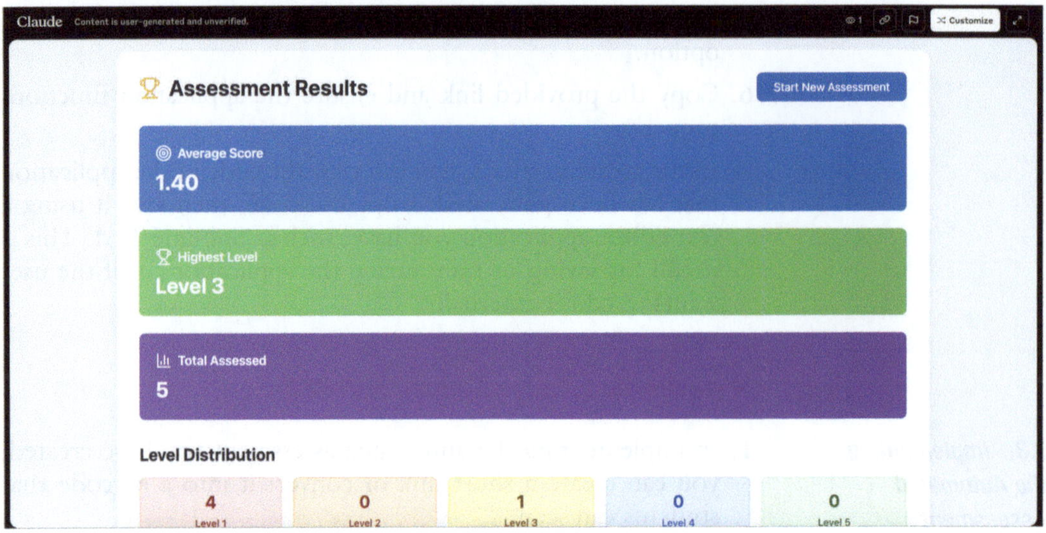

Fig. 3 The main view that students may see after completing the test. Below this, there are explanations and feedback for each question

2. Practitioners need to consider the number of questions given and the allotted time. Considering that these are open-ended essay questions, it is better not to have too many. The author recommends five questions.
3. In making arguments, students are greatly influenced by their position in society; therefore, asking them to take on a specific role (e.g., business owner, government official, or NGO) in the presented issue can lead to different arguments and may be more engaging for them.
4. The created application cannot save the data that has been entered. By refreshing the webpage, the data may be lost; therefore, students need to be informed of this.

Acknowledgments

We extend our gratitude to the Ministry of Higher Education, Science, and Technology of the Republic of Indonesia for their support in funding this research through the PMDSU Program.

Conflicts of Interest The authors have no conflicts of interest to declare that are relevant to the content of this chapter.

References

1. Sadler TD (2009) Socioscientific issues in science education: labels, reasoning, and transfer. Cult Stud Sci Educ 4:697–703. https://doi.org/10.1007/s11422-008-9133-x
2. Sadler TD, Zeidler DL (2005) Patterns of informal reasoning in the context of socioscientific decision making. J Res Sci Teach 42:112–138. https://doi.org/10.1002/tea.20042
3. Ware SA (2001) Lecture presented at the 8 th international chemistry conference in Africa (8 th ICCA). Pure Appl Chem 73:1147–1223
4. Marks R, Eilks I (2009) Promoting scientific literacy using a sociocritical and problem-oriented approach to chemistry teaching: concept, examples, experiences. Int J Environ Sci Educ 4:231–245
5. Sadler TD, Donnelly LA (2006) Socioscientific argumentation: the effects of content knowledge and morality. Int J Sci Educ 28:1463–1488. https://doi.org/10.1080/09500690600708717
6. Venville GJ, Dawson VM (2010) The impact of a classroom intervention on grade 10 students' argumentation skills, informal reasoning, and conceptual understanding of science. J Res Sci Teach 47:952–977. https://doi.org/10.1002/tea.20358
7. Walker KA, Zeidler DL (2007) Promoting discourse about socioscientific issues through scaffolded inquiry. Int J Sci Educ 29:1387–1410. https://doi.org/10.1080/09500690601068095
8. Khishfe R (2022) Nature of science and argumentation instruction in socioscientific and scientific contexts. Int J Sci Educ 44:647–673. https://doi.org/10.1080/09500693.2022.2050488
9. Toulmin SE (2003) The uses of argument: updated edition. Cambridge University Press, Cambridge
10. Sampson V, Schleigh S (2012) Scientific argumentation in biology: 30 classroom activities. National Science Teachers Association, Arlington
11. Erduran S, Simon S, Osborne J (2004) TAPping into argumentation: developments in the application of Toulmin's argument pattern for studying science discourse. Sci Educ 88:915–933. https://doi.org/10.1002/sce.20012
12. Lindblom-ylänne S, Pihlajamäki H, Kotkas T (2006) Self-, peer- and teacher-assessment of student essays. Act Learn High Educ 7:51–62. https://doi.org/10.1177/1469787406061148
13. Hussein MA, Hassan H, Nassef M (2019) Automated language essay scoring systems: a literature review. PeerJ Comput Sci 5:e208. https://doi.org/10.7717/peerj-cs.208
14. Ramesh D, Sanampudi SK (2022) An automated essay scoring systems: a systematic literature review. Artif Intell Rev 55:2495–2527. https://doi.org/10.1007/s10462-021-10068-2
15. Anthropic (2025) Build and host AI-powered apps with Claude – No Deployment Needed \Anthropic. https://www.anthropic.com/news/claude-powered-artifacts. Accessed 15 Aug 2025
16. OpenAI Memperkenalkan GPT-5 | OpenAI. https://openai.com/id-ID/index/introducing-gpt-5/. Accessed 15 Aug 2025
17. xAI Grok | xAI. https://x.ai/grok. Accessed 15 Aug 2025
18. Google Gemini Gemini Canvas—menulis, membuat kode, & berkreasi dengan AI di satu tempat. https://gemini.google/overview/canvas/. Accessed 15 Aug 2025
19. Khishfe R (2012) Relationship between nature of science understandings and argumentation skills: a role for counterargument and contextual factors. J Res Sci Teach 49:489–514. https://doi.org/10.1002/tea.21012
20. von Aufschnaiter C, Erduran S, Osborne J, Simon S (2008) Arguing to learn and learning to argue: case studies of how students' argumentation relates to their scientific knowledge. J Res Sci Teach 45:101–131. https://doi.org/10.1002/tea.20213
21. Zeidler DL, Sadler TD, Simmons ML, Howes EV (2005) Beyond STS: a research-based framework for socioscientific issues education. Sci Educ 89:357–377. https://doi.org/10.1002/sce.20048
22. Çalık M, Wiyarsi A (2021) A systematic review of the research papers on chemistry-focused socio-scientific issues. J Balt Sci Educ 20:360–372. https://doi.org/10.33225/jbse/21.20.360

Chapter 13

Sodium Dodecyl Sulfate–Polyacrylamide Gel Electrophoresis (SDS–PAGE) with Activity Staining for Polyphenol Oxidase (PPO)

Lalla Fall, Aa'ishah Riaz, and Monika Sommerhalter

Abstract

Sodium dodecyl sulfate–polyacrylamide gel electrophoresis (SDS–PAGE) is a popular technique for the separation and characterization of proteins. Whereas most proteins are denatured by the anionic detergent SDS, the enzyme polyphenol oxidase (PPO) is activated by low SDS concentrations, provided that the samples are not heated and reduced. Direct activity staining for PPO is therefore possible on SDS–PAGE gels. Color development from ortho-quinone reaction products can be enhanced with coupling reagents such as 4-amino-N,N-diethylaniline (ADA). As demonstrated with *Agaricus bisporus* (white and brown button mushroom) extracts, multiple PPO forms can be clearly resolved on SDS–PAGE gels. The method described here encompasses sample preparation, electrophoresis, and activity staining. It is a rapid, simple, and high-resolution technique that enables the comparison of PPO forms across several samples without the need for protein purification.

Key words Protein electrophoresis, SDS–PAGE, Enzyme, Polyphenol oxidase, Activity staining

1 Introduction

Sodium dodecyl sulfate polyacrylamide gel electrophoresis (SDS–PAGE) is a well-established laboratory technique that separates proteins based on their size [1]. The anionic detergent SDS denatures proteins and coats them with a negative charge, which forces them to migrate to the positive pole of the electric field applied in electrophoresis [2]. The crosslinked polyacrylamide acts like a sieve. Coomassie or silver staining is commonly used to detect proteins irrespective of their function [3]. Functional staining with reaction mixtures specific for enzyme activity is only possible after non-denaturing (native) PAGE or after renaturation of the protein within the gel after SDS–PAGE [4]. The enzyme polyphenol oxidase (PPO) is activated by low concentrations of SDS [5, 6], and is

Robert E. Dempski (ed.), *Transforming Molecular Biology with Emerging Technologies*, Methods in Molecular Biology, vol. 3043, https://doi.org/10.1007/978-1-0716-5320-3_13,

therefore an ideal candidate for function-specific staining of SDS–PAGE gels.

PPO is a ubiquitous enzyme found in animals, plants, and fungi [7]. It catalyzes the oxidation of o-diphenols to o-quinones in the presence of oxygen, a process described as diphenolase or catechol oxidase activity (EC 1.10.3.1). In some cases, the enzyme can also act as a tyrosinase, catalyzing the oxidation of monophenols to diphenols (EC 1.14.18.1). Plants produce multiple PPO isoforms, with expression patterns depending on tissue type, developmental stage, and environmental conditions, including abiotic and biotic stress [8]. Since high PPO activity can spoil the visual appeal, taste, and nutritional value of vegetables and fruit, the characterization of PPO isoforms is important for agriculture and the food industry [9, 10].

With the protocol described here, the presence of different PPO isoforms in plant tissues can be directly visualized by loading a crude plant extract onto a polyacrylamide gel, performing electrophoresis, incubating the gel in buffer, and then staining for PPO activity. Notably, the crude plant extract is mixed with an SDS-containing loading buffer, but is neither heated nor reduced. Many of the o-quinone products of PPO-catalyzed reactions are sufficiently colorful for detection. In solution-based assays, secondary coupling reactions using 3-methyl-2-benzothiazolinone hydrazone (MBTH) or 4-amino-N,N-diethylaniline (ADA) are often employed to enhance color development [11, 12]. Building on this principle, Rescigno and coworkers developed a rapid detection method for PPO on polyacrylamide gels after non-denaturing PAGE, using 4-tert-butyl-catechol as substrate and ADA as the coupling reagent [13]. As demonstrated below, this coupled staining procedure can also be performed after SDS–PAGE. As an alternative to 4-*tert*-butyl-catechol, other PPO substrates (e.g., dopamine, L-DOPA, catechol, 4-methyl-catechol, and pyrogallol) can be used.

The SDS–PAGE gel shown in Fig. 1 provides an example of the outcome of the protocol presented below. The gel was loaded with crude extracts prepared from white and brown button mushrooms, which belong to the species *Agaricus bisporus*. After electrophoresis, the gel was stained with dopamine and then ADA. PPO catalyzes the oxidation of dopamine into aminochrome, which forms a blue adduct with ADA. The blue bands on the gel in Fig. 1 signify the presence of several PPO forms. The banding patterns between samples from white and brown button mushrooms differ. High molecular PPO bands above 100 kDa are more visible in brown button mushroom extracts. Despite having overall lower protein content per gram tissue, crude extracts prepared from stalks produced more intense bands on the gel than crude extracts prepared from mushroom caps. The most intense set of PPO bands was found around 35 kDa for the extract prepared from the stalk of

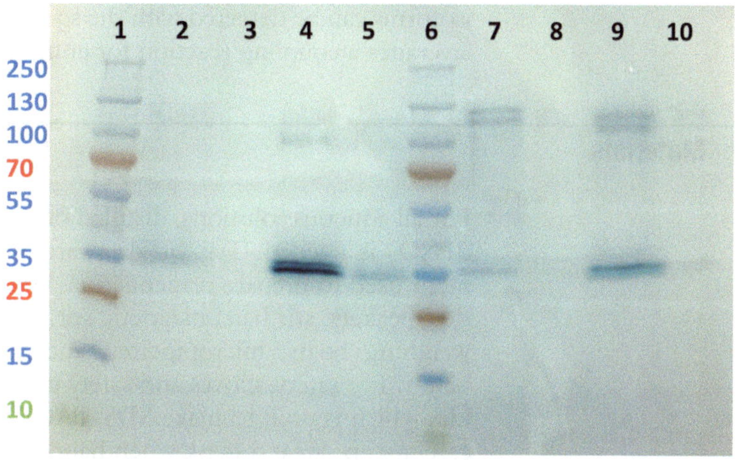

Fig. 1 SDS–PAGE gel (4–20% Tris–glycine) stained with dopamine (10 mM in 0.1 M sodium phosphate buffer, pH 6.5) followed by 4-amino-N,N-diethylaniline (0.6 mM). Lanes 1 and 6: Prestained PageRuler Plus protein ladder. Lanes 2–3: Crude cap extracts from white button mushroom. Lanes 4–5: Crude stalk extracts from white button mushroom. Lanes 7–8: Crude cap extracts from brown button mushroom. Lanes 9–10: Crude stalk extracts from brown button mushroom. Every other mushroom extract sample was diluted ten-fold before loading, resulting in total protein amounts of 20 μg, 2 μg, 10 μg, 1 μg, 20 μg, 2 μg, 10 μg, and 1 μg for lanes 2, 3, 4, 5, 7, 8, 9, and 10, respectively

white button mushrooms. The presence of multiple tightly spaced bands most likely originates from similar PPO isoforms and post-translational modifications (e.g., glycosylation). Other studies have shown that *A. bisporus* has six PPO genes (PPO1–PPO6) with high levels of expression for PPO3 and PPO4 in the stalk and cap [14]. Two subunits of PPO3 and two subunits of lectin-like subunits are known to form a large tyrosinase tetramer [15]. Heated SDS–PAGE samples of purified PPO from *A. bisporus* (100 °C for at least 5 min) showed multiple tightly spaced bands on silver-stained SDS–PAGE gels around 47 kDa [14]. Because our samples were not heated, incomplete denaturation might have resulted in preserving higher oligomerization states. In addition, monomeric PPO forms might have remained more compact, resulting in faster migration and, therefore, lower apparent molecular weights around 35 kDa (Fig. 1).

The method presented here is not suitable for exact protein size determinations, but provides for easy and rapid visualization of PPO isoforms for many samples in parallel. In contrast to native PAGE, the presence of SDS results in higher resolution with tighter and crisper bands and minimizes the effects of intrinsic protein charge. The samples do not require any purification. Despite the presence of numerous other proteins in crude extracts, PPO isoforms can be detected with the specificity of an activity assay that leverages a coupling reaction for enhanced sensitivity.

2 Materials

For all aqueous solutions, distilled or deionized water (abbreviated dH_2O) is needed. Typical glassware and small laboratory equipment used to prepare reagents (i.e., volumetric flasks, storage bottles, beakers, stir bars, magnetic stir plates, scales, weighing paper, weighing boats, micropipettes, and pipette tips) are not listed below. For safety, gloves and safety goggles are required. Acrylamide, which is used to make SDS–PAGE gels, is a neurotoxin and a carcinogen. Wear a mask when handling solid SDS. The fine powder is a health hazard. The chemicals used for staining, dopamine hydrochloride and ADA, are irritants, and dopamine is an environmental hazard. Follow local waste disposal regulations when discarding the gel and staining solutions.

2.1 Sample Preparation

1. Tissue containing PPO (e.g., white or brown button mushroom) (*see* **Note 1**).
2. Buffer used for sample extraction, dilutions, and activity staining: 0.1 M sodium phosphate buffer, pH 6.5. Prepare 250 mL of a 0.1 M sodium phosphate buffer, pH 6.5, by weighing out 1.47 g of $NaH_2PO_4{\cdot}H_2O$ and 3.85 g of $Na_2HPO_4{\cdot}7H_2O$. Dissolve both salts in ~240 mL d H_2O. Check the pH and adjust with a few drops of 1 M HCl or NaOH if the pH deviates from 6.5 (*see* **Note 2**). Transfer the solution to a 250 mL volumetric flask and bring the final volume up to 250 mL with dH_2O. After thorough mixing, transfer the buffer to a storage bottle. Sodium phosphate buffers can be stored for several months at 4 °C.
3. Extraction buffer: 0.1 M sodium phosphate buffer, pH 6.5, 2 mM L-ascorbic acid, 1% (w/v) polyvinylpolypyrrolidone (PVPP) (*see* **Note 3**). Shortly before sample preparation, weigh out 10 mg L-ascorbic acid and 0.25 g PVPP and transfer both solids into a 25 mL volumetric flask. Add 0.1 M sodium phosphate buffer, pH 6.5, and shake the slurry before filling up to the 25 mL mark with 0.1 M sodium phosphate buffer, pH 6.5. Keep the extraction buffer at 4 °C.
4. Sample loading buffer (5×): 0.25 M Tris–HCl buffer, pH 6.8, 10% (w/v) SDS, 30% (v/v) glycerol, 0.02% (w/v) bromophenol blue (*see* **Notes 4** and **5**). To make 1 mL of 5× SDS–PAGE loading buffer, combine the following solutions and solids in a microcentrifuge tube: 0.125 mL of 2 M Tris–HCl buffer, pH 6.8, 0.1 g SDS (solid), 0.30 mL glycerol, 0.50 mL of 0.04% (w/v) bromophenol blue, and 0.075 mL of

dH_2O. The 0.04% (w/v) bromophenol blue solution is prepared by dissolving 0.040 g bromophenol blue in 100 mL of dH_2O. The bromophenol blue solution and the loading buffer can be stored at room temperature for several months.

5. Positive control sample: *A. bisporus* tyrosinase preparation (T3824 from Sigma-Aldrich, St. Louis, MO, USA, *see* **Note 6**). Dissolve the contents of the T3824 vial (3.8 mg of tyrosinase powder) in 2 mL of 100 mM sodium phosphate buffer, pH 6.5. To make a 10-fold dilution, mix 100 μL stock solution and 900 μL of buffer in a new microcentrifuge tube by pipetting up and down.
6. Negative control sample: 1 mg/mL bovine gamma globulin (BGG; *see* **Note 7**).
7. Ice bucket filled with ice.
8. Mortar and pestle.
9. Microcentrifuge tubes.
10. Microcentrifuge.

2.2 SDS–Page

1. Polyacrylamide Tris–glycine gel (*see* **Note 8**).
2. Prestained protein ladder (*see* **Note 9**).
3. Running buffer: 25 mM Tris, 192 mM glycine, 0.1% w/v SDS (*see* **Note 10**). To make 500 mL running buffer, dissolve 1.51 g of Tris base, 7.22 g of glycine, and 0.5 g of SDS in less than 500 mL of dH_2O in a larger beaker equipped with a magnetic stir bar. Stir on a magnetic stirring plate at room temperature until all solids are dissolved. Transfer the solution to a graduated cylinder and fill up to the 500 mL mark with distilled or deionized water. Transfer the solution into a 500 mL glass bottle and store at room temperature for several months.
4. Electrophoresis apparatus.
5. Power supply.

2.3 PPO Activity Staining

1. Staining buffer: 0.1 M sodium phosphate buffer, pH 6.5 (*see* **Notes 2** and **11**).
2. Substrate stock solution: 200 mM dopamine hydrochloride (*see* **Notes 12** and **13**). Dissolve 0.379 g dopamine hydrochloride (MW 189.64 g/mol) in dH_2O using a 10 mL volumetric flask. Prepare fresh for each experiment.
3. Activator: 10% (w/v) SDS (*see* **Notes 14** and **15**). To prepare 10 mL of a 10% (w/v) SDS solution, dissolve 1.0 g of SDS (solid) in 10 mL dH_2O. The solution will be viscous and might require gently heating on a magnetic stir plate.
4. Enhancer solution: 25 mM 4-amino-N,N-diethylaniline sulfate (ADA sulfate), 10 mM HCl (*see* **Note 16**). Weigh out 0.0328 g

of ADA sulfate. Dissolve the solid in less than 5.0 mL of 10 mM HCl in a beaker with magnetic stirring. Transfer the solution to a 5.0 mL volumetric flask and fill up to the 5.0 mL mark with 10 mM HCl. Prepare fresh for each experiment and avoid light exposure.

5. Staining tray.
6. Rocking platform.
7. Camera or imaging system.

3 Methods

Unless otherwise noted, carry out all steps at room temperature. Rinse all glassware with distilled or deionized water before use. Four to five hours are needed to prepare all reagents and crude extracts, perform the electrophoresis run, and stain the gel for PPO activity. The high voltage (~225 V) applied to the gel within the electrophoresis apparatus is a safety hazard. Make sure that the apparatus and all its connections are properly set up.

3.1 Sample Preparation

1. Precool the extraction buffer on ice.
2. Cut or tear the starting tissue into small pieces (e.g., the cap and/or stalk of a white button mushroom).
3. Using a pestle, homogenize ~2 g precut tissue in a mortar with cold extraction buffer. Add the buffer in small portions until you have added 8 or 10 mL, reaching a 1:4 or 1:5 w/v ratio of tissue to extraction buffer. Work swiftly to minimize browning.
4. Transfer most of the liquid from the mortar into an even number of microcentrifuge tubes.
5. Centrifuge at 10,000–15,000 rpm for 10–15 min at 4 °C.
6. Collect the supernatant as crude extract and store what is not processed right away in small 0.1–1 mL portions at −80 °C (*see* **Note 17**).
7. To prepare SDS–PAGE samples, mix the crude extract or control samples with an appropriate amount of loading buffer (i.e., 40 μL of sample with 10 μL 5× sample loading buffer; *see* **Note 5**). Mix these SDS–PAGE samples by pipetting up and down. Do not heat these samples, as the final gel is later stained for enzyme activity!

3.2 SDS–PAGE

1. Remove the comb and tape from the cassette containing the precast SDS–PAGE gel.
2. Rinse the cassette with distilled or deionized water.
3. Place and secure the gel in the electrophoresis apparatus.

4. Fill the running buffer into both chambers of the electrophoresis apparatus. Ensure that the wells of the gel are filled with the running buffer.
5. Rinse the wells of the gel by pipetting the running buffer into the wells of the gel.
6. Load the samples (10–40 μL) and the pre-stained protein ladder (5–10 μL) into the wells of the gel (*see* **Note 18**).
7. Connect the lid with the electrodes to the core unit of the electrophoresis apparatus.
8. Connect the electrodes to the power supply. Switch on the power supply.
9. Set the power supply to a constant voltage of 225 V for 40 min (*see* **Note 19**).
10. Press the start key on the power supply. The resulting current will be approximately 30–40 mA/gel, and "bubbles" start to form in the chamber as the dye front migrates down the gel.
11. At the end of the electrophoresis run, the pre-stained protein ladder should already be visible as separated bands on the gel. Turn off the power, unplug the power supply, and the electrodes before disassembling the electrophoresis apparatus to retrieve the gel cassette.

3.3 PPO Activity Staining

1. Remove the gel from the cassette and let it fall into a tray filled with 190 mL of 0.1 M sodium phosphate buffer, pH 6.5. Incubate the gel for 5 min at room temperature by placing the tray on a rocking platform (*see* **Note 20**).
2. Add 10 mL of a 200 mM dopamine hydrochloride solution, agitate the tray gently, place the tray on a rocking platform, and wait for 5–20 min. If large amounts of active PPO are present, orange-colored bands will appear (Fig. 2a). To activate PPO, add 0.5–1 mL of 10% (w/v) SDS (*see* **Note 14**).
3. Add 5 mL of a 25 mM solution of ADA (*see* **Note 21**). The ADA solution is acidic as it was prepared with 10 mM HCl. The reaction is completed within 5 to 20 min. Previously orange-colored bands will turn blue (Fig. 2b). Bands with very low PPO activity might only be visible after the addition of ADA.
4. Place the gel into a clear plastic tray and take a picture with a camera. If available, the gel can also be placed on an imager for documentation.

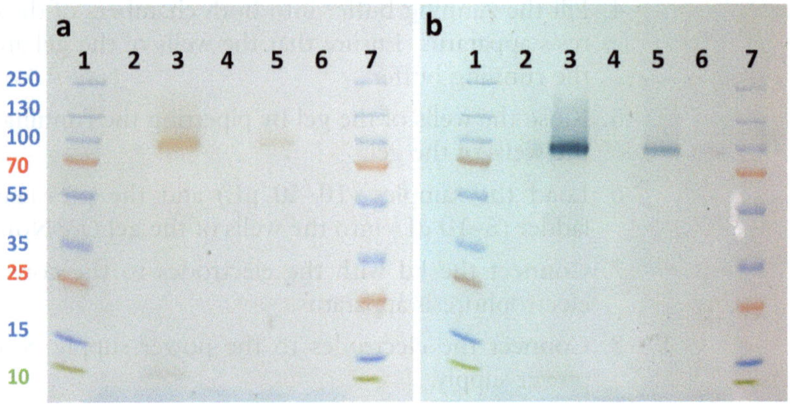

Fig. 2 SDS–PAGE gel (4–20% Tris–glycine) photographed after staining with dopamine (10 mM in 0.1 M sodium phosphate buffer, pH 6.5, panel **a**) and after addition of 4-amino-N, N-diethylaniline (0.6 mM, panel **b**). Lanes 1 and 7: Prestained PageRuler Plus protein ladder. Lanes 2 and 4: Negative control: BGG (5.0 μg protein in each lane). Lanes 3, 5, and 6: Positive control: commercially available tyrosinase preparation (6.0 μg, 0.6 μg, and 0.06 μg protein)

4 Notes

1. Any plant or animal tissue sample containing PPO can be used for this experiment. Mushrooms from the genus *A. bisporus* are known to be rich in PPO, and an enriched protein preparation can be purchased for method testing purposes (*see* **Note 6**).
2. Sodium phosphate buffers within a pH range of 5.8 and 8.0 can also be prepared by mixing two stock solutions of monobasic and dibasic sodium phosphate. The mixing ratio between these two stock solutions determines the final pH value. To prepare 250 mL of a 0.1 M sodium phosphate buffer, pH 6.5, combine 171 mL of 0.1 M NaH_2PO_4 and 79 mL of 0.1 M Na_2HPO_4. If the pH value deviates from the desired value, more monobasic stock solution can be added to lower the pH, and more dibasic stock solution can be added to increase the pH. To make 1 L of 0.1 M NaH_2PO_4 (monobasic stock solution), weigh out 13.8 g of $NaH_2PO_4 \cdot H_2O$ and dissolve it in less than 1 L of dH_2O using a magnetic stir plate. Transfer the solution into a graduated cylinder and fill up to the 1-L mark. To make 1 L of 0.1 M Na_2HPO_4 (dibasic stock solution), weigh out 26.8 g of $Na_2HPO_4 \cdot 7\ H_2O$ and dissolve it in less than 1 L of dH_2O using a magnetic stir plate. Transfer the solution into a graduated cylinder and fill up to the 1-L mark. After transfer into two separate bottles, both stock solutions can be stored for several months at 4 °C. The time invested in preparing two separate stock solutions can pay off later, for example, if different staining conditions are to be tested (*see* **Note 11**).

3. To reduce browning during sample preparation, the extraction buffer is enriched with 2 mM ascorbic acid and 1% (w/v) PVPP. Ascorbic acid acts as a reductant that gradually degrades through air oxidation. At a low concentration, it will not interfere with PPO activity staining. Also, any other small inhibitor molecules present in the plant tissue will be removed from PPO during the electrophoresis procedure. PVPP is a polymer that binds phenolic compounds during extraction. Since PVPP is not water-soluble, the extraction buffer will be a slurry, and the PVPP will be removed with the pellet during the centrifugation step. Other components of the extraction buffer may include protease inhibitors, such as 0.1 mM phenylmethylsulfonyl fluoride (PMSF). Avoid the addition of ethylenediamine tetraacetic acid (EDTA) or other metal chelators, as they might strip the copper ions from the PPO active site.
4. The SDS–PAGE loading buffer does not contain any reducing agent. The reducing agent would interfere with the detection of PPO.
5. It is also possible to use a commercially available loading buffer, for example, 4× Laemmli loading buffer from Bio-Rad Laboratories, Inc. (Hercules, CA, USA; product number #1610747). This product contains no reductant. The ratio of sample to loading buffer needs to be adjusted depending on the concentration of the loading buffer. For SDS–PAGE samples prepared with 4× loading buffer, mix 30 μL sample with 10 μL 4× loading buffer. For SDS–PAGE samples prepared with 5× loading buffer, mix 40 μL sample with 10 μL 5× loading buffer.
6. The commercially available tyrosinase preparation (T3824 from Sigma–Aldrich, St. Louis, MO, USA) is best prepared as a concentrated stock solution and then diluted 10-fold. The total protein content of the stock solution can be determined via a Bradford assay if needed [16].
7. BGG or bovine serum albumin (BSA) can serve as a negative control. Prediluted solutions of BGG and BSA at 1 mg/mL or 2 mg/mL are readily available from various vendors. BSA and BGG are well-characterized, stable proteins that do not have PPO activity. They can help rule out artefacts, such as binding of quinones to proteins other than PPO. Also, filling all lanes of a gel during protein electrophoresis can help ensure even migration and prevent distortion of band patterns.
8. The polyacrylamide gel needs to be compatible with the electrophoresis apparatus. The following gel can be used with the XCell SureLock Mini-Cell: Novex® Tris–Glycine Mini Protein Gel, 4–20%, 1.0 mm, WedgeWell format (Invitrogen, Carlsbad, CA, USA). Purchasing precast SDS–PAGE gels reduce the

health risk of exposure to acrylamide. The number of wells and acrylamide percentage of the gel depend on the experiment and the expected molecular size of the PPO isoforms. Using a gradient gel (4–20% acrylamide) provides for a broad molecular weight range, which is especially useful if the molecular weight range of expected PPO isoforms is not yet known.

9. To avoid additional steps to visualize the protein ladder via Coomassie or silver staining, it is highly recommended to use a pre-stained protein ladder, such as the Thermo Scientific™ PageRuler™ Plus Prestained Protein Ladder, 10 to 250 kDa (product number PI26619 from Thermo Fisher Scientific, Waltham, MA, USA).
10. The running buffer can either be purchased as a 10× stock solution from various vendors (e.g., 10× Tris/Glycine/SDS #1610732 from BioRad) or be prepared in the lab.
11. The pH value of the staining buffer should be acidic and close to the pH optimum of the investigated PPO. Working in a range of pH 5.5–6.5 is a good starting point. Here we use a 0.1 M sodium phosphate buffer at pH 6.5. At neutral to alkaline pH, nonenzymatic background reactions (i.e., autoxidation of diphenolic compounds) occur alongside PPO activity, producing streaks or diffuse banding patterns on the gel.
12. Autoxidation causes dopamine solutions to change color from clear to pink within several hours and to brown overnight. The dopamine solution is best prepared while performing electrophoresis, as it should be used fresh. The final concentration of dopamine in the staining tray will be 10 mM after adding the full 10 mL of the 0.2 M substrate stock solution to 190 mL of staining buffer. Lower final dopamine concentrations, 5 mM, for example, also give good results.
13. Alternative PPO substrates, such as L-DOPA and catechol, can also be used. However, not all substrates are as soluble as dopamine hydrochloride (~500 mg/mL in water) and catechol (~450 mg/mL in water). L-DOPA, for example, has a water solubility of only ~5 mg/mL. If a substrate solution can only be prepared in lower concentrations (1–25 mM), the staining method needs to be adjusted. For renaturation, the gel is first incubated in 200 mL buffer. The buffer is then removed and exchanged for a substrate solution, which was prepared in the same buffer.
14. PPO is activated by the anionic detergent SDS. The addition of SDS can enhance color development and decrease the latency of monophenolase reactions. The optimum concentration of SDS, often reported to lie between 0.5 and 3 mM, depends on the PPO source. Adding 1 mL of a 10% (w/v) SDS solution

Table 1
Pipetting scheme to record loading of SDS–PAGE gel

Lane/well	1	2	3	4	5	6	7	8	9	10
Sample										
Loading volume (μL)										

results in a final concentration of 0.05% (w/v) SDS, equivalent to 1.7 mM SDS in the staining tray.

15. SDS solutions (10% w/v) can be kept at room temperature for several months. If the 10% (w/v) SDS solution is stored cold, precipitation of SDS occurs, and the solution needs to be reheated.

16. The diluent to prepare the ADA solution is 10 mM HCl. Typically, 1 M HCl is available in a Chemistry lab. To dilute this mineral acid, add less than 100 mL of dH_2O into a 100 mL volumetric flask. Pipette 1 mL of 1 M HCl into this flask. Use dH_2O to fill up to the 100 mL mark. Swirl the flask gently to ensure the solution is mixed. This 10 mM HCl solution can be stored at room temperature for many months. The ADA solutions, however, cannot be stored and are best prepared during the electrophoresis run. ADA solutions turn pink if exposed to light.

17. Performing a Bradford assay with a portion of the crude extract is highly recommended to keep track of the total amount of protein that is loaded onto the gel [16].

18. The protein ladder can be loaded into the first lane as well as into another lane on the other side of the gel to monitor for uneven sample progression during the electrophoresis run. It is best to load a gel in a non-symmetric manner, so that the orientation of the gel can always be identified even if the transparent gel was flipped while staining. Keeping a record of the pipetting scheme (*see* Table 1) is essential.

19. A high voltage setting of 225 V can be used for most commercially available precast gels. High voltage leads to shorter running times, but can also heat up the gel, resulting in gel deformation and uneven migration. For a voltage setting of 125 V, the running time increases to 90 min.

20. Soaking the gel after SDS–PAGE in an SDS-free buffer is important for the renaturation of PPO. Immediately before loading, the sample contains 2% SDS ($\approx$70 mM), whereas the running buffer contains 0.1% SDS ($\approx$3.5 mM). Optimal PPO activity is usually observed at lower SDS levels ($\approx$0.5–3 mM).

21. In the standard protocol, the ADA enhancer solution is directly added to the staining solution containing the PPO substrate and the staining buffer. To reduce any background staining, a washing step can be introduced. The first staining solution, containing buffer, the PPO substrate, and the optional activator SDS, is decanted, and a fresh portion of 200 mL buffer is poured into the staining tray before adding the ADA enhancer solution.

Conflicts of Interest

The authors have no conflicts of interest to declare that are relevant to the content of this chapter.

References

1. Laemmli UK (1970) Cleavage of structural proteins during the assembly of the head of bacteriophage T4. Nature 5259:680–685. https://doi.org/10.1038/227680a0
2. Gallagher SR (2008) SDS-Polyacrylamide Gel Electrophoresis (SDS-PAGE). Curr Protoc Essent Lab Tech:7.3.1–7.3.25. https://doi.org/10.1002/9780470089941.et0703s00
3. Winkler C, Denker K, Wortelkamp S, Sickmann A (2007) Silver-and Coomassie-staining protocols: detection limits and compatibility with ESI MS. Electrophoresis 28:2095–2099. https://doi.org/10.1002/elps.200600670
4. Bischoff KM, Shi L, Kennelly PJ (1998) The detection of enzyme activity following sodium dodecyl sulfate polyacrylamide gel electrophoresis. Anal Biochem 260:1–17. https://doi.org/10.1006/abio.1998.2680
5. Moore BM, Flurkey WH (1990) Sodium dodecyl sulfate activation of a plant polyphenoloxidase. Effect of sodium dodecyl sulfate on enzymatic and physical characteristics of purified broad bean polyphenoloxidase. J Biol Chem 265:4982–4988. https://doi.org/10.1016/S0021-9258(19)34072-4
6. Espín JC, Wichers HJ (1999) Activation of a latent mushroom (*Agaricus bisporus*) tyrosinase isoform by sodium dodecyl sulfate (SDS). Kinetic properties of the SDS-activated isoform. J Agric Food Chem 47:3518–3525. https://doi.org/10.1021/jf981275p
7. Mayer AM (2006) Polyphenol oxidases in plants and fungi: going places? A review. Phytochemistry 67:2318–2331. https://doi.org/10.1016/j.phytochem.2006.08.006
8. Zou H, Xiao Q, Li G, Wei X, Tian X, Zhu L, Ma F, Li M (2025) Revisiting the advancements in plant polyphenol oxidases research. Sci Hortic 341:113960. https://doi.org/10.1016/j.scienta.2025.113960
9. Sui X, Meng Z, Dong T, Fan X, Wang Q (2023) Enzymatic browning and polyphenol oxidase control strategies. Curr Opin Biotechnol 81:102921. https://doi.org/10.1016/j.copbio.2023.102921
10. Tilley A, McHenry MP, McHenry JA, Solah V, Bayliss K (2023) Enzymatic browning: the role of substrates in polyphenol oxidase mediated browning: mechanisms of enzymatic browning. Curr Res Food Sci 7:100623. https://doi.org/10.1016/j.crfs.2023.100623
11. Rodriguez-Lopez JN, Escribano J, Garcia Canovas F (1194) A continuous spectrophotometric method for the determination of monophenolase activity of tyrosinase using 3-methyl-2-benzothiazolinone hydrazone. Anal Biochem:205–212. https://doi.org/10.1006/abio.1994.1026
12. Rescigno A, Sanjust E, Pedulli GF, Valgimigli L (1999) Spectrophotometric method for the determination of polyphenol oxidase activity by coupling of 4-tert-butyl-o-benzoquinone and 4-amino-N,N-diethylaniline. Anal Lett 32:2007–2017. https://doi.org/10.1080/00032719908542948
13. Rescigno A, Sollai F, Rinaldi AC, Soddu G, Sanjust E (1997) Polyphenol oxidase activity staining in polyacrylamide electrophoresis gels. J Biochem Biophys Methods 34:155–159. https://doi.org/10.1016/S0165-022X(96)01201-8
14. Lopez-Tejedor D, Palomo JM (2018) Efficient purification of a highly active H-subunit of tyrosinase from *Agaricus bisporus*. Protein

Expr Purif 145:64–70. https://doi.org/10.1016/j.pep.2018.01.001

15. Ismaya WT, Rozeboom HJ, Weijn A, Mes JJ, Fusetti F, Wichers HJ, Dijkstra BW (2011) Crystal structure of *Agaricus bisporus* mushroom tyrosinase: Identity of the tetramer subunits and interaction with tropolone. Biochemistry 50:5477–5486. https://doi.org/10.1021/bi200395t
16. Bradford MM (1976) A rapid and sensitive method for the quantitation of microgram quantities of protein utilizing the principle of protein-dye binding. Anal Biochem 72:248–254. https://doi.org/10.1016/0003-2697(76)90527-3

Chapter 14

Integrating Maker Technology with Research: DIY Continuous Flow Colorimetric Detector with Integrated Data Acquisition and Processing

Adam Ee Xian Loh and Kheng Soo Tay

Abstract

Continuous flow analysis (CFA) offers a robust platform for real-time chemical monitoring and analysis, but the high cost and complexity of commercial systems limit their usage. This chapter describes a practical, low-cost, and reproducible approach for developing a continuous flow colorimetric detection system using open-source electronics, 3D-printed components, and custom data processing software. The setup incorporates a peristaltic pump, an Arduino-controlled solenoid valve, and an AS7341 visible light sensor aligned with a flow-through cuvette. A key innovation of this system is the integration of a Python-based algorithm to process real-time absorbance data and automatically identify and exclude interference-induced signal artifacts. This method enables low-cost detection of colorimetric changes in flowing samples and is particularly suited for biological and chemical analyses involving colorimetric detection, as well as educational settings. The system's modular design allows for easy customization, and its reproducibility makes it useful for dissemination across teaching laboratories or low-resource research environments.

Key words Flow injection analysis, Colorimetric detection, Arduino, AS7341 sensor, Solenoid valve, Peristaltic pump, Bubble correction, Open-source hardware, DIY analytical instruments

1 Introduction

Flow injection analysis (FIA) is an automated analytical technique that streamlines chemical and biomolecular detection, offering chemists and biologists a powerful tool for high-throughput analysis. In the biological field, FIA is used in immunoassays [1, 2], enzymatic assays [3–5], DNA analysis [6], and monitoring of biological processes [7, 8]. In chemistry, FIA is widely used for a variety of applications, particularly in environmental [9, 10], food [11], and pharmaceutical analysis [12, 13]. In FIA, a sample is injected into a continuously flowing carrier stream, where it reacts with reagents to induce a physicochemical change, such as a color shift or electrochemical signal. This change is measured in real-time

Robert E. Dempski (ed.), *Transforming Molecular Biology with Emerging Technologies*, Methods in Molecular Biology, vol. 3043, https://doi.org/10.1007/978-1-0716-5320-3_14,

as the sample passes through a detector, enabling precise quantification of the analyte.

Photometric (colorimetric) detection is particularly well-suited to FIA due to its simplicity, broad applicability, and compatibility with low-cost hardware [14]. In recent years, the emergence of open-source, do-it-yourself (DIY) hardware platforms has lowered the barrier to entry for various system development, empowering laboratories to construct affordable, customizable instruments for diverse applications [15]. However, most recent advancements in DIY-based FIA have centered on microfluidic chips and low-cost pumping systems [16, 17] while the development of equally accessible, sensitive photometric detectors and integrated data processing software has lagged behind. This gap constrains the broader utility of FIA in field-deployable and resource-limited settings. In this chapter, we describe a practical and reproducible approach to building a low-cost continuous-flow colorimetric detector, coupled with custom software for real-time signal acquisition and analysis. One persistent challenge in FIA systems is the formation of air bubbles during reagent or sample flow, which can introduce signal artifacts and compromise measurement reliability [18]. Commercial systems often address this issue with advanced degassing modules [19], which are impractical for low-cost setups. In this work, we present a software-integrated solution that detects and accounts for signal disruptions caused by bubbles, enabling more accurate and robust data acquisition even in basic hardware configurations.

2 Materials

2.1 Chemicals and Standard Solutions

1. All reagents should be of analytical grade unless otherwise specified. Prepare all solutions using distilled or deionized water. The following reagents are required for detector calibration and cleaning.
2. Accurately weigh 0.200 g of methylene blue and dissolve it in 1 L of deionized water using a volumetric flask to prepare a 200 ppm stock solution. From this stock, prepare working standards by pipetting 12.5, 25.0, 37.5, 50.0, and 75.0 mL into separate 100 mL volumetric flasks and diluting to the mark with deionized water to obtain standard concentrations of 25, 50, 75, 100, and 150 ppm, respectively (*see* **Note 1**).
3. Deionized Water—Used for reagent preparation, dilution, and system flushing.
4. 70% Isopropyl Alcohol—Used for cleaning optical surfaces and degreasing flow components if needed.

2.2 Fluidic Flow System

1. DC 4.5 V mini micro-one-way normal close solenoid air valve (*see* **Note 2**).
2. 17HS4401 Bipolar Stepper Motor.
3. ZK-SMC01 stepper motor controller.
4. TB6600 stepper motor driver.
5. Capillary silicone tube transparent (2 mm I.D. × 4 mm O.D.).
6. Polypropylene Y-connector (Connector size: 3.5 mm).
7. Polypropylene spring-type one-way valve (Connector size: 3 mm).
8. Ball bearing (3 × 8 × 4 mm).
9. Bolts and nuts (M2.5 × 0.35).
10. DC power supply (12 V).

2.3 Optical Detection System

1. AS7341 11-channel visible light sensor (*see* **Note 3**).
2. Kingbright 5 mm through-hole common cathode RGB LED (forward voltages: red 2.5 V, green 4.0 V, and blue 4.1 V. Peak wavelengths for red, green, and blue are 645, 515 nm, and 460 nm).
3. Flow Cell Cuvette: Quartz flow-through cuvette (volume: 0.018 mL; dimensions: 45 × 12.5 × 12.5 mm; optical path length: 10 mm; internal (flow path width: 1.5 mm).
4. Arduino UNO R3 board (ATmega328P-based).
5. Breadboard and jumper wires.
6. Type-C USB cable for Arduino boards.
7. External 9 V 1A DC power adapter.

2.4 3D Printing

1. Polylactic acid (PLA) 3D filament.
2. Engineering photopolymer UV SLA resin.
3. CAD software (such as Autodesk Fusion 360).
4. Ultimaker Cura 5.5.0 or any slicer software.
5. FDM 3D printer.
6. SLA 3D printer.

2.5 Software, Libraries, and Scripts

1. Arduino IDE—Used for uploading control logic to the Arduino UNO. Compatible with Windows, macOS, and Linux.
2. Python IDE—Any Python development environment (e.g., Thonny, VS Code, or command-line interpreter) can be used to run data acquisition and processing scripts.
3. AS7341 Library—Enables I2C communication with the AS7341 sensor module.
4. Pandas Library—For data handling and manipulation.

5. Pyserial Library—For serial communication with Arduino.
6. Matplotlib Library—For real-time plotting and visualization.
7. *CFA.ino*—Arduino script for controlling the RGB LED, solenoid valve, and AS7341 data acquisition.
8. data_acquisition.py—Python script for real-time reading and visualization of spectral intensity.
9. data_extraction.py—Python script for processing time-based absorbance data and extracting peak values.

3 Method

3.1 Fabrication of the Cuvette Holder for Optical Flow Cell Integration

The cuvette holder is designed to securely hold a flow-through cuvette and ensure stable optical alignment between the light source and light sensor.

1. Use CAD software to create a cuvette holder model with the following features (Fig. 1):
   (a) A recessed slot to accommodate the 10 mm pathlength cuvette.

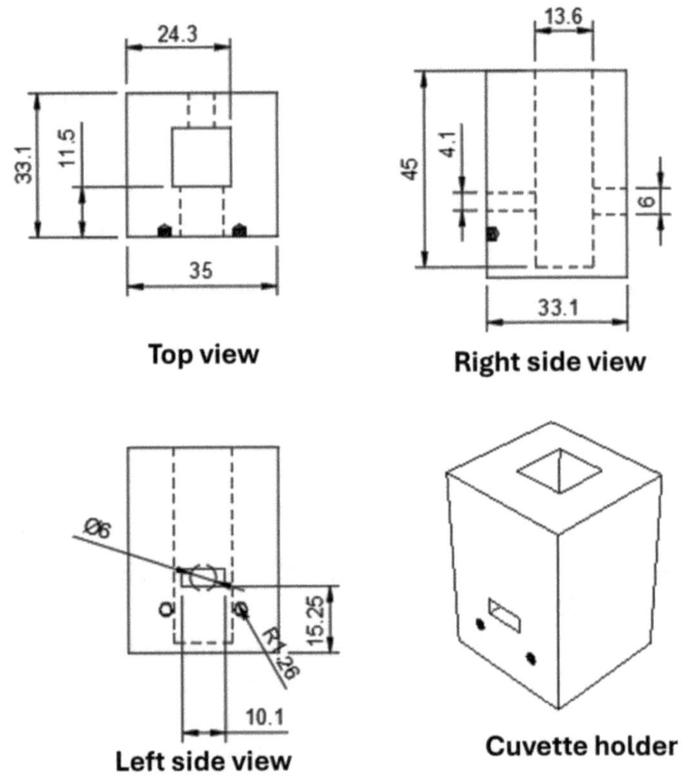

Fig. 1 CAD rendering of the 3D-printed cuvette holder with optical windows and mounting positions

(b) Two opposing optical windows for horizontal light transmission.

(c) Mounting for the RGB LED (light source) and AS7341 visible light sensor, aligned to the cuvette's optical axis.

2. Export the CAD model to a slicer-compatible file format (e.g., .*stl*) and load the file into a slicing software and slice. Use the following main print settings:

   (a) Nozzle diameter: 0.2 or 0.4 mm.

   (b) Layer height: 0.2 mm.

   (c) Infill density: 30% for mechanical stability.

3. 3D Print the design using black PLA filament to minimize stray light and improve signal-to-noise ratio during optical measurements.

4. Insert the cuvette into the holder. Ensure a snug fit; if excessive resistance is observed, modify the design tolerances and reprint as needed.

5. Attach the RGB LED and AS7341 light sensor to the designated mounting. Ensure that the LED beam travels horizontally through the cuvette's optical window and is centered on the AS7341 sensor aperture.

3.2 Fabrication of DIY Peristaltic Pump

The peristaltic pump is designed to deliver a flow of reagent or sample by mechanically compressing flexible tubing using a rotating rotor fitted with ball bearings.

1. Design the DIY peristaltic pump component, which consists of the pump casing (Fig. 2) and rotor (Fig. 3), using CAD software.

2. Print the designed components with an SLA 3D printer using engineering-grade UV SLA resin. Set the layer height to 50 μm (*see* **Note 5**).

3. Assemble the rotor by securing 4 units of ball bearings using a screw onto the rotor disc. Attach the rotor to the shaft of a stepper motor and secure it using a screw (Fig. 4).

4. Assemble the casing and the rotor on the stepper motor. Ensure the inner cavity allows smooth rotation of the bearing-loaded rotor.

5. Set up the circuit controller of the stepper motor using the TB6600 stepper motor driver, ZK-SMC01 stepper motor controller, and a 12 V power supply. Wire the system as shown in Fig. 5:

   (a) Connect ENA–, ENA+, DIR+, and PUL+ of the TB6600 stepper motor to GND, EN, DIR, and PUL of ZK-SMC01.

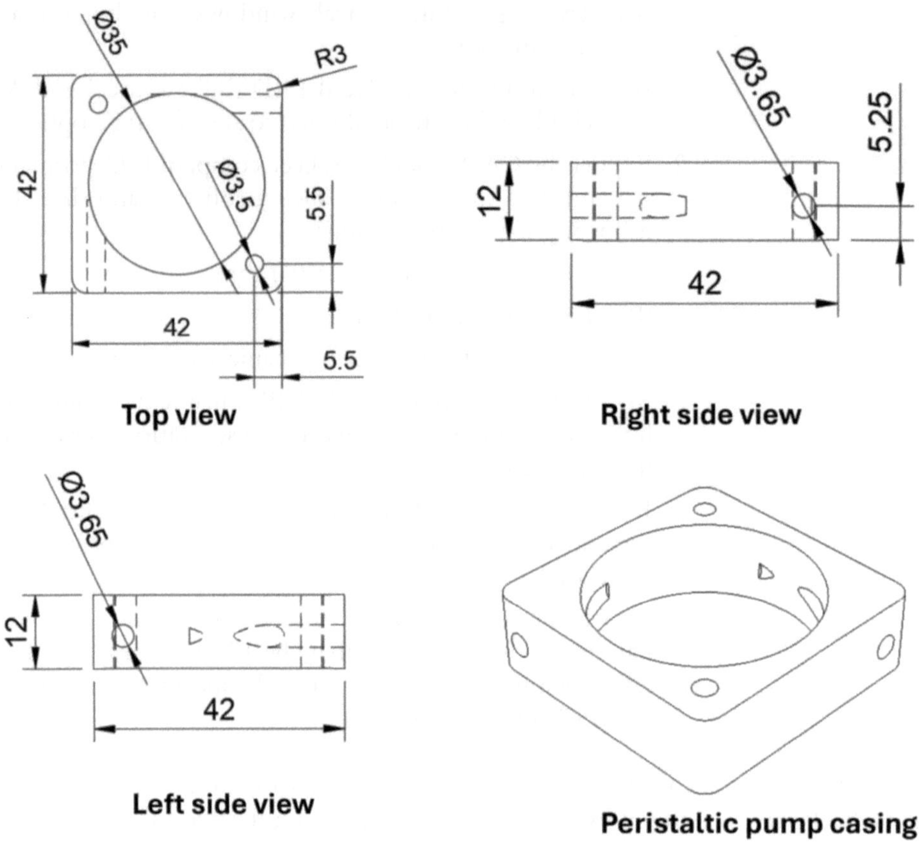

Fig. 2 Model and dimensions of the casing of the DIY peristaltic pump. (unit = mm)

(b) Connect ENA–, DIR–, and PUL– on the TB6600 stepper motor driver together with wires.

(c) For the connection of the stepper motor to the TB6600 stepper motor driver, connect one coil pair to A+/A– and the other to B+/B–.

(d) Connect the 12 V positive to VCC and the negative to GND of the TB6600 stepper motor driver.

(e) Power on and adjust the ZK-SMC01 speed knob. Swap one coil pair if the motor vibrates but does not rotate.

6. Test the rotor motion by connecting the stepper motor to the motor controller and verifying consistent rotation. Check for uniform flow and absence of tubing slippage. If the rotor does not rotate smoothly or fails to compress the tubing adequately, inspect for misalignment, loose bearings, or mechanical friction. Disassemble and reassemble the rotor components as needed to ensure proper function. Modify the design and reprint if needed.

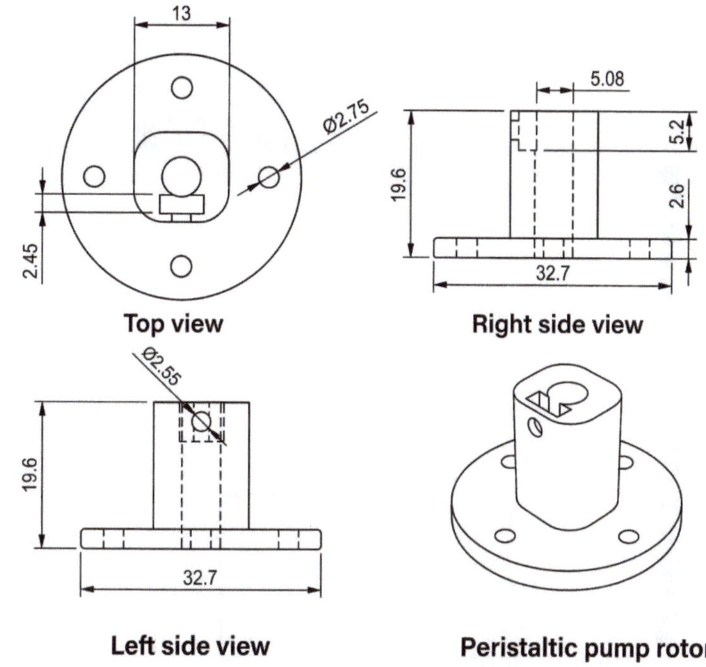

Fig. 3 Model and dimensions of the rotor of the DIY peristaltic pump (unit = mm)

Fig. 4 Assembled DIY peristaltic pump

3.3 Assembly of the Fluidic Flow System

The fluidic flow system consists of sample and reagent inlet lines, a mixing coil, a peristaltic pump, and an optical detection system (Fig. 6). Follow the steps below for proper assembly:

1. Secure the cuvette into the 3D-printed cuvette holder (as detailed in Subheading 3.1). Ensure that the optical window of the cuvette is properly aligned with the RGB LED (light source) on one side and the AS7341 light sensor on the opposite side.
2. Construct the fluidic inlet system using silicone tubing. Connect the following components:

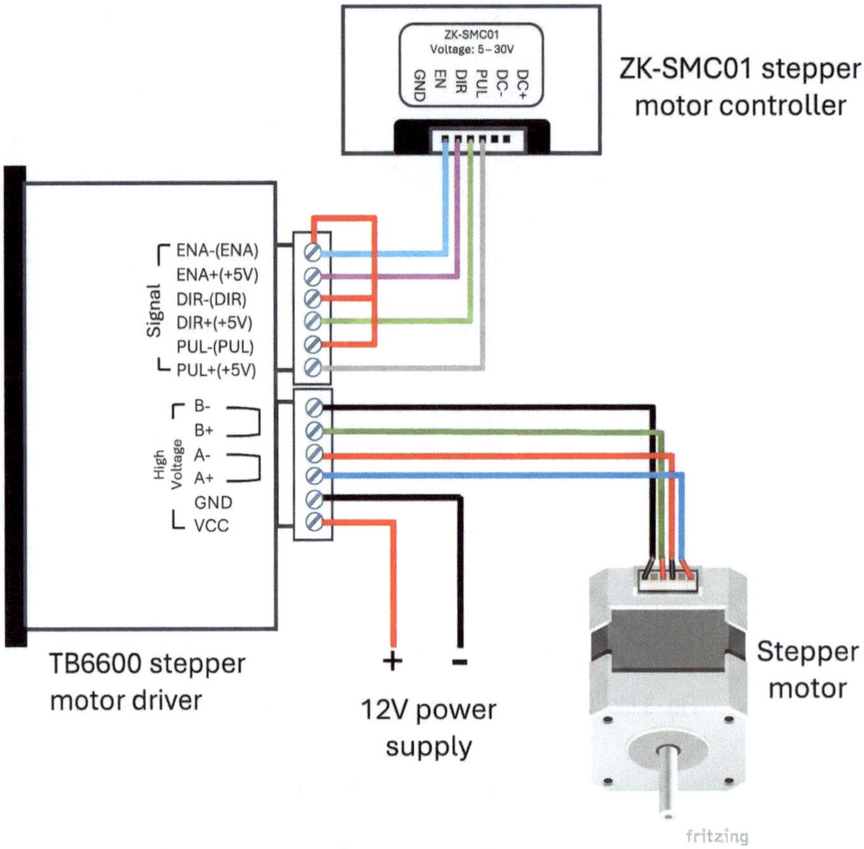

Fig. 5 Wiring diagram showing the connection of the stepper motor controller to the TB6600 driver, stepper motor, and 12 V DC power supply

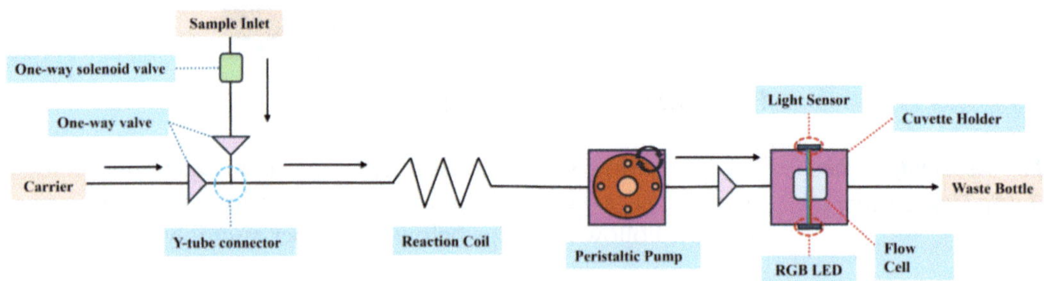

Fig. 6 Schematic of the integrated flow system

(a) One silicone tubing from a sample reservoir to a solenoid valve as an inlet.

(b) One silicone tubing from a reagent or carrier reservoir.

Each line should be routed through a one-way valve to prevent backflow and to a Y-connector that merges into a

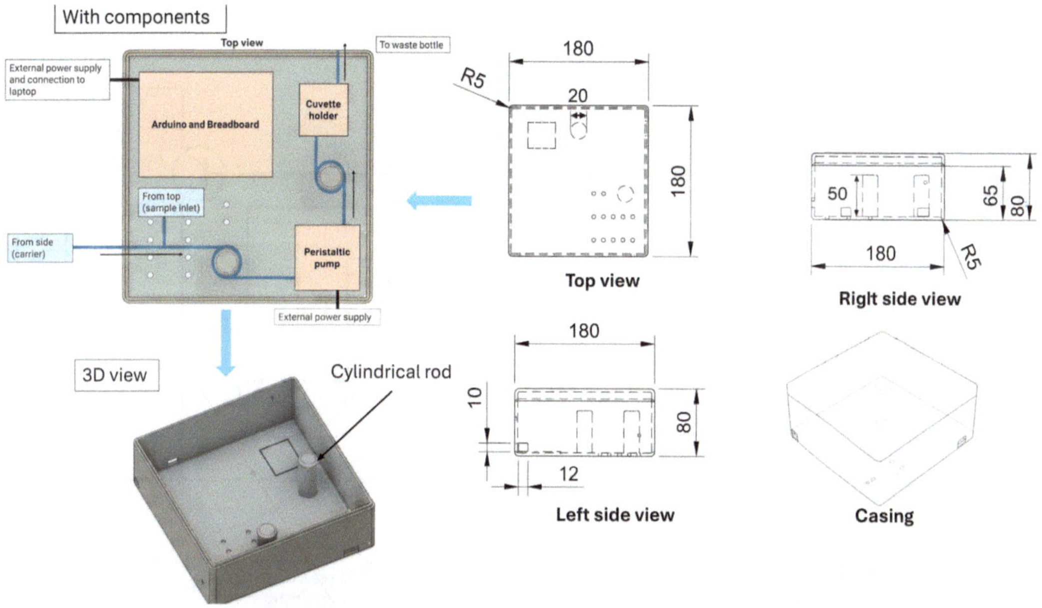

Fig. 7 Customized box with portable lid to assemble all the components. (unit = mm)

single channel (Fig. 6). The solenoid valve will be electronically controlled (*see* **Note 6**) to allow the sample solution to flow unidirectionally.

3. From the Y-connector outlet, direct the fluid into a 20 cm length of silicone tubing wrapped around a 3D-printed cylindrical rod to form a helical mixing coil (Fig. 7).
4. Insert the outlet of the reaction loop into the DIY peristaltic pump (Fig. 4). Ensure that the tubing is properly aligned within the pump's rotor assembly to avoid slippage or pinching.
5. Attach the outlet tubing from the peristaltic pump to the inlet of the cuvette. Ensure the connection is tight and leak-free.
6. Attach another silicone tubing to the cuvette outlet and direct it into a sealed waste container. This ensures safe disposal of reagents if needed.
7. Assemble and enclose all components, including the Arduino board, cuvette holder, tubing, and peristaltic pump, within a custom-designed box to enhance portability and ease of use. This setup allows for a more compact and organized system. An example of the fabricated custom-designed box is shown in Fig. 7.

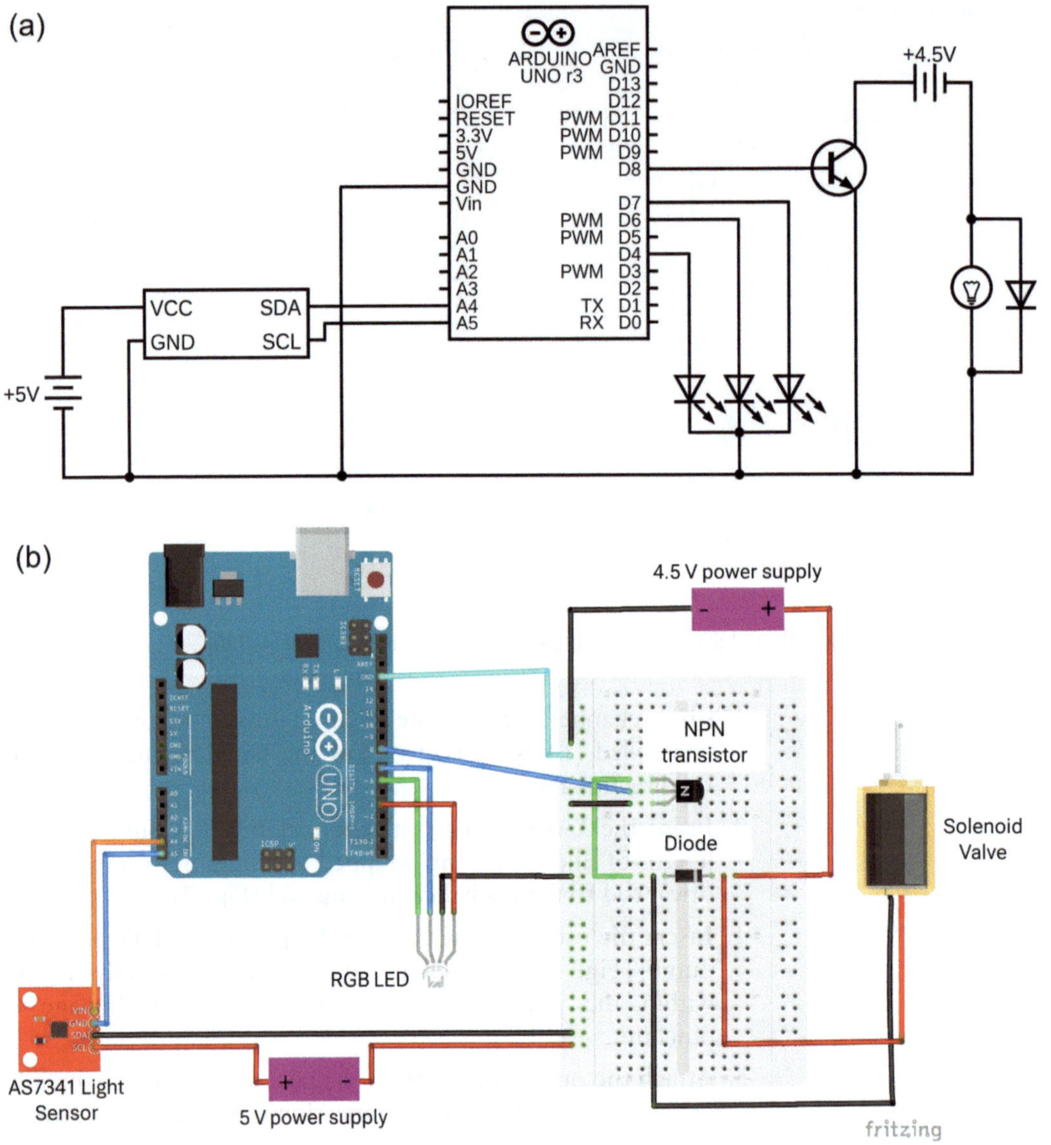

Fig. 8 Circuit for RGB LED, AS7341 visible light sensor, and solenoid valve (**a**) Schematic diagram and (**b**) Circuit diagram

3.4 Setting Up Data Acquisition Via AS7341 Light Sensor and Arduino Uno Board

1. Connect the SDA, SCL, VCC, and GND pins of the AS7341 light sensor to the A4, A5, 5 V, and GND pins of the Arduino Uno board, respectively, using jumper wires (Fig. 8). By default, the F8 channel (670–690 nm) of the AS7341 light sensor is activated via the *CFA.ino* script for the development and verification stage. This channel can be changed according to specific needs (*see* **Note** 7).
2. Connect the red, green, and blue anodes of the RGB LED (pins 1, 4, and 3, respectively) to digital pins D4, D6, and D7 on the Arduino Uno board via jumper wires. Connect the

common cathode (pin 2) to GND on the Arduino Uno board. By default, full red illumination is enabled using the *CFA.ino* script. The color of the emitted light of RGB can be changed according to specific needs (*see* **Note 8**).

3. Connect the positive terminal of the solenoid valve to the positive rail of the external 4.5 V power supply.
4. Connect the negative terminal of the solenoid valve to the collector (middle pin) of the 2 N2222 transistor.
5. Connect the emitter pin of the transistor to the GND rail on the breadboard.
6. Insert a 1 N4007 diode across the solenoid valve terminals for flyback protection, with the cathode (striped end) connected to the positive terminal and the anode connected to the negative terminal (collector side).
7. Connect the base of the transistor to the digital output pin D8 on the Arduino board. This pin will control the valve's ON/OFF state.
8. Ensure all components are connected via a solderless breadboard, allowing clean distribution of power and logical connections.
9. Connect the Arduino Uno board to a computer via a Type-C USB cable and upload the code via the *CFA.ino* script (*see* **Note 9**).
10. Adjust the solenoid valve activation parameters in the Arduino sketch as needed (*see* **Note 10**). This allows you to control the time between activations and the duration for which the valve remains open.

3.5 Priming and Initialization of the Fluidic System

Prior to analysis, the entire fluidic pathway must be primed to eliminate air bubbles and ensure uninterrupted flow through the detection cell.

1. Ensure all tubing connections are secured, and the cuvette is properly inserted in the holder.
2. Fill both the sample and carrier reservoirs with deionized water for initial system flushing.
3. Set the initial peristaltic pump speed using the ZK-SMC01 controller. A 20 mL/min flow rate is recommended.
4. Activate the peristaltic pump to initiate flow. On the Arduino serial monitor interface (*see* **Note 11**), enter the command "Purge" (*see* **Note 12**) to activate the solenoid valve for 2 min, allowing deionized water to flow through both the sample and carrier lines. Continue priming until no visible bubbles remain in the lines and cuvette tubing. Repeat the purging process as necessary.

5. Once the system is primed and the cuvette is filled with water, verify that no bubbles are trapped in the optical path.
6. On the serial monitor interface, monitor the intensity readings. Once the readings are stable, the system is now physically ready for electronic control and optical measurement.

3.6 Data Acquisition and Calibration

1. Immerse the sample inlet tubing in the vial containing the standard methylene blue solution with the lowest concentration (25 ppm). Allow the standard solution to fill up the sample line (*see* **Note 13**) before running the *data_acquisition.py* script in a Python IDE.
2. Allow the program to capture five consecutive stable intensity readings and establish the time-based absorbance plot.
3. Allow the standard solution to flow into the system for 3 min (*see* **Note 14**) and monitor the absorbance plot.
4. Transfer the sample inlet tubing to another vial or container containing deionized water for 2 min to create a buffer zone between the current and subsequent standard solutions (*see* **Note 15**).
5. Pass on to the next standard solution with a higher concentration (50 ppm).
6. Repeat **steps 3–5** for all standard solutions to obtain the data required (Fig. 9) for constructing the calibration curve.
7. Once the analysis of all standard solutions is completed, replace the standard solutions with the sample solutions and repeat **steps 3–5** to collect absorbance data for the unknown samples.
8. After all target solutions have passed through the system and the corresponding absorbance plots are generated, power off the Arduino UNO to terminate data acquisition. Close the data visualization interface displaying the time-based absorbance graph. A comma-separated values file (.csv) containing the absorbance readings will be automatically generated in the same folder as the script, and the data is ready for processing.

3.7 Data Processing

1. Create a .csv file that specifies the time intervals corresponding to each concentration of standard and sample solutions (*see* **Note 16**).
2. Execute *data_extraction.py* script (*see* **Note 17**) to process the time-based absorbance data. A new .csv file containing the corresponding concentration versus absorbance values (*see* **Note 18**) will be automatically generated in the same folder as the script.
3. With the absorbance values extracted, a calibration curve can be plotted by graphing absorbance against concentration for the standard solutions (Fig. 9). This curve can then be used to

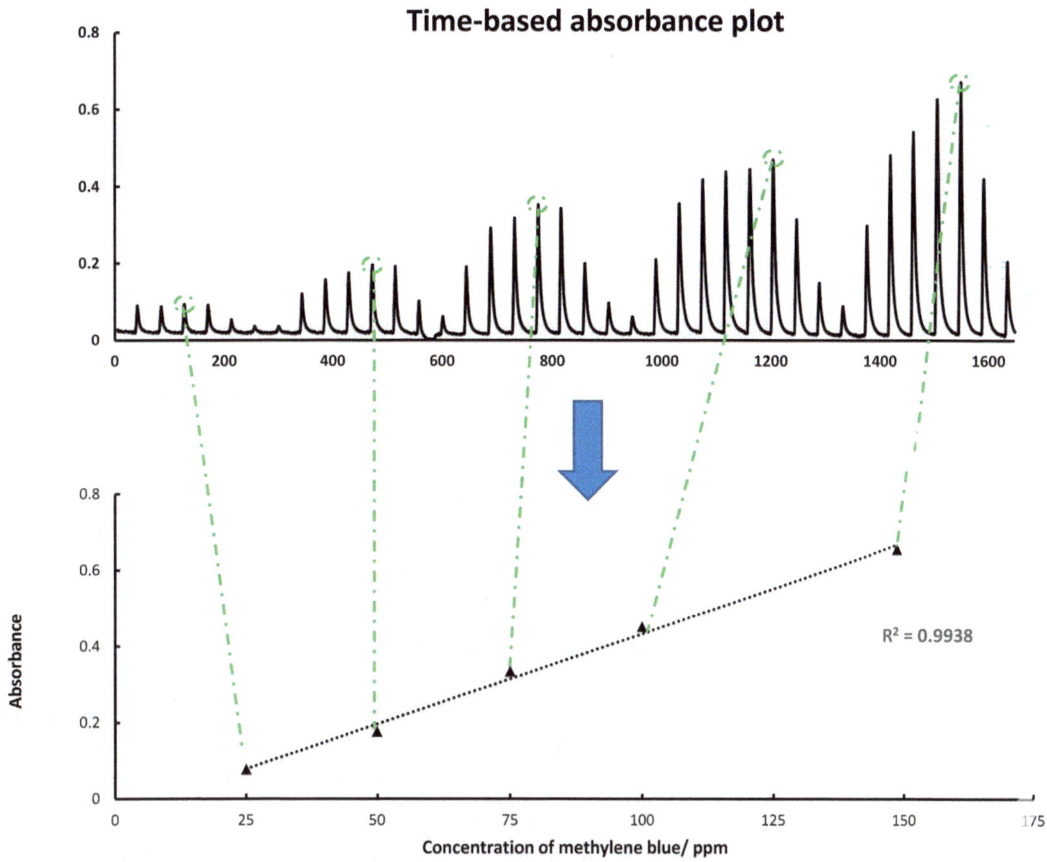

Fig. 9 Illustration of calibration curve using the time-based absorbance plot via Python

determine the concentration of an unknown sample by locating its absorbance value on the curve and finding the corresponding concentration.

4 Notes

1. Methylene blue is used for the detector development and verification due to its well-defined colorimetric response and suitability for visible light detection.
2. The solenoid valve has inlet and outlet diameters of 3 mm and 3.5 mm, respectively (Fig. 10). It is normally closed; when powered, the inlet and outlet become connected, allowing the fluid flow.
3. The AS7341 visible light sensor features eight visible channels, one near-IR channel, and one unfiltered channel. Visible channels include F1 (405–425 nm), F2 (435–455 nm), F3 (470–490 nm), F4 (505–525 nm), F5 (545–565 nm), F6 (580–600 nm), F7 (620–640 nm), and F8 (670–690 nm).

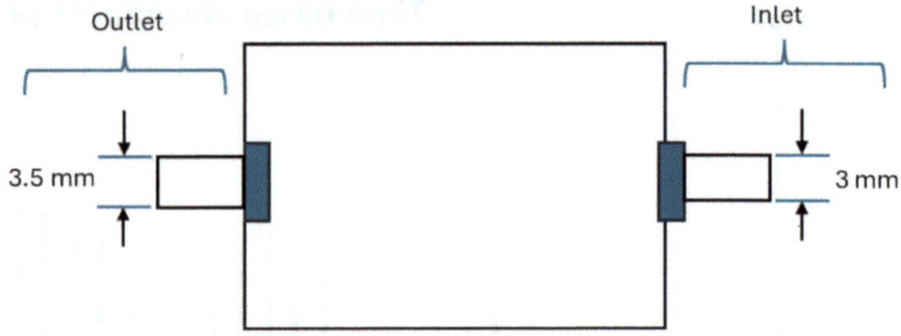

Fig. 10 Schematic diagram of the one-way solenoid valve

For targeted detection, any of these channels can be selectively activated, as shown in **Note 6**.

4. All required libraries and scripts for system operation, data visualization, and analysis are available via the GitHub repository (https://github.com/khengsoo/DIY-Flow-Photometric-Analyzer). The AS7341 Arduino library enables full communication and functionality with the sensor.

5. PLA filament is not suitable for peristaltic pump components as the PLA has lower heat resistance and often softens when exposed to heat. Heat is generated during peristaltic pump operation. More durable engineering-grade resin is recommended. SLA resin printing is recommended as these components require higher precision.

6. The Arduino triggers the solenoid valve via the *CFA.ino script*. By default, the valve opens for 1 s every 45 s. This allows controlled injection of the sample solution. To adjust sample introduction timing, refer to **Note 8**.

7. A specific AS7341 light sensor channel can be changed in the *CFA.ino* script using the following code:

```
as7341.startMeasure(as7341.eF5F8ClearNIR);
data2 = as7341.readSpectralDataTwo();
intensity = data2.ADF8;
```

To switch to another channel (e.g., F1), modify the code as follows:

```
as7341.startMeasure(as7341.eF5F8ClearNIR);
data2 = as7341.readSpectralDataTwo();
intensity = data2.ADF1;
```

8. The RGB LED output can be modified in *CFA.ino* script using *setColor(R, G, B)* function, where each value (0–255) represents the red, green, and blue intensity. For example:

(a) *setColor(255, 0, 0)* for red
(b) *setColor(0, 255, 0)* for green
(c) *setColor(0, 0, 255)* for blue
(d) *setColor(0, 255, 255)* for cyan (green + blue)

9. The CFA.ino script controls the RGB LED, AS7341 sensor, and solenoid valve. It operates the valve periodically (default: 1 s ON for every 45 seconds) and continuously records light intensity from the F8 channel. A manual purge mode is also included. By typing *purge* in the Serial Monitor, it opens the valve for 2 min to flush the system.
10. Solenoid valve activation parameters are controlled in the sketch via:

```
const long totalInterval = 45000; // time between activations
const long openDuration = 1000; // valve ON duration
```

Adjust these values to match flow requirements.

11. The Arduino Serial Monitor can be opened via Tools > Serial Monitor in the Arduino IDE.
12. Typing purge in the Serial Monitor activates continuous flow mode for 2 minutes. The purge duration is set by:

```
const long purgeDuration = 120000;
```

Modify this value to change the flush time if needed.

13. Ensure that the sample (methylene blue solution) fills the inlet tubing before script initiation. This step ensures precise alignment between timing and flow introduction. If the line contains only water, delayed signal response and data misalignment may occur.
14. Adjust the flow time for each standard based on observed absorbance behavior. The default is 3 min, but actual timing can be optimized via real-time monitoring.
15. A buffer zone (e.g., deionized water) between standards helps prevent carryover. It also creates a visible baseline dip in the absorbance–time plot, marking the transition between samples (Fig. 11).
16. To create a .csv file, open any text editor (like Notepad) or a spreadsheet tool (like Microsoft Excel). If using Notepad, type your data with commas separating each value (e.g., Concentration, start_time, and end_time). Then save the file with a .csv extension, such as concentration_range.csv. If using Excel, enter your data into rows and columns, then go to File > Save As, and choose CSV (Comma-delimited) (*.csv) as the

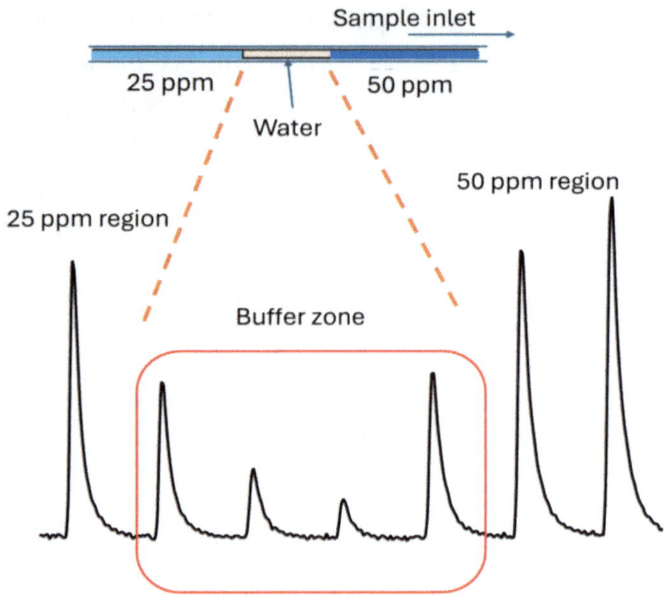

Fig. 11 Illustration of the buffer zone on the absorbance–time plot

Table 1
Example input CSV file used to define time-steps for absorbance extraction

Concentration	start_time	end_time
25	1	300
50	301	600
75	601	900
100	901	1200
150	1201	1500
200	1501	1800

file format. The .csv file defines the time intervals (time-steps) corresponding to each standard and sample concentration, enabling the Python script to map segments of the absorbance–time data to their respective solutions. This facilitates accurate extraction of absorbance values during data processing. Table 1 presents an example of a time-step configuration for a protocol consisting of 3 min of standard solution flow followed by 2 min buffering interval in csv file.

17. Ensure that the name of the target data file to be analyzed matches the filename specified in the *data_extraction.py* script, allowing the script to correctly locate and process the data. For example, the following line in the script defines the input file:

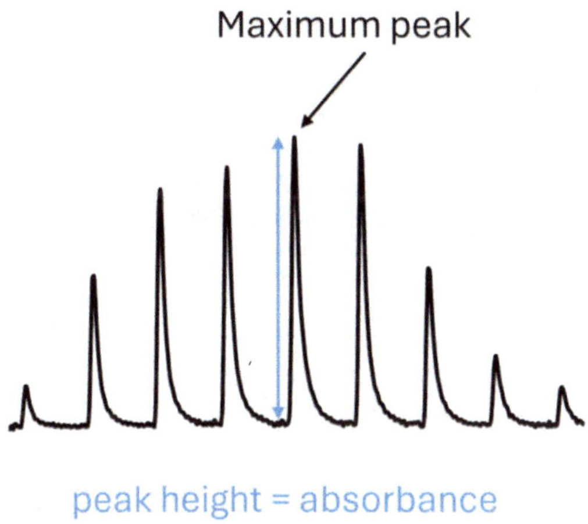

Fig. 12 Maximum peak detected and peak height saved as absorbance

```
file_path = "absorbance_data.csv"
```

If the actual file name differs, this line must be updated accordingly to avoid file-not-found errors during execution.

18. The *data_extraction.py* processes the absorbance–time data based on the time intervals defined in the input CSV file. Within each specified time segment, the script uses the find_-peaks() function to identify absorbance peaks. To ensure accurate results, a classification step is included to filter out invalid peaks, such as those caused by air bubbles, which typically appear as abrupt spikes in the signal. After filtering, the script selects the maximum valid peak height as the absorbance value corresponding to that specific standard or sample solution (Fig. 12). The final output, concentration values, and their corresponding absorbance (peak height) readings are saved as a new .csv file in the same directory as the script for further analysis or documentation.

Conflicts of Interest

The authors have no conflicts of interest to declare that are relevant to the content of this chapter.

References

1. Raju CV, Rani GM, Haribabu J, Kumar SS (2022) Flow injection analysis-based electrochemiluminescence: an overview of experimental design and its biosensing applications. ECS Sens Plus 1:031604

2. Rhee JI (2010) Immuno-flow injection analysis (FIA), process monitoring and control. In: Flickinger MC (ed) Encyclopedia of industrial biotechnology: bioprocess, bioseparation, and cell technology. Wiley, London

3. Hansen EH, Miró M (2009) Flow injection analysis in industrial biotechnology. In: Flickinger MC (ed) Encyclopedia of industrial biotechnology: bioprocess, bioseparation, and cell technology. Wiley, London
4. Ogawa K, Stöllner D, Scheller F, Warsinke A, Ishimura F, Tsugawa W, Ferri S, Sode K (2002) Development of a flow-injection analysis (FIA) enzyme sensor for fructosyl amine monitoring. Anal Bioanal Chem 373:211–214
5. Thiruppathi M, Tsai CY, Wang TW, Tsao Y, Wu TH, Ho JAA (2020) Simple and cost-effective enzymatic detection of cholesterol using flow injection analysis. Anal Sci 36: 1119–1124
6. Chen XW, Chen ML, Chen SA, Wang JH (2008) Flow-based analysis: a versatile, powerful platform for DNA assays. Trac-Trend Anal Chem 27:762–770
7. Albanese D, Liguori C, Paciello V, Pietrosanto A (2011) Winemaking process monitoring based on a biosensor automatic system. IEEE Trans Instrum Meas 60:1909–1916
8. Rhee IL, Ritzka A, Scheper T (2004) On-line monitoring and control of substrate concentrations in biological processes by flow injection analysis systems. Biotechnol Bioprocess Eng 9: 156–165
9. Trojanowicz M, Pyszynska M (2022) Flow-injection methods in water analysis-recent developments. Molecules 27:1410
10. Xu WH, Sandford RC, Worsfold PJ, Carlton A, Hanrahan G (2005) Flow injection techniques in aquatic environmental analysis: recent applications and technological advances. Crit Rev Anal Chem 35:237–246
11. Ruiz-Capillas C, Nollet LML (2015) Flow injection analysis of food additives. CRC Press, Boca Raton
12. Kurbanoglu S, Unal MA, Ozkan SA (2018) Recent developments on electrochemical flow injection in pharmaceuticals and biologically important compounds. Electrochim Acta 287: 135–148
13. Tzanavaras PD, Themelis DG (2007) Review of recent applications of flow injection spectrophotometry to pharmaceutical analysis. Anal Chim Acta 588:1–9
14. Salimi M, Nouroozi S (2022) Photometric flow injection analysis of As(III) by using a homemade, LED-based flow-cell device and methyl orange reagent. Spectrochim Acta A Mol Biomol Spectrosc 283:121713
15. Cokley JA, Jones KR, Delp NM, Rogers KR, Davis BW (2024) A 3D-printable, modular absorption spectrophotometer and fluorimeter for use in education. J Chem Educ 101:2665–2671
16. Sadegh-cheri M (2022) Using the recycled parts of a computer DVD drive for fabrication of a low-cost Arduino-based syringe pump. J Chem Educ 99:521–525
17. Wu YT, Chen YX, Cheng Y (2024) Building an Arduino-based open-source programmable multichannel syringe pump: a useful tool for fluid delivery in microfluidics and flow chemistry. J Chem Educ 101:1951–1958
18. de Oliveira LH, Trindade MAG, Angnes L (2025) Chronoamperometric flow-cell designed to be unaffected by air bubbles. Talanta 281:126875
19. Ruzicka J, Chocholous P (2024) Next generation of flow analysis is based on flow programming. Talanta 269:125410

Chapter 15

Open-Sourced In Silico Drug Screening

Audrey G. Fikes and Melissa C. Srougi

Abstract

Advances in structure-based computer-aided drug design have yielded robust computational techniques now pivotal for chemical biologists. These tools predict small molecule–protein interactions, significantly accelerating the drug discovery process. Unlike other computational techniques, structure-based design requires prior knowledge of a target molecule's three-dimensional structure, often leveraging molecular docking to predict target–ligand interactions. The primary goal is to design or identify small molecules that can bind to the target with high affinity and modulate its activity. Once a target's binding site(s) are determined, high-throughput in silico screens of large chemical libraries can predict binding energies, non-covalent interactions, and intermolecular forces crucial for target–ligand binding. These powerful methods accelerate the bench-to-bedside process, aiding in the repurposing of FDA-approved drugs, identifying novel ligands, and informing rational drug design, optimization, and/or elucidation of mechanism(s) of action. In this chapter, we describe a structure-based computational approach to perform high-throughput ligand screens of chemical libraries using open-source software programs. We illustrate this workflow with the enzymatic molecular target NAD(P)H:quinone oxidoreductase1 (NQO1), which is overexpressed in a number of human solid tumors. Importantly, this adaptable workflow requires minimal coding knowledge.

Key words Molecular docking, Structure–activity relationship modeling, In silico, Drug screening, Open-source, Drug discovery

1 Introduction

Computer-aided drug design (CADD) approaches can expedite the time, scale, and cost of traditional drug discovery and development experimental processes. CADD uses computational methods and simulations to analyze biological data and predict how potential drugs will interact with their intended biological target. CADD may include virtual screening of large chemical libraries for novel drug candidates, identification of drugs with higher efficacy, new drug targets, optimization of candidate compounds, and further assessment of their potential cytotoxicity. Additionally, it has been employed in drug repurposing approaches to identify FDA-approved drugs with the potential to modulate druggable

Robert E. Dempski (ed.), *Transforming Molecular Biology with Emerging Technologies*, Methods in Molecular Biology, vol. 3043, https://doi.org/10.1007/978-1-0716-5320-3_15,

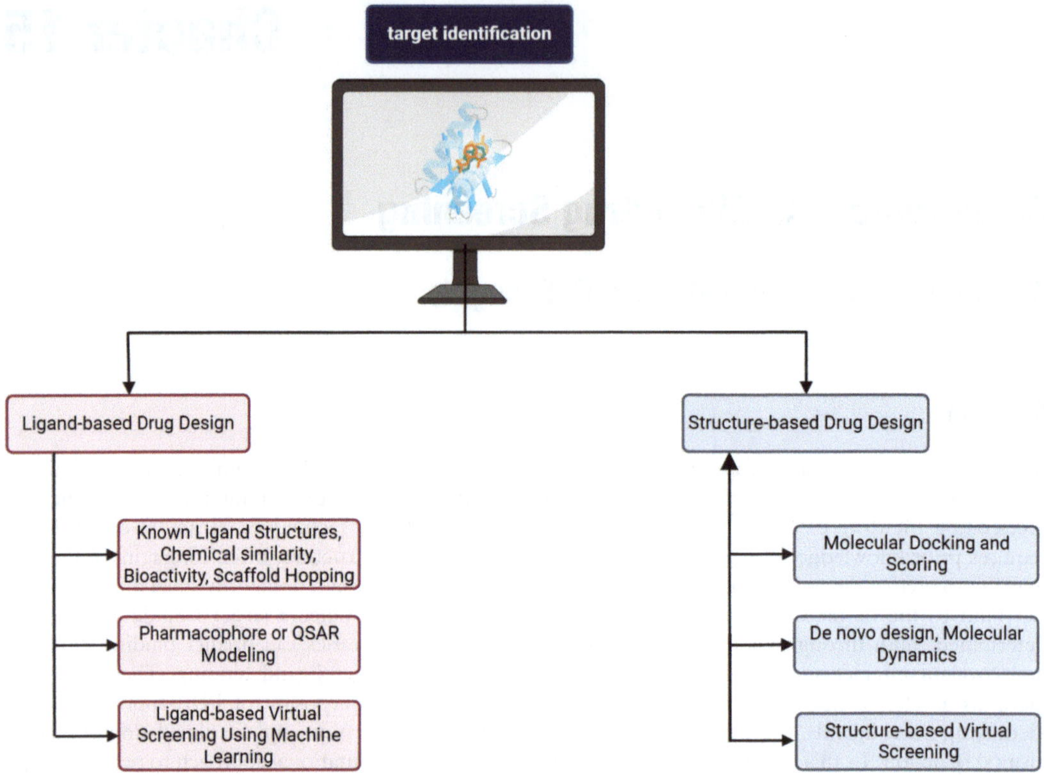

Fig. 1 Diagram summarizing general workflows for the two types of computer-aided drug design approaches. Ligand-based drug design (*left side*) and structure-based drug design (*right side*). (Image created in BioRender)

proteins that affect disease phenotypes ranging from cancer to heart disease. Key features of CADD processes include target identification, virtual screening, molecular modeling, absorption, distribution, metabolism, excretion, and toxicity (ADMET) predictions. Using available biological information, particularly structural information on the biological target and in some cases relevant antitargets (nontargeted subtypes of the target or drug transporters) and mechanisms of disease pathology, CADD has been used to predict putative drug targets.

There are two main types of CADD approaches: structure-based and ligand-based, which are briefly described herein (Fig. 1).

1.1 Structure-Based

One type of CADD approach utilizes structure-based drug design (SBDD) strategies. These strategies require previous knowledge of the target molecule's molecular structure (typically proteins and in some cases RNA molecules) and may employ molecular docking tools which enable researchers to predict target–ligand interactions based on the crystal structure of the target molecule [1]. The aim of this approach is to design and/or identify small molecules that can bind to specific sites on the target with high affinity and modulate its activity, leading to a therapeutic effect [2, 3]. When the structure of the target is known, potential binding sites on the target can be identified if they are otherwise unknown. There are a number of computational methods that can assist in binding site identification by analyzing the target's binding site(s), topology and cavities [2]. Once the binding site(s) are determined, virtual screening of large chemical libraries can be performed via molecular docking to predict small molecules expected to fit into the binding site. There are many docking programs available, including commercially available options such as GOLD, MOE-Dock, and Glide as well as open-source programs like AutoDock Vina [4, 5], DOCK, and rDock. Each of these programs use algorithms or combinations of algorithms (e.g., Monte Carlo simulations and genetic algorithms) to predict multiple binding poses of each molecule within the binding site and estimate the strength of the interaction [6]. After screening, molecules are ranked by predicted binding affinities and intermolecular forces evaluated to formulate the top-ranked hit molecules. It is important in this process to also account for the molecular dynamics of the protein target to include different protein conformations and flexibility [7], although this workflow treats the target protein as rigid and focuses solely on ligand flexibility and conformations. This decreases the computational demand and time needed to carry out high-throughput screening. Additionally, if the structure of the target is known in complex with its native ligand, structure-based pharmacophore models can be generated. These models can use key steric and intermolecular features necessary for ligand binding to screen small molecule libraries for molecules with similar electrochemical properties [8].

However, these methods are not without their limitations. A number of challenges exist in correctly predicting binding affinity due to accurately modeling the flexibility of the protein, as well as the effect of solvent, protonation states, and the complexity of proteins that contain multiple domains. Moreover, high-quality 3D structural data are required in this approach [7].

1.2 Ligand-Based

Ligand-based drug design (LBDD), on the other hand, is often used when a high-quality, well-defined 3D molecular structure of the biological target is not available [9, 10]. Instead, ligand-based drug design uses the similarity principle, which assumes compounds with similar chemical structures or physicochemical properties are likely to exhibit similar biological activities due to common molecular targets. The method focuses on analyzing known ligands that interact with the target of interest. The properties of these compounds can then be used to derive information to guide the design and identification of novel molecules with the desired activity [11]. A number of techniques are utilized in LBDD. These techniques include quantitative structure–activity relationship modeling (QSAR) [12]. Using the structure–function relationship of a known ligand–target, mathematical models can be generated that quantify their physiochemical or structural properties. When combined with regression analysis and machine learning, they can predict the activity of novel compounds. Typically, this method works by selecting a diverse set of known ligands to the target, programming the chemical descriptors of these ligands into the model, then investigating the correlation between these descriptors and activity while assessing the model's predicted capabilities [13–16].

Another commonly employed LBDD method involves pharmacophore modeling. Here, the pharmacophore establishes the 3D arrangement of steric and electronic properties necessary for the ligand to interact with its target and cause a biological response [17]. The 3D structures of active molecules are analyzed to ascertain common features that are crucial for binding. These models can include qualitative and/or quantitative data, such as activity data to train virtual screening models [18]. This is especially useful for identification of new compounds or those with structurally diverse chemical scaffolds [19, 20].

LBDD can be a faster approach, especially when used to train large datasets and identify structurally diverse hits [21]. It is an often-used approach to develop structure–activity relationships (SAR). However, LBDD is limited by the known ligand data used to build the models. Typically, integrated approaches using elements of both LBDD and SBDD are often employed.

This chapter details an SBDD approach, leveraging known 3D molecular crystal structure and ligand binding site information of a target protein. We will demonstrate how to perform a high-throughput in silico screen using open-source software. This methodology aims to identify promising lead compounds by evaluating their putative binding energies and characterizing crucial non-covalent interactions within the binding site. Ultimately, this chapter provides a robust computational workflow for accelerating lead compound discovery against a defined molecular target.

1.3 AutoDock Vina

AutoDock Vina is a free, open-source molecular docking program well known for its speed and ease of use [4, 5]. It was developed by Oleg Trott and Arthur Olson at The Scripps Research Institute based on the need for molecular docking software for novice users. AutoDock Vina uses a combined Monte Carlo-BFGS based algorithm that predicts the probability of outcomes by using repeating rounds of random sampling and nonlinear optimization to find global energy minima [5]. Although originally designed for the docking of individual ligands, the open-source nature of AutoDock Vina means that it can be easily manipulated to perform large high-throughput screens with minimal coding knowledge by the user.

1.4 Target and Lead Compound Identification

To model the high-throughput drug screening workflow, we will focus on an enzymatic molecular target, NAD(P)H:quinone oxidoreductase1 (NQO1), also known as DT-diaphorase. NQO1 is a stress-inducible two-electron oxidoreductase whose normal function is to protect cells against oxidative stress [22]. One of its protective functions is to detoxify quinones into less toxic hydroquinones that can be excreted from the body [22]. However, in the case of unique rare quinone-containing compounds such as beta-lapachone [23–25], dexoynyboquinone (DNQ) [26], mitomycin C (MMC) [27], and others [28], NQO1-mediated metabolism results in their bioreductive activation resulting in the rapid production of reactive oxygen species [24, 29] and cell death [24, 30, 31]. Interestingly, NQO1 is highly over-expressed in a number of solid tumors, including breast [32], non-small cell lung carcinomas [29], pancreas [33], etc., which makes it a viable target for personalized cancer therapy as it is often overexpressed compared to levels in associated normal tissues. Therefore, identifying novel NQO1-bioactivated compounds through methods such as in silico high-throughput drug screening is of high clinical interest.

To begin with the docking procedure, the protocol described below utilizes five well-characterized NQO1 substrates: beta-lapachone, menadione, mitomycin C (MMC), diaziquone (AZQ), and DNQ. These compounds serve as a benchmark for comparing novel compounds identified from in vitro drug screens. Researchers can, however, use alternative experimental starting points for their SAR studies. This initial setup is designed to assist researchers in becoming familiar with the docking procedure before applying their own experimental conditions.

2 Materials

For this protocol, you will need to download the open-source software programs listed in Table 1.

The instructions for downloading and installing these five programs (i.e., Autodock Vina, MGLTools, PyMOL, Strawberry Perl, and Open Babel) can be found below. The workflow described was adapted and reprinted with permission from Design and Implementation of an Accessible and Open-Sourced *In Silico* Drug Screening Activity for Cancer Drug Discovery, by Fikes and Srougi *J. Chem. Educ.* 2023, 100, 10, 4125–4130. Copyright 2023 American Chemical Society [34].

Note, if you plan on using known ligand structures from online databases, you do not need ChemDraw or ChemSketch. However, if you plan on performing molecular docking with novel ligands, you will need access to one of these programs to make the ligand files. ChemDraw requires a subscription license. ChemSketch is free and is available at https://www.acdlabs.com/resources/

Table 1
Software programs used for the high-throughput molecular docking protocol. Download instructions are provided below

Software	Usage
AutoDock Tools[a] (Part of the MGL Tools software suite) [36]	Preparation of protein receptor file
AutoDock Vina [4, 5]	Performs molecular docking
Open Babel [37]	Convert chemical file formats
Strawberry Perl [38]	Runs scripts to dock and rank multiple ligands at once
PyMOL [39]	Visualization of results

[a]Note that the 1.5.7 version of MGLTools is not compatible with Mac OS; Windows or Linux OS must be used.

free-chemistry-software-apps/chemsketch-freeware/ (accessed 31 July 2025). Five programs required for molecular docking are listed below.

2.1 AutoDock Vina

The AutoDock Vina download can be found at: https://github.com/ccsb-scripps/AutoDock-Vina/releases (accessed 31 July 2025).

1. Download *v1.2.3* for Windows.
2. Open the downloaded file. A warning may appear about running an unrecognized app. Select *More Info* and then choose *Run Anyway.*

2.2 MGLTools, Including AutoDockTools

The download for MGLTools can be found at: https://ccsb.scripps.edu/mgltools/downloads/ (accessed 31 July 2025).

1. Download the version for Windows (latest version is 1.5.7).
2. Open the downloaded file. The *MGLTools Setup* (*InstallJammer Wizard*) dialog box will appear. Select *Next.*
3. Read and accept the terms of the license, agree and select *Yes.*
4. The next window will be the *Choose Destination Location* window. The default location should be *C:\Program Files (x86)\MGLTools-1.5.7.* If this is not the default location, then click *Browse* and choose *C:\Program Files (x86)\MGLTools-1.5.7.* Select *Next.*
5. A dialog box about commercial usage and citing software will appear. Read the dialog box and select *OK.* Select *Finish.*
6. The *Python Molecular Viewer* may open, but it can be closed.

2.3 PyMOL

The purchase and download instructions for PyMOL can be found at: https://www.pymol.org/buy.html (accessed 31 July 2025).

2.4 Strawberry Perl

The download for Strawberry Perl can be found at: https://strawberryperl.com/ (accessed 31 July 2025).

1. Download the appropriate version (32 bits or 64 bits) for the computer. To determine the type of processor the computer has, go to *Start* > *Settings* > *System* > *About*. The information should be listed under *System Type*.
2. Open the downloaded file. The *Setup Wizard* should appear. Select *Next*.
3. Check the box next to "I accept the terms in the License Agreement" and select *Next*.
4. Choose the *Destination Folder*. Choose the default location in the C:\ drive. If the program needs to be installed in a different location, click *Change* to choose the desired location. Select *Next*.
5. Select *Install*. A dialog box may appear asking to allow this app from an unknown publisher to make changes to your device. Select *Yes*. Select *Finish*.
6. A README file will appear with more information about *Strawberry Perl*. It is not necessary to read this, and the file can be closed.

2.5 Open Babel

1. Go to https://openbabel.org (accessed 31 July 2025). Select *User Guide* on the lower right side of the screen.
2. Under the *Section Navigation Menu*, select *Install Open Babel*.
3. Under the *Windows* heading, select the link for the latest version on *GitHub*. Select *OpenBabel-3.1.1.exe* (current version at time of publication).
4. Open the downloaded file. A dialog box will appear asking, "Do you want to allow this app from an unknown publisher to make changes to your device?" Select *Yes*.
5. The OpenBabel 3.1.1 Setup dialog box will appear. Select *yes*. The *License Agreement* will appear. Select *I Agree*.
6. *Choose the Install Location* window will appear. The default location *(C:\Program Files (x86)\OpenBabel-3.1.1\)* does not need to be changed. Select *Next*.
7. *Choose the Start Menu Folder* Window will appear. The text box should say *OpenBabel 3.1.1*. If not, type that in. Select *Install*.
8. The *Completing OpenBabel 3.1.1 Setup* window will appear with the *Run OpenBabel 3.1.1* option selected. Select *Finish* to run the program.

3 Methods

For this procedure, we will be using a few known ligands of our molecular target as an example on how to apply the screening procedure. However, it is common to run up to hundreds of ligands during a given computational run. The five NQO1 ligands that will be used here are: beta-lapachone, menadione, MMC, AZQ, and DNQ.

Each individual screening run should have its own dedicated working directory (i.e., a specified folder on the computer that is the default location for project files) that contains the input/output files generated for the screening. In addition to keeping the results organized, this allows the use of the same file naming conventions (protein.pdbqt, ligand.txt, etc.) each time a new screening is performed. Even though the files have the same name, they are saved in different folders. This approach allows the user to apply the same command for each screening session, without the need for modification. The only change required for subsequent screenings on the *same protein* is to specify a different working directory. In this case, if additional screenings are employed the following files should be copied to the new working directory location:

1. protein.pdbqt
2. grid.txt
3. conf_vs.txt
4. vina.exe
5. vina_license.rtf
6. vina_split.exe
7. Vina_windows.pl
8. rank_vina_results.pl

The original files can be reused for virtual screening because there is no need to re-configure the protein file or find the Grid Box coordinates, since the molecular target remains the same. However, if carrying out a screening procedure on a *different target protein*, only the following files should be copied over to the new working directory:

1. vina.exe
2. vina_license.rtf
3. vina_split.exe
4. Vina_windows.pl
5. rank_vina_results.pl

The original protein files in this case cannot be reused because the new screening procedure will require the new configured protein file and respective Grid Box coordinates.

The high-throughput virtual screening procedure consists of six main steps:

1. Configuring the protein files.
2. Configuring the ligand files.
3. Selecting the docking site on the protein.
4. Performing the screening.
5. Ranking docked ligands.
6. Visualizing the results.

3.1 Configuring the Protein File

1. Go to the *Protein Data Bank* (https://www.rcsb.org/) and search for the molecular target protein using the 4-digit code *5EA2* (human NQO1). Select the blue *Download Files* button and select *Legacy PDB Format.* Move the file to the working directory and rename it ***protein.***
2. Open AutoDockTools and go to *File* > *Read Molecule* and select the file *Protein.* This should open the file downloaded from the *Protein Data Bank.*
3. Several modifications must be made to the protein structure before converting it to the PDBQT format for docking, including deleting water molecules. Go to *Edit* > *Delete Water.*
4. It is often useful or necessary to delete any unnecessary or redundant structures, such as currently bound ligands or identical monomers. This example does not need any structures deleted. *See* **Note 1** for instructions on how to delete unneeded structures.
5. Hydrogen atoms and appropriate charges also need to be added to the protein. Go to *Edit* > *Hydrogens* > *Add.* Select *Polar Only* and select *OK.* Go to *Edit* > *Charges* > *Add Kollman Charges* and select *OK.*
6. To ensure that there are no nonbonded atoms in the protein, in the left-hand window, under *All Molecules* and *Current Selection*, select the + icon next to *protein.* Select all the chains in the protein by selecting the ovals directly to the right, under the letter S. Selected chains will have their respective ovals highlighted yellow (*see* Fig. 2). Selected chains will also be highlighted in yellow in the visualizer. Go to *Edit* > *Misc* > *Check for Missing Atoms.* A dialog box will appear that lists any residues that contain missing atoms. There should be zero in this case. Select *Dismiss* and deselect the protein chains. When working with a different protein, there may be atoms missing due to

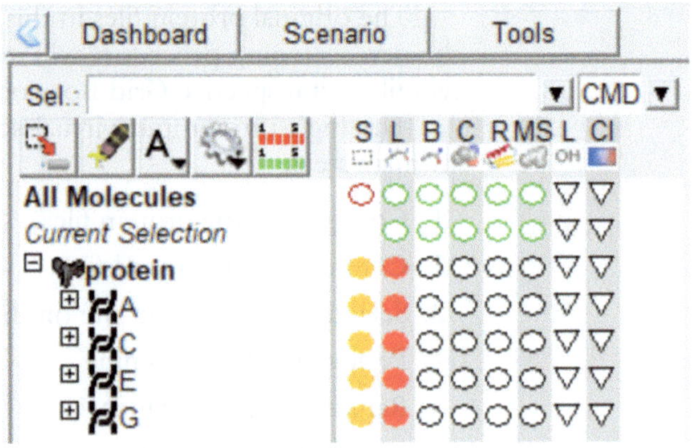

Fig. 2 AutoDockTools Menu for selecting protein chains, amino acids, or ligands within a crystal structure

incomplete data from the crystal structure. For instructions on how to address this, *see* **Note 2**.

7. To save the file in the PDBQT format needed for docking, go to *Grid* > *Macromolecule* > *Choose*. Select *Protein* and choose *Select Molecule*. A dialog box will appear that states "*contains no non-bonded atoms*." Select *OK*. Choose the appropriate working directory for the file location. Select the *PDBQT files (*.pdbqt)* as the *Save As* type. Name the file *Protein* and select *Save*

3.2 Configuring the Ligand Files

The next step is configuring the ligand files. This protocol uses known and commercially available ligands that can be downloaded from databases such as PubChem or Zinc. The files used in this example will come from PubChem. The files downloaded from PubChem are not in the correct file format to carry out molecular docking. They must first be converted from *SDF* files to *PDBQT* files. There are multiple ways to convert file types; however, the easiest is to use Open Babel, a program specifically designed for the conversion of chemical file formats. If the ligand files are not available on PubChem, Zinc, or a comparable database, *see* **Note 3.**

1. Go to *PubChem* (https://pubchem.ncbi.nlm.nih.gov/) and search for the first compound, *beta-lapachone*. Choose *beta-lapachone* from the list of search results and select it. Select *Download*. In the 3D conformer section of the dialog box, select *Save* next to *SDF*. Rename the file *beta-lapachone* and move it to the working directory. File names should contain no spaces. *See* **Note 4** about ligand file naming convention.
2. Repeat the previous step with the remaining four ligands: menadione, mitomycin C (MMC), deoxynyboquinone (DNQ), and diaziquone (AZQ). Name each file after its respective compound.

3. Open the *Open Babel GUI*. Under *INPUT FORMAT* on the left side of the window, use the dropdown menu to select *sdf—MDL MOL* format. Select the *More Options* button and navigate to the working directory where the ligand files are located. Choose all of the *SDF* ligand files and select *Open*.
4. On the right side of the window under *OUTPUT FORMAT*, use the dropdown menu to select *pdbqt—AutoDock PDBQT* format. In the text box below *Output file* type **.pdbqt* , and then select *CONVERT*. All of the *SDF* ligand files should now be in the working directory as PDBQT files.

3.3 Selecting the Docking Site on the Protein

1. Open *AutoDockTools*. Go to *File* > *Read Molecule*. Select *protein.pdbqt* from the working directory and select *Open*. Go to *Grid* > *Macromolecule* > *Choose*. Select *protein* and select *Molecule*.
   (a) A dialog box will appear that says "Protein may already have per atom partial charge. Do you want to preserve these charges instead of adding Gasteiger charges?" Select *No*. There is no need to preserve the charges as this step is only being used to identify the docking site on the protein.
   (b) A warning dialog box may appear indicating that there is an unbonded atom. Select *OK*. As with the charges, there is no need to address this warning because this step is only for selecting the docking site on the protein.
2. Go to *Grid* > *Grid Box*. A *Grid Options* dialog box will appear along with a red, green, and blue cube in the molecule visualization area. This represents the area where the docking will occur. Adjustments can be made to the location and size of the cube by selecting the red, green, and blue dials in the dialog box and moving the mouse left or right. The top three dials change the box size. The bottom three dials change the box location. Use these dials to center the box over the active site of the protein. Left click and move the mouse in the visualization window to move the protein and make sure the Grid Box is in the appropriate location. To change the view of the protein, left click to rotate, right click to move. If using a track pad, moving two fingers away from each other or toward each other will zoom in and out, respectively. Alternatively, if using a mouse, the scroll wheel can be used zoom in and out.
3. To ensure that the Grid Box is centered in the desired search location the Grid Box view can be changed from a solid cube to transparent cube. The transparent Grid Box has a yellow crosshair that denotes the center of the Grid box. This yellow crosshair can be helpful in centering the Grid box (*see* Fig. 3)

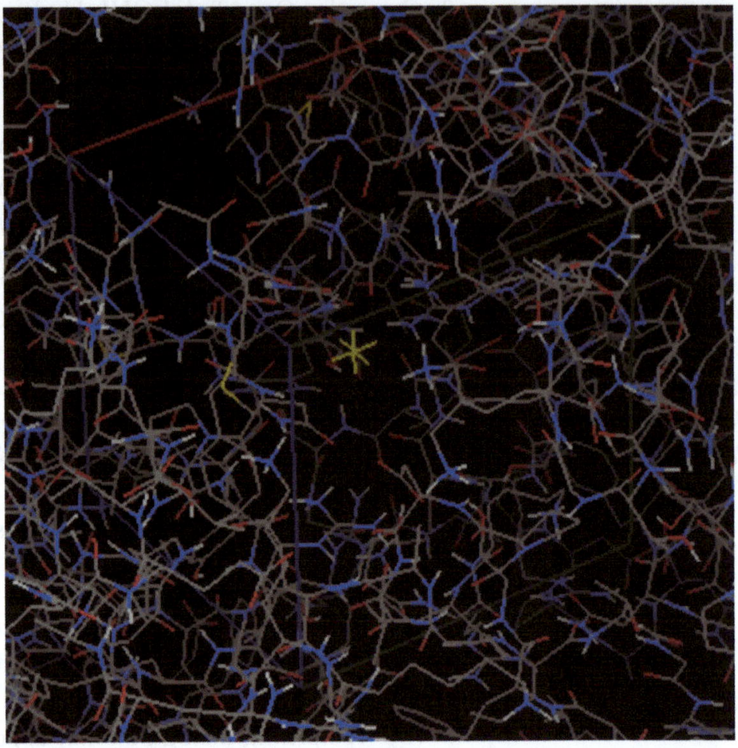

Fig. 3 The Grid Box is represented as a cube with edges of red, green, and blue. The grid box sets the search area for docking. The yellow cross-hairs represent the center of the grid box and can be useful for determining the correct coordinates for the search area

4. In order to get consistent results, the same Grid Box dimensions (i.e., search area) should be used every time docking is conducted on a particular protein. For NQO1 (PDB 5EA2) use the following values:

Number of points in the x-dimension: **30**

Number of points in the y-dimension: **26**

Number of points in the z-dimension: **30**

X center: **5.848**

Y center: **-5.736**

Z center: **9.701**

5. Please note that these values are optimized for the protein NQO1 (PDB 5EA2). The values will be different for different proteins. For instructions on how to find the Grid Box for a different protein, *see* **Note 5**.

6. Once the box is in the correct location, in the *Grid Options* dialog box go to *File > Output grid dimension file*. Select the appropriate working directory location. Select *.txt (*.txt)* as the *Save as* type. Name the file *Grid* and select *Save*. This will save the grid box dimension values so that there is an easily accessible record of the dimensions for future reference. Close *AutoDockTools*.
7. Open the *Notepad* application and type (or copy and paste) the following:

```
receptor = protein.pdbqt

size_x = 30
size_y = 26
size_z = 30

center_x = 5.848
center_y = -5.736
center_z = 9.701

energy_range = 4

exhaustiveness = 16
```

The center and size values should correspond with the Grid Box values from Subheading 3.3, **step 4**. Recall that these values can always be found in the *grid.txt* file from Subheading 3.3, **step 6**. The values in the *grid.txt* file are listed in order x, y, and z. Be sure to leave a space between the equal sign and the number values. Go to *File > Save As* and select the working directory. Select *Text Documents (*.txt)* as the *Save as* type. Name the file *conf_vs* and select *Save* and close the *Notepad* application. For information about energy range and exhaustiveness, *see* **Note 6**.

3.4 Performing the Screening

1. Open *File Explorer* and select the *C:* drive. Open the *Program Files (x86)* folder. Open *The Scripps Research Institute* folder. Open the *Vina* folder. Select all three files in the *Vina* folder and copy and paste them into the working directory. If *The Scripps Research Institute folder* is not present, *see* **Note 7**.
2. In order for AutoDock Vina to dock multiple ligands at once, a script must be written and run. Open *Notepad* and type the following command:

```
#!/usr/bin/perl
print"Ligand_file:\t";
```

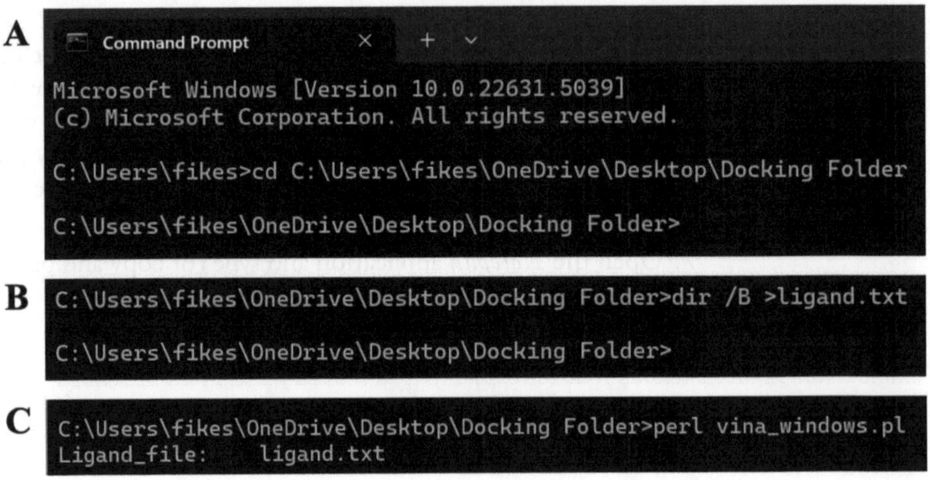

Fig. 4 (**a**) The command line indicating the working directory location for the drug screening. (**b**) The command line indicates that a new text file has been created, which lists all files in the working directory. (**c**) The command line used to initiate the docking program

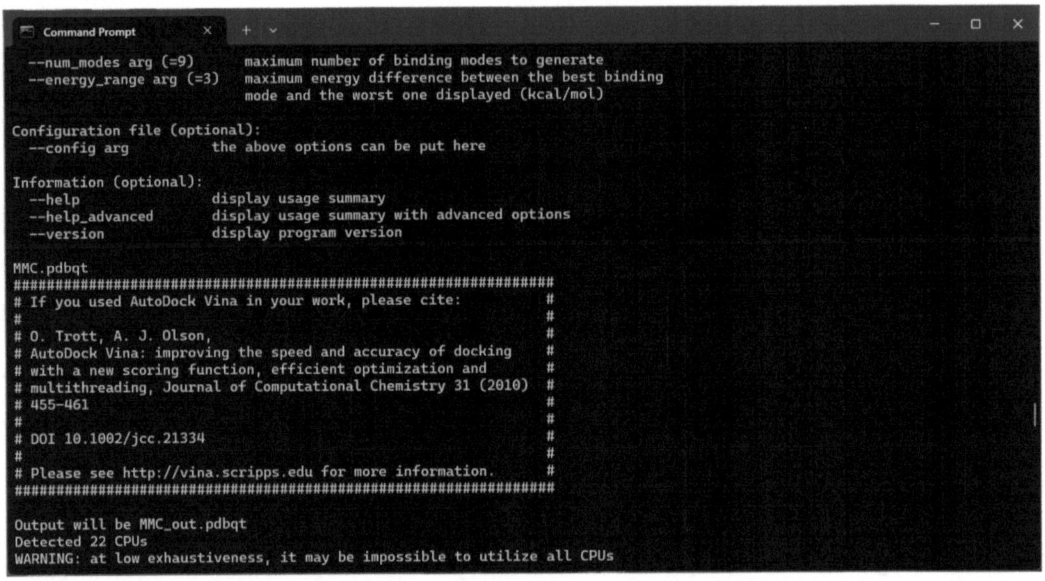

Fig. 5 Screenshot of the Command Prompt window displaying the successful execution of a molecular docking script

```
$ligfile=<STDIN>;
chomp $ligfile;
open (FH,$ligfile)||die "Cannot open file\n";
@arr_file=<FH>;

for($i=0;$i<@arr_file;$i++)
{
```

```
print"@arr_file[$i]\n";
@name=split(/\./,@arr_file[$i]);
}
for($i=0;$i<@arr_file;$i++)
{
chomp @arr_file[$i];
print"@arr_file[$i]\n";
system("vina.exe --config conf_vs.txt --ligand @arr_file
[$i] --log @arr_file[$i]_log.log");
}
```

Go to *File* > *Save as.* Select the working directory as the location to save the file. Under *Save as type* use the dropdown menu to select *All files.* Name the file *vina_windows.pl* and close *Notepad.*

3. Go to the Windows search bar and type *cmd* and select *Command Prompt* to open the *Windows Terminal.*
4. Open the working directory in Windows *File Explorer,* and right-click in the address bar and copy the address of the working directory. In *Command Prompt,* type: *cd* (followed by a space) and then paste the copied address of the working directory and hit the *Enter key* on the keypad. It should look similar to Fig. 4a. Note that there should be no space before *cd,* but there should be one space after it. This step will identify the working directory as the location for all subsequent activities related to this screening.
5. This step serves as a shortcut to make a list of all of the ligand files that will be used for the screening. Type *dir/B > ligand.txt* and hit the *Enter key* on the keypad (*see* Fig. 4b). A new file named *ligand.txt* should appear in the working directory. Open the file with *Notepad* and delete all the files that are not *ligand_name.pdbqt* files. *Save* the file and *close* it.
6. Return to the *Windows Terminal* and type *perl vina_windows.pl* and hit the *Enter key* on the keypad. Type *ligand.txt* and hit the *Enter key* (*see* Fig. 4c). To confirm the docking is running successfully, your *Windows Terminal* should look like Fig. 5. If an error message appears, *see* **Note 8**.
7. The screening is now running. After each compound has been run, a table of results, including binding energies, will appear in the *Windows Terminal.* The results for each ligand are also saved as two files in the working directory: (1) a log file that contains the table of binding energies shown in the *Windows terminal,* and (2) an output file (.out) that contains the molecular conformations of the compound in relation to the protein.

The table ranks the different compound conformations and the associated binding energy for each pose. The most energetically favorable pose will be the first one listed in the table and will have the lowest binding energy. Each pose can be further visualized and evaluated using *PyMOL*.

3.5 Ranking Docked Ligands

AutoDock Vina does not automatically rank the results of different ligands that have been docked simultaneously. It only ranks the different conformations/binding poses for each individual ligand. To compare and rank all of the ligands from the screening, a script must be written.

1. Open *Notepad* application and type (or copy and paste) the following command:

```
#!/usr/bin/perl

use strict;
use warnings;

# --- Configuration ---
# You can change the name of the output file here if you like.
my $output_filename = 'ranked vina results.txt';
# --- End Configuration ---

# --- Main Script ---

 # This version uses glob() to handle wildcards (*.log)
internally.
 # This makes it work correctly on all operating systems,
including Windows cmd.exe.
my @files_to_process;
if (!@ARGV) {
      # If the user provides no arguments, assume they want to
process *.log
       @files_to_process = glob("*.log");
} else {
       # If they do provide arguments, glob each one
       foreach my $pattern (@ARGV) {
               push @files_to_process, glob($pattern);
       }
}

if (!@files_to_process) {
       print "\n--- No Log Files Found ---\n";
       print "No files matching your input ('*.log' or other)
were found in this directory.\n";
```

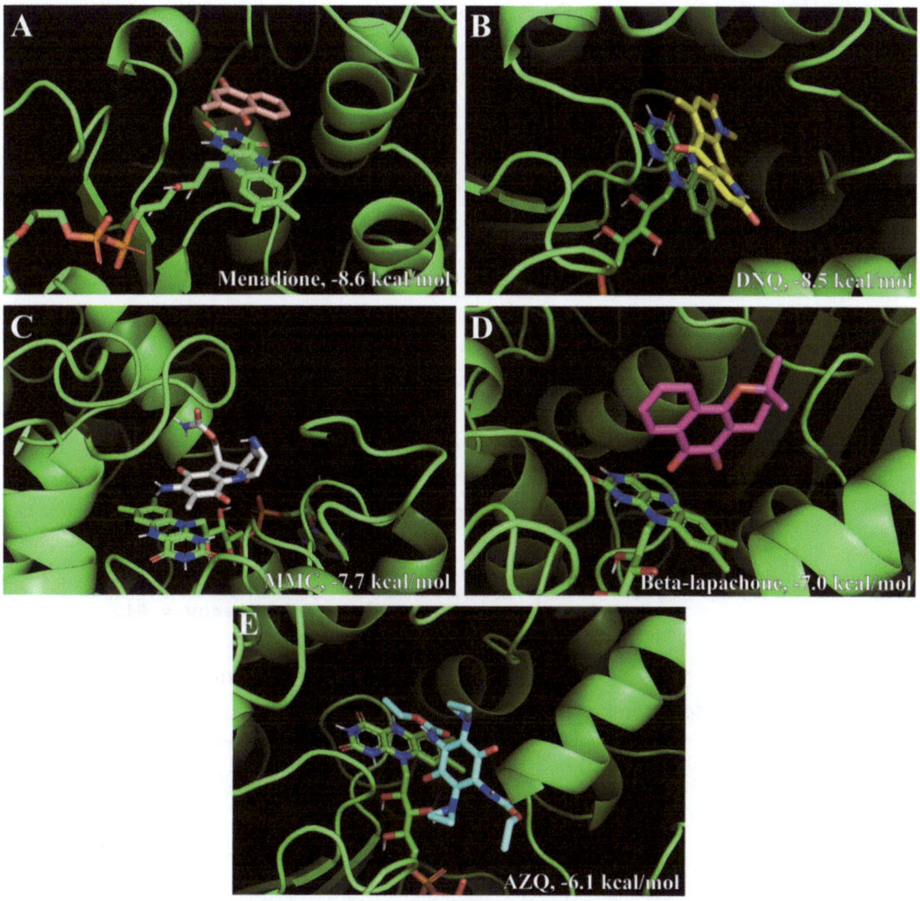

Fig. 6 Representative screening results of five ligands with NQO1 and their respective binding energy values. (**a**) Menadione, −8.6 kcal/mol; (**b**) DNQ, −8.5 kcal/mol; (**c**) MMC, −7.7 kcal/mol; (**d**) Beta-lapachone, −7.0 kcal/mol; (**e**) AZQ, −6.1 kcal/mol

```
            print "Please make sure you are in the correct
directory.\n";
        print "Usage: perl rank_vina_results.pl *.log\n";
        exit;
 }

 print "--- Starting Vina Log Ranker ---\n";
 print "Found ". scalar(@files_to_process). " file(s) to
process.\n\n";

 my %results; # Hash to store filename => best_energy

 # Process each log file found by glob()
 foreach my $logfile (@files_to_process) {
        my $best_energy = undef;
        my $fh; # Declare the filehandle
```

```
        # Attempt to open and read the log file
        unless (open($fh, '<', $logfile)) {
              warn "Warning: Could not open file '$logfile': $!
\n";
               next;
        }

       print "Processing file: $logfile\n"; # This still prints
to screen

        while (my $line = <$fh>) {
               chomp $line;

               # The regex looks for a line starting with '1'
(for mode 1)
                # and captures the floating point number that
follows.
               if ($line =~ /^\s*1\s+([-\d\.]+)/) {
                      my $current_energy = $1;

                                 if (!defined $best_energy ||
$current_energy < $best_energy) {
                               $best_energy = $current_energy;
                             last; # Found mode 1, stop reading
this file.
                      }
               }
        }
        close $fh;

        if (defined $best_energy) {
               $results{$logfile} = $best_energy;
        } else {
              warn "Warning: Could not find binding energy for
mode 1 in '$logfile'.\n"; # This still prints to screen
        }
 }

 # Check if any results were successfully extracted
 if (!%results) {
        print "\n--- No Results Found ---\n";
        print "The script ran and processed files, but did not
find any valid binding energies to rank.\n";
          print "Please check that your log files contain a
results table with a line for 'Mode 1'.\n";
        exit;
 }
```

```
# --- NEW: Open a file handle for writing the output ---
open(my $out_fh, '>', $output_filename) or die "FATAL: Cannot
open '$output_filename' for writing: $!";

# --- MODIFIED: These print statements now write to the file
via $out_fh ---
print $out_fh "--- Ranked Vina Results (Lower energy is
better) ---\n";
foreach my $logfile (sort { $results{$a} <=> $results{$b} }
keys %results) {
        printf $out_fh "%-50s: %8.3f kcal/mol\n", $logfile,
$results{$logfile};
}

# --- NEW: Close the output file ---
close $out_fh;

# --- MODIFIED: Print a final confirmation message to the
screen ---
print "\n--- Processing Complete ---\n";
print "Ranked results have been saved to: $output_filename
\n";

exit 0;
```

Go to *File* > *Save as*. Select the working directory as the location to save the file. Under *Save as type*, use the dropdown menu to select *All files*. Name the file *rank_vina_results.pl* and close *Notepad*.

2. In the *Windows Terminal*, type *perl rank_vina_results.pl* and hit the *Enter Key*. The ranked results will appear in the *Windows Terminal*. The results will also be saved as a text file named *ranked vina results.txt* in the working directory. If an error appears, go to **Note 9**. The ranked results for the screening are shown with images from *PyMOL* in Fig. 6.

3.6 Visualizing the Results

PyMOL is one of the most commonly used molecular visualization programs. Many tutorials exist for PyMOL [35]. Thus, the section below only gives brief descriptions and instructions on PyMOL and its user interface (Fig. 7).

1. Open PyMOL. Go to *File* > *Open* and choose the protein. pdbqt file from the working directory and select *Open*. Be sure to open the PDBQT file and not the PDB file.
2. Go to *File* > *Open* and choose the *beta-lapachone_out.pdbqt* file from the working directory and select *Open*. Each compound

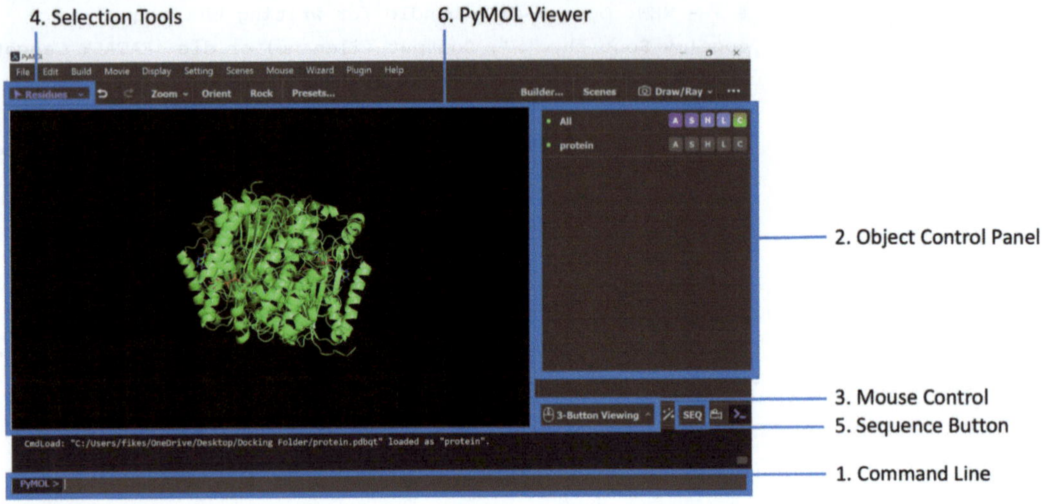

Fig. 7 The PyMOL user interface

can be analyzed individually in its own PyMOL session, or all of the compounds can be opened in one PyMOL session with different compounds shown or hidden as needed. If you are not familiar with PyMOL, it is easiest to view each compound in its own separate session.

3. Both the protein and the docking results for the ligand (s) should now be visible in the PyMOL viewer window. The different poses (i.e., binding conformations) for the ligand (s) can be viewed by using the *left and right arrow keys* on the keyboard. The following section will go over a brief introduction to PyMOL, enabling further analysis of docking results.
4. To save the PyMOL session at any point, go to *File* > *Save Session As*. Select the working directory. Name the file and select *Save*.

3.7 Key Features of the PyMol User Interface

Key features of the PyMol user interface (Fig. 7) are described in detail in Subheadings 3.7, 3.8, 3.9, and 3.10 below.

1. *Command Line*: This is where you can type in commands. The command line is generally most useful for experienced users. This brief tutorial focuses on using the mouse-initiated functions instead of the command line.
2. *Object Control Panel*: This contains the list of objects that are currently being visualized. Selecting object names in the *Object Control Panel* will hide them in the PyMOL viewer. Selecting again will cause the object to reappear. After each object is a set of buttons that can be used to adjust viewing modes, change colors, add labels, etc. The buttons and their functions are briefly summarized below:

(a) *A—Actions*: Rename, duplicate, remove, apply presets (like "ball-and-stick" or "publication"), and perform computations.

(b) *S—Show*: Change the way things appear, e.g., change to stick or cartoon view.

(c) *H—Hide*: Things that are shown using *S* accumulate, and don't automatically replace the last view. *H* is the opposite of *S* and hides unwanted representations.

(d) *L—Label*: Label atoms, residues, etc.

(e) *C—Color*: Change the color of atoms and groups.

3. *Mouse Control Legend*: This is the guide for using the mouse. The default is for a 3-button mouse. These settings can be changed for different mouse types by selecting this area, but that is generally unnecessary. The most useful mouse controls for visualizing the protein are listed below.

(a) Left Click: Select (the default is to select residues).

(b) Hold Left Click: Rotate protein.

(c) Ctrl Left Click: Move.

(d) Hold Right Click: Zoom in and out.

(e) Scroll Wheel: Add or remove foreground and background details.

4. *Selection Tools*: Clicking here will change what is selected when clicking the left mouse button. The default is the selection of *Residues*, but other options include *Chains*, *Molecules*, and *Atoms*. Leaving it on *Residues* is sufficient for the purposes of this protocol.

5. *Sequence Viewer*: Selecting this button will bring up the amino acid sequence of the protein. Clicking the button again will hide the amino acid sequence. While the amino acid residue sequence is visible, clicking the one-letter abbreviations in the sequence will select them in the PyMOL viewer. Click them again to deselect.

6. *PyMOL Viewer*: This is where all of the visualization takes place. Left-clicking on objects in the PyMOL viewer selects them. Clicking on the objects again deselects them. Objects that are selected are listed as *(sele)* in the *Object Control Panel.*

3.8 Visualizing Protein Surface

One of the first things that is helpful for visualizing protein–ligand interaction(s) is viewing the protein surface.

1. Select the *S* (Show) button for the *Protein* in the *Object Control Panel* and select *Surface.*

2. To hide the protein surface, select the *H* (Hide) button for the *Protein* in the *Object Control Panel* and select *Surface.*

3.9 Visualizing Amino Acid Side Chains in the Active Site

1. Select the *S* (Show) button for the *Protein* in the *Object Control Panel*. Select *side chain* > *sticks*. All of the protein amino acid side chains should now be visible.
2. In the PyMOL viewer, select the side chains of the amino acid residues that are relevant to the docking results.
3. Select the *A* (Action) button for *(sele)*. Go to *modify* > *invert* > *within molecule(s)*. Click on the *H* (Hide) button for *(sele)*. Select *Side chain*. Now only the relevant amino acid side chains are visible. Side chains can also be labeled by selecting *L* (Label) > *residue*.

3.10 Measuring Distances Between Atoms

1. Go to *Wizard* > *Measurement*. This will bring up the Measurement Wizard below the Object Control Panel.
2. Click on the first atom of interest and then on the second atom of interest. The distance in Ångstroms will appear between the two atoms.
3. To delete the measurement, select *Delete Last Object*.
4. Selecting *Done* will close the Measurement Wizard but leave any remaining measured values in the PyMOL viewer.

4 Notes

1. From Subheading 3.1, **step 4**, in AutoDock Tools on the left-hand window, under *All Molecules* and *Current Selection*, select the "+" icon next to the file name *protein*. Select any chains in the protein that are redundant or unnecessary by selecting the ovals directly to the right, under the letter S. Selected chains will have their respective ovals highlighted yellow (see Fig. 2). Selected chains will also be highlighted in yellow in the visualizer. Go to *Edit* > *Delete* > *Delete Selected Atoms*. A dialog box will appear that gives the following warning: "This command cannot be undone. If you choose to continue the undo list will be reset. Do you want to continue?" Select *Continue*.
2. From Subheading 3.1, **step 6**, when the Check for Missing Atoms Dialog box appears, choose *Select All Residues* and then select *Dismiss*. Go to *Edit* > *Repair Missing Atoms*.
3. From Subheading 3.2, if screening ligands that are novel compounds and do not have files available for download from PubChem or Zinc, the structures can be drawn using Chem3D (a part of ChemDraw and requires a license) or ChemSketch found at: https://www.acdlabs.com/resources/free-chemistry-software-apps/chemsketch-freeware/. The steps below are the instructions for making ligand files using

Chem3D and should replace **step 2**: Configuring the ligand files below.

(a) Open *Chem3D*. Click in the white *ChemDraw* box to bring up the *ChemDraw* tools. Use the *ChemDraw* tool to draw the ligand. A 3D representation of the ligand appears on the left-hand side of the viewer window.

(b) Go to *File > Save as*. Choose the working directory for the file location. Select *SDFile (*.sdf; *.sd)* as the *Save as* type. Name the file and select *Save*.

(c) Optional Step: If you want to save the ligand structure to use it again, as in the case of similar ligand structures, go to *File > Save As*. Choose *Chem3D XML (*.c3xml)* as the *Save as* type. Choose a file name and location to save it. Close *Chem3D*.

(d) Repeat steps *a* and *b* above for any other novel ligands to be included in the screening.

(e) Open *Open Babel GUI*. Under *INPUT FORMAT* on the left side of the window, use the dropdown menu to select *sdf — MDL MOL* format. Select the *More Options* button and navigate to the working directory where the ligand files are located. Choose all of the *SDF* ligand files and select *Open*.

(f) On the right side of the window under *OUTPUT FORMAT*, use the dropdown menu to select *pdbqt — AutoDock PDBQT* format. In the text box below Output file type **.pdbqt*. Select *CONVERT*. All of the *SDF* ligand files should be in the working directory as *PDBQT* files.

4. From Subheading 3.2, **step 1**, ligand file names should not contain any spaces. This will result in an error when performing the molecular docking in Subheading 3.2, **step 4**. Additionally, ligand file names should be easily recognizable. This is especially important when screening hundreds of ligands.

5. From Subheading 3.3, **step 5**, when using a protein other than NQO1, it is necessary to determine the coordinates of the binding site (also referred to as the search area or Grid Box). This is potentially the most time-consuming process. Finding the correct area and the correct area size (the Grid Box location and size) on the protein for the docking to occur often requires trial and error if no reported dimensions are found in the literature. The Grid Box size should not exceed 40 by 40 by 40. It can be helpful to use a PDB file that has a known ligand bound in the active site. The Grid Box can be centered around that ligand in the protein active site and used to determine the center and size values needed. The steps below show how to

find the binding site (Grid Box) on a protein using a PDB file with a known ligand.

(a) Go to *Protein Data Bank* (https://www.rcsb.org/) and search for the protein. If there are multiple structures available, choose one that has a known *Unique Ligand*. After navigating to the webpage for that specific crystal structure, select the blue *Download Files* button and select *Legacy PDB Format*. Move the file working directory and rename it *protein*.

(b) Open AutoDockTools and go to *File* > *Read Molecule* and select the file *Protein*. This should open the file downloaded from the *Protein Data Bank*.

(c) In AutoDock Tools on the left-hand window, under *All Molecules* and *Current Selection*, select the "+" icon next to the file name *protein*.

(d) In the menu tree, locate *Select* the + icon. This will bring up a list of amino acid residues in the protein. If the protein has multiple polypeptide chains, a list of chains will appear instead. In this scenario, choose the appropriate polypeptide chain and select the + icon to bring up the list of amino acid residues.

(e) Scroll to the bottom of the list. Any ligands should be listed at the bottom. Select the first oval (under the letter) next to the ligand on the menu tree. The ligand should now be highlighted in yellow in the crystal structure.

(f) Follow Subheadings 3.3, **steps 2** and **3** to find the grid box dimensions. The Grid Box should be centered around the location of the ligand. Use these values for Subheading 3.3, **step** 7. If there is no PDB file with a known ligand in the crystal structure, the location of a known residue in the protein binding site can be used. Follow the above steps *a* through *e*, but select the appropriate residue and center the Grid Box around that residue.

6. From Subheading 3.3, **step 7,** the exhaustiveness value indicates how thoroughly AutoDock Vina will explore the possible ligand conformations within the receptor Grid Box. Increasing the exhaustiveness value increases the likelihood of a more accurate pose at the cost of increased computational time. A value of 8 is the default value. This protocol has increased the value to 16 for increased accuracy. Values above 25 offer diminishing returns for accuracy for the increased amount of time. Energy range determines the maximum energy difference between the lowest energy conformation and other possible conformations calculated during the docking.

7. From Subheading 3.4, **step 1**, if *The Scripps Research Institute* folder is not present in the *Program Files (x86)* folder, it could be because AutoDock Vina was not installed in the default folder location. In this case, navigate to the folder drive or folder where AutoDock Vina was installed. Open *The Scripps Research Institute* folder. Open the *Vina* folder. Select all three files in the *Vina* folder and copy and paste them into the working directory. Continue the protocol at Subheading 3.4, **step 2**.
8. From Subheading 3.4, **step 6**, it is common for an error message to appear at this step if there are any mistakes in the *vina_windows.pl* script. Mistakes may arise due to copying and pasting the code from a web page or pdf which can result in unrecognized characters in the script. The most reliable way to fix this, or any error, is to locate the indicated numbered line in the *vina_windows.pl* script file, delete the entire line, and retype it manually. Large language models (LLMs) like Gemini, ChatGPT, and Copilot are also useful tools for diagnosing simple code errors. To use one, start by telling the LLM what your code is meant to do. For example, you might use the prompt, "I'm running simultaneous docking on multiple ligands with AutoDock Vina using Perl script and receiving the following error." After that, copy and paste both the error message and the code from your *vina_windows.pl* script into the chat window. The LLM should be able to pinpoint any errors or provide a corrected version of the code that can be copied and pasted directly into the *vina_windows.pl* file.
9. From Subheading 3.5, **step 2**, similar to **Note 8**, it is common for an error message to appear at this step if there are any mistakes in the *rank_vina_results.pl* script. Mistakes may arise due to copying and pasting the code from a web page or pdf which can result in unrecognized characters in the script. The most reliable way to fix this, or any error, is to locate the indicated numbered line in the *rank_vina_results.pl* script file, delete the entire line, and retype it manually. Large language models (LLMs) like Gemini, ChatGPT, and Copilot are also useful tools for diagnosing simple code errors. To use one, start by telling the LLM what your code is meant to do. For example, you might use the prompt, "I am writing a Perl script to rank the results of multiple.log files from an AutoDock Vina virtual screening." After that, copy and paste both the error message and the code from your *rank_vina_results.pl* script into the chat window. The LLM should be able to pinpoint any errors or provide a corrected version of the code that can be copied and pasted directly into the *rank_vina_results.pl* file and replace the old code.

Acknowledgments

The authors gratefully acknowledge the NC State Biotechnology Program and the Chemistry of Life Program for supporting this work. Financial support was generously provided by NC State University's Provost's Office.

Conflicts of Interest The authors have no conflicts of interest to declare that are relevant to the content of this chapter.

References

1. Wei H, McCammon JA (2024) Structure and dynamics in drug discovery. npj Drug Discov 1(1):1. https://doi.org/10.1038/s44386-024-00001-2
2. Yu W, MacKerell AD (2017) Computer-aided drug design methods. In: Sass P (ed) Antibiotics: methods and protocols. Springer New York, New York, pp 85–106
3. Anderson AC (2003) The process of structure-based drug design. Chem Biol 10(9):787–797. https://doi.org/10.1016/j.chembiol.2003.09.002
4. Trott O, Olson AJ (2010) AutoDock Vina: improving the speed and accuracy of docking with a new scoring function, efficient optimization, and multithreading. J Comput Chem 31(2):455–461. https://doi.org/10.1002/jcc.21334
5. Eberhardt J, Santos-Martins D, Tillack AF, Forli S (2021) AutoDock Vina 1.2.0: new docking methods, expanded force field, and python bindings. J Chem Inf Model 61(8): 3891–3898. https://doi.org/10.1021/acs.jcim.1c00203
6. Forli S et al (2016) Computational protein–ligand docking and virtual drug screening with the AutoDock suite. Nat Protoc 11(5): 905–919. https://doi.org/10.1038/nprot.2016.051
7. Lavecchia A, Giovanni CD (2013) Virtual screening strategies in drug discovery: a critical review. Curr Med Chem 20(23):2839–2860. https://doi.org/10.2174/09298673113209990001
8. Agu PC et al (2023) Molecular docking as a tool for the discovery of molecular targets of nutraceuticals in diseases management. Sci Rep 13(1):13398. https://doi.org/10.1038/s41598-023-40160-2
9. Mavromoustakos T, Kellici TF (ed) (2018) Rational drug design: Methods and protocols. Humana Press. https://doi.org/10.1007/978-1-4939-8630-9
10. Jahn A, Hinselmann G, Fechner N, Zell A (2009) Optimal assignment methods for ligand-based virtual screening. J Chem 1(1): 14. https://doi.org/10.1186/1758-2946-1-14
11. Bacilieri M, Moro S (2006) Ligand-based drug design methodologies in drug discovery process: an overview. Curr Drug Discov Technol 3(3):155–165. https://doi.org/10.2174/157016306780136781
12. Acharya C, Coop A, Polli JE, MacKerell AD (2011) Recent advances in ligand-based drug design: relevance and utility of the conformationally sampled pharmacophore approach. Curr Comput Aided Drug Des 7(1):10–22. https://doi.org/10.2174/157340911793743547
13. Fang C, Xiao Z (2016) Receptor-based 3D-QSAR in drug design: methods and applications in kinase studies. Curr Top Med Chem 16(13):1463–1477. https://doi.org/10.2174/1568026615666150915120943
14. Wang T, Wu M-B, Lin J-P, Yang L-R (2015) Quantitative structure–activity relationship: promising advances in drug discovery platforms. Expert Opin Drug Discov 10(12): 1283–1300. https://doi.org/10.1517/17460441.2015.1083006
15. Tamay-Cach F et al (2016) In Silico studies Most employed in the discovery of new antimicrobial agents. Curr Med Chem 23(29): 3360–3373. https://doi.org/10.2174/0929867323666160210141912
16. Ouassaf M, Bourougaa L, Bahaz F, Alhatlani BY (2024) Exploring the antiviral potential of Artemisia annua through JAK-STAT pathway targeting: a network pharmacology approach. Pharmaceuticals 17

17. Dror O, Shulman-Peleg A, Nussinov R, Wolfson HJ (2004) Predicting molecular interactions in silico: I. A guide to pharmacophore identification and its applications to drug design. Curr Med Chem 11(1):71–90. https://doi.org/10.2174/0929867043456287
18. Guner OF (2002) History and evolution of the pharmacophore concept in computer-aided drug design. Curr Top Med Chem 2(12): 1321–1332. https://doi.org/10.2174/1568026023392940
19. Babu S et al (2022) Identification of potent and selective JAK1 lead compounds through ligand-based drug design approaches. Front Pharmacol 13-2022
20. Zhang W et al (2024) Identification of mIDH1 R132C/S280F inhibitors from natural products by integrated molecular docking, pharmacophore modeling and molecular dynamics simulations. Pharmaceuticals 17
21. Shim J, MacKerell JAD (2011) Computational ligand-based rational design: role of conformational sampling and force fields in model development. MedChemComm 2(5):356–370. https://doi.org/10.1039/C1MD00044F
22. Ross D, Siegel D (2018) NQO1 in protection against oxidative stress. Curr Opin Toxicol 7: 67–72. https://doi.org/10.1016/j.cotox.2017.10.005
23. Bey EA et al (2013) Catalase abrogates beta-lapachone-induced PARP1 hyperactivation-directed programmed necrosis in NQO1-positive breast cancers. Mol Cancer Ther 12(10):2110–2120. https://doi.org/10.1158/1535-7163.MCT-12-0962
24. Bentle MS et al (2006) Calcium-dependent modulation of poly(ADP-ribose) polymerase-1 alters cellular metabolism and DNA repair. J Biol Chem 281(44):33684–33696. https://doi.org/10.1074/jbc.M603678200
25. Bentle MS et al (2006) New tricks for old drugs: the anticarcinogenic potential of DNA repair inhibitors. J Mol Histol 37(5):203–218. https://doi.org/10.1007/s10735-006-9043-8
26. Huang X et al (2012) An NQO1 substrate with potent antitumor activity that selectively kills by PARP1-induced programmed necrosis. Cancer Res 72(12):3038–3047. https://doi.org/10.1158/0008-5472.CAN-11-3135
27. Gibson NW, Ross D (1993) Mitomycin C. Cancer Chemother Biol Response Modif (0921-4410 (Print))
28. Ross D et al (2000) NAD(P)H:quinone oxidoreductase 1 (NQO1): chemoprotection, bioactivation, gene regulation and genetic polymorphisms. Chem Biol Interact 129(1–2):77–97. https://doi.org/10.1016/s0009-2797(00)00199-x
29. Bey EA et al (2007) An NQO1- and PARP-1-mediated cell death pathway induced in non-small-cell lung cancer cells by β-lapachone. Proc Natl Acad Sci 104(28):11832–11837. https://doi.org/10.1073/pnas.0702176104
30. Huang X et al (2016) Leveraging an NQO1 bioactivatable drug for tumor-selective use of poly(ADP-ribose) polymerase inhibitors. Cancer Cell 30(6):940–952. https://doi.org/10.1016/j.ccell.2016.11.006
31. Bentle MS et al (2007) Nonhomologous end joining is essential for cellular resistance to the novel antitumor agent, β-Lapachone. Cancer Res 67(14):6936–6945. https://doi.org/10.1158/0008-5472.CAN-07-0935
32. Yang Y et al (2017) β-lapachone suppresses tumour progression by inhibiting epithelial-to-mesenchymal transition in NQO1-positive breast cancers. Sci Rep 7(1):2681. https://doi.org/10.1038/s41598-017-02937-0
33. Li LS et al (2011) Modulating endogenous NQO1 levels identifies key regulatory mechanisms of action of beta-lapachone for pancreatic cancer therapy. Clin Cancer Res 17(2): 275–285. https://doi.org/10.1158/1078-0432.CCR-10-1983
34. Fikes AG, Srougi MC (2023) Design and implementation of an accessible and open-sourced in Silico drug screening activity for cancer drug discovery. J Chem Educ 100(10): 4125–4130. https://doi.org/10.1021/acs.jchemed.3c00307
35. Vertrees J (2018) Practical Pymol for beginners. In: SBGrid Consortium, Harvard Medical School
36. Morris GM et al (2009) AutoDock4 and AutoDockTools4: automated docking with selective receptor flexibility. J Comput Chem 30(16): 2785–2791. https://doi.org/10.1002/jcc.21256
37. O'Boyle NM et al (2011) Open babel: an open chemical toolbox. J Chem 3(1):33. https://doi.org/10.1186/1758-2946-3-33
38. https://www.perl.org/get.html. https://www.perl.org/get.html. Accessed 17 Mar 2022
39. PyMOL The PyMOL Molecular Graphics System. In, Version 3.0 edn. Schrodinger, LLC

Chapter 16

Exploring 3D Models Via Extended Reality Tools Using Common Smart Devices

Clara A. von Randow and Günther Thiele

Abstract

Extended reality (XR) tools have been embellishing our real-life experiences increasingly since the late twentieth century. With augmented reality applications being suitable for common smart devices and with technological advances becoming more and more commercially viable at a faster pace, XR is continuously growing in visibility and its applications. One of its potentials lies in the context of teaching, in particular in relation to concepts that require high spatial awareness—such as structural biology and chemistry. We hereby present examples and guides to generating and exploring 3D models via augmented reality and mixed reality using common smart devices. In addition, we present our fully immersive and interactive virtual science education platform *VeRidium*.

Key words Augmented Reality, Virtual Reality, Extended Reality, Mixed Reality, Structural Chemistry, 3D Models, Hands-On Learning

1 Introduction

1.1 A Brief Guide to Extended Reality

Expanding our real-life experiences to a world including virtual depictions and interactions is becoming more and more prominent. The concept of this so-called extended reality (XR) was first explored by Morton Heilig in 1957. The "Sensorama Simulator" enabled a multisensory theater for the user, combining 3D images with stereo audio, vibrations, scents, and feedback to the handlebars [1, 2]. A significant advance in XR was achieved by Ivan Sutherland in 1968 by means of a head-mounted display, effectively a helmet with screens placed in front of each eye. Movement of the head translates into the images being reconstructed at a new position, creating the illusion of the synthetic world remaining static, while the user is fully immersed and exploring the three-dimensional space [1, 3, 4]. Jaron Lanier coined the term "virtual reality" in 1989 and developed the "AudioSphere," the "EyePhone," and the "DataGlove," which, in combination, enhanced

Robert E. Dempski (ed.), *Transforming Molecular Biology with Emerging Technologies*, Methods in Molecular Biology, vol. 3043, https://doi.org/10.1007/978-1-0716-5320-3_16,

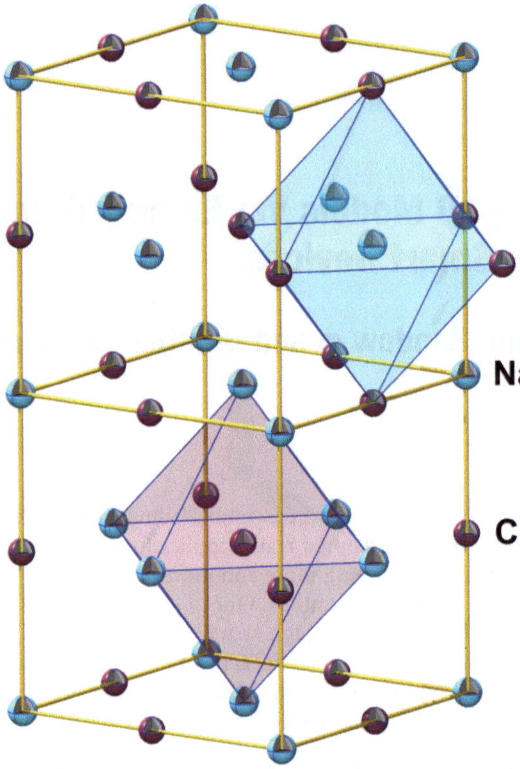

Fig. 1 Excerpt from the rock salt (NaCl) structure depicting the face-centered cubes and the octahedral sites in either blue or purple

the user's sensory experience in such a way to create an immersive experience [5, 6].

XR includes augmented reality (AR), virtual reality (VR), and mixed reality (MR) [7]. While our real-life environment is defined by the laws of physics, AR describes a real-life physical reality which is enhanced—augmented—with virtual objects that are projected onto our natural surroundings. MR describes a subset of augmented reality, which includes interaction between the real physical world and augmented features that have been projected into the real-world surroundings. This can involve projected objects hiding behind physical items, or a physical interaction with a real-life object, such as a dummy steering wheel, by the user directly translating into a change in motion of a projection [7, 8]. In VR, the participant is immersed in a fully computed environment that replaces the natural reality [1, 9].

Besides being a playful endeavor, the technology of XR was further developed with the goal of providing military training opportunities with minimal risk [1, 2, 5]. A second huge motivation for the realization of VR was to enable social interaction between different users [10, 11]. However, with time, the scope of applications became broader and more accessible, including multiple applications in training and teaching environments. It was seen to be particularly helpful in the comprehension of complex concepts that require high spatial awareness [2], such as chemistry or structural biology. Examples of educational applications using XR in these fields include the virtual exploration of the comprehensive database *DrugBank* [12], the VR software *Nanome* for the visualization of proteins from the *Protein Data Bank* [13, 14], interactive molecular dynamics between organic molecules [15], the virtual interaction with a 3D model of the potassium channel [16], or a VR tool named *Peppy,* developed to help teach the principles of peptide and protein structures to undergraduate students [17].

Many fields of chemistry rely on high spatial awareness of the individual in order to fully grasp the behavior and reactivity of the molecules. Structural chemistry includes complex structures of periodically ordered atoms, such as the rock salt structure. This structure can be described by alternating sodium (Na) and chlorine (Cl) ions in a cubic arrangement (Fig. 1). However, in order to systematically derive this structure type, structural chemistry describes the rock salt structure as a "face-centered cubic" arrangement of one atom type, with the other atom type occupying so-called "octahedral sites" (highlighted in blue for Na and purple for Cl, respectively, in Fig. 1) [18]. Without going into the details and advantages of these descriptions, it becomes clear that studying these structures requires a high degree of spatial understanding. Typically, these spatially demanding structures are introduced by means of 2D representations and physical models, which are passed around during classes and lectures. As an alternative, XR tools can be applied as a powerful aid to visualize these complex structures to further spatial awareness and deeper understanding [2].

Our approach to XR provides a freely accessible online database with supplementary information for 3D models and AR representations of chemical structures. Our goal is to complement and enhance chemistry courses by providing alternative tools for interacting with complex concepts from the curriculum, encouraging each individual's learning experience [19, 20]. In addition to our freely accessible online database and step-by-step guides to the AR realizations, we have developed the virtual reality science education platform *VeRidium,* which allows for a fully immersive experience of chemical structures. The user can explore various metal structure types in their own time. Following a tutorial, the user is prompted to partake in a guided interactive lesson, which concludes in a quiz in order to reiterate the learning content and to foster retention [9]. Comparative studies of students' learning achievements by means of either conservative teaching methods, such as 2D representations and physical models, or alternatively via extended reality tools, have confirmed the potential of AR, MR, and VR applications. The surveys have shown that students display a higher level of motivation for engaging with and learning complex concepts, which is strongly conducive to the retention of the taught contents [9, 16, 19, 21–23].

1.2 The Ins and Outs of AR, MR, and VR in Education

While VR tools are becoming increasingly popular, AR remains the choice for many applications. Its strongest asset is its accessibility: Most common smart devices with a camera are perfectly suitable for augmented reality projections. Making AR freely available on a device as widespread as a smart device has caused a significant increase in awareness for this technique in the wider public [24]. Viewing chemical structures via AR allows for the simultaneous exploration of complex structures at an independent and individual pace, even among a large group of people. This poses one of the most critical advantages compared to, for example, physical 3D models, which typically are very limited in numbers and are passed around in class. Depending on the number of students, with physical models, there will be a significant time-lapse between the introduction of the model by the teacher and the student investigating the model, which compromises the act of having a teaching aid for demanding concepts in the first place [20]. Expanding AR to MR allows for the unique enhancement of the AR technology through interaction with the physical world. Depending on personal preferences, the manipulation of a physical object in the real-world, and the movement thereof directly translating into movement of the virtual projection, can help promote the comprehension of a complex matter immensely.

VR offers a fully immersive experience in an entirely synthetic, computed world for the user, like none of the aforementioned XR tools. With this distinct feature, users can reach a new level of engagement with the topics at hand in order to spur deeper interest and to facilitate learning. The technological playfulness of VR is a massive contributor toward the large appeal of VR applications; captivating its users and increasing motivation for engaging with the learning content [15, 21].

An obvious challenge around implementing virtual reality applications is the necessity of a VR headset. While individuals might have access to the hardware due to personal interests in VR, providing access to a full set of headsets and controllers in the context of education, i.e., for a tutorial class, let alone for a lecture hall full of students, requires funds. In order for funds to be increasingly mobilized toward VR, strategic applications of the technique with the goal of complementing the curriculum are pivotal [9]. The continuous advancement of technology, combined with the development and commercialization of XR tools, results in an ever-growing offer for and visibility of VR in (chemistry) education. However, it is worth noting that due to personal preferences and/or learning predispositions, applying XR tools will not necessarily be the method of choice for everyone and thus should complement but not entirely substitute more traditional teaching methods.

The user learning experience of mixed reality was evaluated in the scope of a third-year inorganic chemistry undergraduate course with the focus on solid-state structures. Forty six percentage of the participants enjoyed the experience, confirmed that studying with this tool helped them better grasp the 3D structures and requested that this learning method be integrated in more classes [19]. We have further evaluated the efficacy of interacting with 3D structures of chemical compounds by means of virtual reality compared to three-dimensional clay models. For both cases, the users displayed a significant learning gain—independent of their learning environment [9]. It became clear that more than the concrete choice of model, varying teaching techniques within academic settings lead to a significant increase in motivation and concentration, which is highly conducive to the learning gain.

2 Materials

The following outlines the respective required hardware and software for exploring 3D models of structural chemistry by means of augmented, mixed, and virtual reality tools. Please note that a connection to the Internet is necessary in order to download the applications and/or object files; no Internet connection is needed for running the tools following successful setup.

2.1 Structural Chemistry via AR and MR

In order to implement AR and MR, a smart device with a camera and access to the Internet such as a mobile phone or a tablet is required. Structures from the database (vide infra) can be loaded into the application *Merge Object Viewer*, which is freely available in common app stores or from https://mergeedu.com/cube (accessed 20 August 2025) and is compatible with all major operating systems. The device's camera recognises the surface of the *Merge Cube* as the projection target for the structures. The *Merge Cube* can either be purchased from *Merge* or *Amazon*, or alternatively be handcrafted from a print-out which is freely available from https://mergecube.com/paper-pdf (accessed 20 August 2025). For the generation of 3D models for projection via augmented and mixed reality, please install the crystal and molecular structure visualization software *Diamond* version 4.6 or later from *Crystal Impact* free of charge from https://www.crystalimpact.com/diamond/download.htm (accessed 20 August 2025), as well as the software *Blender* for 3D modeling, which can be accessed free of charge from https://blender.en.softonic.com/ (accessed 27 August 2025).

2.2 Structural Chemistry via VR

To enter the virtual reality, the participant requires access to an immersive VR headset such as *Meta Quest 2*, *Meta Quest 3*, *Meta Quest 3S*, or *Meta Quest Pro*. The application *VeRidium* can be downloaded free of charge from the *Meta Quest Store* or directly via https://www.meta.com/en-gb/experiences/veridium/7441691572588551/ (accessed 21 August 2025).

3 Methods

The following sub-chapters comprise step-by-step instructions for the generation and projection of 3D models of chemical structures by means of extended reality tools. The AR and 3D structure database (vide infra) contains numerous 3D models for direct use in lectures; however, the creation of novel models or variations of existing models is also possible, as described below. The generation of 3D models results in a link, a QR code or an "Object Code" which you can share with others for access to the respective structure (*see* Subheading 3.1). Projection via the *Merge Object Viewer* allows for visualization of the models by means of AR (*see* Subheading 3.2.1) and MR (*see* Subheading 3.2.2). The platform *p3d.in* poses a second approach to visualizing the 3D models via AR utilizing a smart device of choice. *p3d*.in runs in a browser and does not require the download of an application, rendering this path highly accessible (*see* Subheading 3.3). Finally, please find enclosed a guide to our fully immersive VR science education platform *VeRidium* (*see* Subheading 3.4).

3.1 Generation of 3D Models for Projection via AR or MR

1. In order to generate 3D models for AR depiction, e.g., from experimentally determined structures or from other published structures, please load the respective file into the crystal and molecular structure visualization software *Diamond* version 4.6 or later from *Crystal Impact* (*see* **Note 1**).
2. Export the structure as a 3D model in the *.obj*-file wavefront format after choosing the depiction options of choice: "File" → "Save As" → "Save Graphics As" → Choose "Wavefront OBJ (*.obj)" → "Save" (*see* **Note 2**).

3.1.1 Generation of 3D Models for Projection via Merge Object Viewer

1. Open *Blender* and delete the grey cube that is generated automatically by the software when starting up. Navigate to "File" → "Import" → "Wavefront (.obj)" → "Import" and select the respective *.obj*-file of your model of choice.
2. Select "File" → "Export" → "FBX (.fbx)" → "Export" and choose a folder to save your 3D model as a *.fbx*-file.
3. Sign in to your *Merge EDU* account under https://mergeedu.com/ (accessed 21 August 2025) (*see* **Note 3**).

4. Navigate to "Browse Content and Upload Objects."
5. Upload the generated *.fbx*-file under "Upload" (*see* **Note 4**).
6. Under the menu option "Share Options", you will receive both an "Object Code" and a QR code, which you can share with others to view the respective structure.

3.1.2 Generation of 3D Models for Projection Via p3d.in

1. Sign up for a free account under https://p3d.in/ (accessed 27 August 2025) (*see* **Note 5**).
2. Select "New Model" from the menu bar on the top right of your screen.
3. Drag and drop both the *.obj*-file and the *.mtl*-file of your generated 3D model into the window, which will translate into the respective 3D model on the screen immediately (*see* **Note 6**).
4. You can choose different options for the depiction of the model from the menu bar on the top left of the depiction. Select "Save Changes" once you have made your choice.
5. Navigate back to "My Gallery" from the menu bar on the top right of your screen, and you will see the respective model uploaded successfully.
6. Select the drop-down menu from the arrow that appears when hovering over the model with the cursor. Select "Share Model" in order to choose the preferred level of access and to retrieve the link for sharing the model with others. You can transform the link into a QR code to facilitate sharing access to the structure with others using any common QR code generator.

3.2 Projection of Given 3D Models via Merge Object Viewer

1. Open the application *Merge Object Viewer* on your smart device.
2. Sign up for a free *Merge EDU account* (*see* **Note 3**).
3. Go to the AR and 3D structure database under https://linktr.ee/thielelabs (accessed 27 August 2025) and pick a structure from the database. Available structure depictions include the following:
   (a) Element allotropes.
   (b) Aristotypes of structural chemistry.
   (c) Mineral and silicates.
   (d) Intermetallic compounds.
   (e) Metal–organic frameworks.
   (f) Boranes.
   (g) Noble gas compounds.
   (h) Isomerism and functional groups.

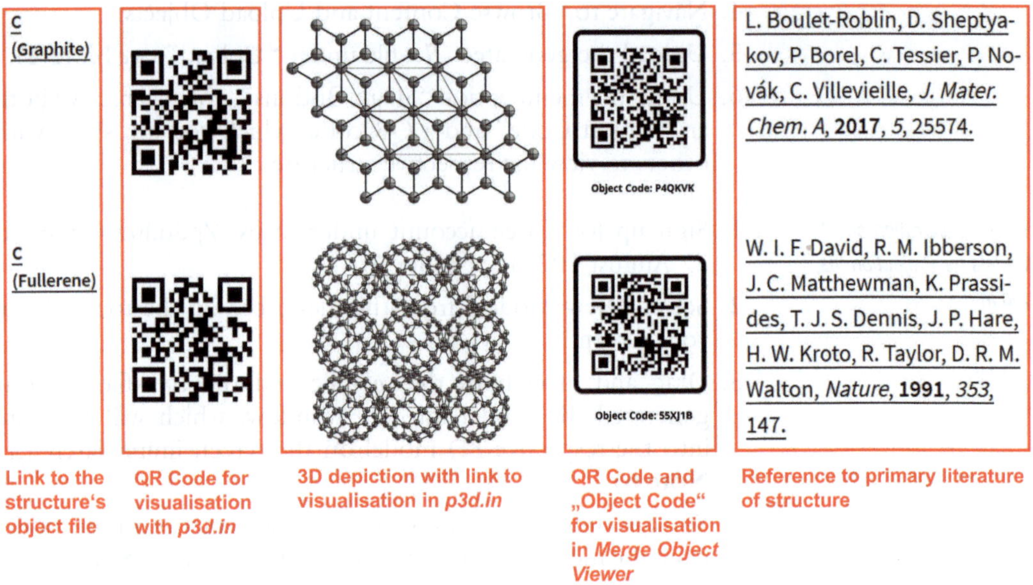

Fig. 2 Excerpt of the freely accessible database under https:/linktr.ee/thielelabs (accessed 27 August 2025) with explanations of the various elements of the database

(i) Coordination numbers.

(j) Bravais lattices and space groups.

(k) Atomic orbitals.

(l) Teaching models.

4. You can now scan the given QR code on the right (Fig. 2) given for the respective structure in the database with your smart device, which should open the *Merge Object Viewer* application automatically and initiate download of the structure (*see* **Note 7**).

   Alternatively, type in the respective "Object Code" from the chosen structure into the "Code Search" bar from the drop-down menu in the application in order to start downloading the structure onto the smart device (*see* **Note 8**).

   To access structures that you created and deposited under your account, type the generated "Object Code" into the "Code Search" bar from the drop-down menu in the application or scan the respective QR Code.

5. The structure will appear as a 3D model on the display of your smart device, which you can interact with via conventional movements on the screen.

3.2.1 Projection via AR

1. After loading a 3D model into the *Merge Object Viewer* (*see* Subheading 3.2), you can change the viewing mode from the bar "Change Mode" displayed at the top of your screen. Choose "WORLD" and scan a surface in front of you by

moving your phone around slowly in order to let the camera scan the surface. Select "Place" from the menu bar at the bottom of the display and tap the screen in order to place the projection of the 3D structure onto the surface in front of you. The projection of the structure can be rotated and changed in size via typical movements with your fingers on the screen.

2. To change to another structure, simply navigate back to the beginning and download another structure to your device. Please note that you can view and select any recently downloaded objects under "Recently Downloaded" from the dropdown menu.

3.2.2 Projection via MR

1. After loading a 3D model into the *Merge Object Viewer* (*see* Subheading 3.2), you can change the viewing mode from the bar "Change Mode" displayed at the top of your screen. Choose "CUBE" and point the camera of your device onto the *Merge Cube* in order to start the projection of the 3D structure onto the cube (Fig. 3).
2. Movement of the cube in your hand is directly translated into movement of the projection visible on the display of your device. In order to zoom in or out of the projection, please move your fingers up and down on the screen.
3. The object can be placed in space with the help of the button "Stamp" on the bottom right of the display. Now you can investigate the 3D structure on your display in a given orientation independent of the *Merge Cube* in the designated space of your choice.
4. To change to another structure, simply navigate back to the beginning and download another structure to your device. Please note that you can view and select any recently downloaded objects under "Recently Downloaded" from the dropdown menu.

3.3 Depiction of 3D Models via AR with p3d.in

1. Go to the structure database under https://linktr.ee/thielelabs (accessed 27 August 2025) and select a structure from the database either by selecting the QR code on the left (Fig. 1) or the depiction of the 3D model. Scanning the QR code with the camera of a smart device will open up the browser-based depiction of the 3D model on the respective device automatically.
2. The object can be moved around in space by use of the mouse (left: rotation; right: translation; scrolling: increase/decrease in size) (*see* **Note 9**). Please note: If you are viewing the browser-based 3D model in a smart device with a camera, the menu bar at the top left of the display will include an option for an AR

Fig. 3 MR visualization of a 3D chemical structure of zeolite using the commercially available *Merge Cube* and the *Merge Object Viewer* on a common smart phone. (Adapted and reprinted with permission from J. R. Schmid, M. J. Ernst, G. Thiele, *J. Chem. Educ.* **2020**, *97*, 4515–4519. Copyright 2025 American Chemical Society)

depiction. Simply select "AR" for projection of the model by the device's camera into the real world (Fig. 4) (*see* **Note 10**).

3.4 Interaction with 3D Models via VR

1. Put on the VR headset and familiarize yourself with it and the two controllers (*see* **Note 11**). Please note that you can adjust the length of the straps and the distance of the lenses from each other for optimal support on your head.
2. Ensure that you are in a location without any objects to bump into within a $(1.5 \times 1.5 \times 1.5)$ m^3 perimeter. When prompted in the virtual space, please set the boundaries around you using the controllers (*see* **Note 12**).
3. To start the application *VeRidium*, please navigate to the app library on the far right of the menu bar at the bottom of your screen, and select *VeRidium* from the list.
4. You will find yourself in a virtual space with a dashboard in front of you (Fig. 5). Please choose the respective language you

Fig. 4 Visualization of a chemical structure via *p3d.in* in the browser (left) and projected into the real world via the AR mode in *p3d.in* (right)

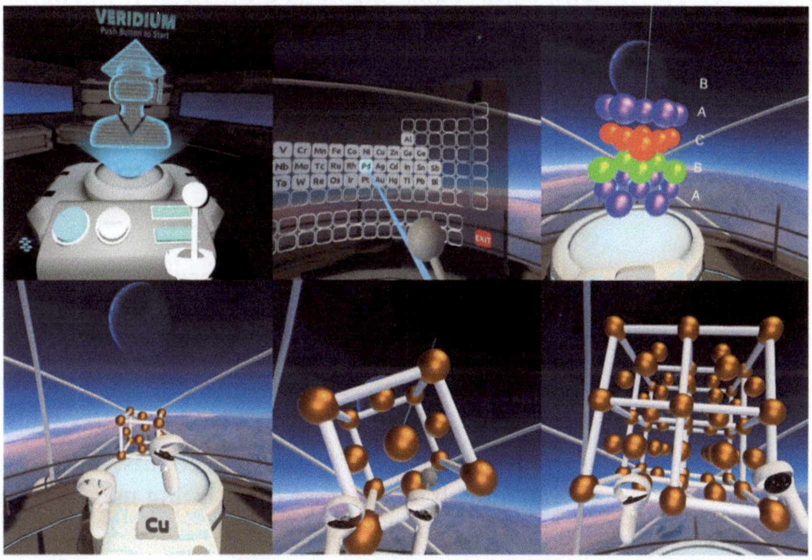

Fig. 5 Excerpts from the lessons on chemical structures on the VR science education platform *VeRidium* showing the user interacting with the virtual structure using the controllers

would like to proceed in with your right controller (*see* **Note 13**).

5. In the next step, please select "Tutorial" with your right controller in order to familiarize yourself with the virtual space (*see* **Note 14**).
6. After the tutorial is finished, you will once again find yourself in front of the dashboard. Please select "Lab" with your right controller.
7. You will now see a periodic table of elements on your left side. To get started, grab one of the highlighted elements and drop it into the slit of the podium in front of you. Follow the audio prompts (*see* **Note 15**) to ensure the maximum learning experience from this lesson (*see* **Note 16**). Every lesson will end with an interactive quiz. If you encounter any type of problems while using the application, please *see* **Note 17**.

4 Notes

1. Please note that a selection of file formats containing 3D models is supported by *Diamond* version 4.6 or later: *.cif*, *.cry*, *.csd*, *.dat*, *.pdb*, *.ent*, *.ins*, *.res*, *.idf*, *.kpl*, *.xyz*, *.emo*, *.mol*, *.mdl*, *.mol2*, and *.cssr*. Additionally, many databases, such as the *Protein database* (PDB, https://www.rcsb.org/, accessed 28 August 2025), the *Protein Data Bank in Europe* (PDBe, https://www.ebi.ac.uk/pdbe/, accessed 28 August 2025), or the *Protein Data Bank Japan* (PDBj, https://pdbj.org/, accessed 28 August 2025) have online 3D structure viewers that allow direct export of the models as *.obj*-file (typically via "Export Geometry").
2. Please note that saving the structure as an *.obj*-file will always generate both an *.obj*-file including the geometric arrangement of the 3D model and an *.mtl*-file containing details on the coloring, etc.
3. If you have one of the following accounts, it is possible for you to sign in with them rather than setting up a new account with *Merge EDU*: *Microsoft*, *Google*, *Classlink*, *Clever*, or *Apple*.
4. Each user can upload up to 50 objects in their personal account under https://mergeedu.com/ (accessed 20 August 2025).
5. If you have one of the following accounts, it is possible for you to sign in with them rather than setting up a new account with *p3d.in*: *Google* and *Facebook*.
6. Please note that you can also proceed by uploading the respective *.fbx*-file created by the software *Blender*.

7. For those structures in the database for which no Object Code or QR code has been provided for direct implementation as a 3D model in AR, download the .zip file from the respective structure containing the necessary object files. For upload and AR projection, follow the steps described under Subheading 3.1.
8. Depending on the structure, downloading the file onto your smart device might take up to a few minutes and can require a large battery capacity—please ensure that your smart device is fully charged.
9. The menu at the top left corner of the browser-based depiction provides various display options for you to choose from.
10. If you encounter difficulties, please ensure that you have granted your browser permission to access the camera on your device. For very large models, the AR option may not be available due to AR viewer limitations of the respective operating system (for more details, see model requirements at https://p3d.in/faq/augmented-reality, accessed 28 August 2025).
11. Please note that while each controller has five buttons and a joystick, you will only be using two of the buttons. Please refer to the tutorial within *VeRidium* in order to familiarize yourself with the controllers.
12. You will be asked to confirm the level of the floor by touching your controller to the floor and pressing "Confirm." In the next step, you will need to switch and then confirm your stationary boundary.
13. For now, users can choose between English and German. Further languages are planned in the future.
14. VR can cause dizziness and nausea in some users. If users feel disoriented, dizzy, nauseous, or experience any other form of feeling unwell, the headset should be removed immediately and other means of visualization, e.g., via screencast, should be made available. When used in a teaching setting, students should be made aware of potential risks *before* launching the tool.
15. Users will have the chance to interact with and to examine the structures in a detailed manner at their own leisure before moving on with the single segments of the lesson.
16. The VR learning platform is best experienced in smaller groups in a setting allowing for sufficient space for the users to move around in the virtual space.
17. If you encounter problems within the application, please proceed through the following steps in order to resolve the issue:

(a) Close and restart the application.
(b) Switch off and restart the headset.
(c) Change the batteries in the controllers.

Acknowledgments

This work was facilitated by financial support from the Verband der Chemischen Industrie e.V. and a fellowship for postdoctoral research by the Leverhulme Trust.

Conflicts of Interest The authors have no conflicts of interest to declare that are relevant to the content of this chapter.

References

1. Arena F, Collotta M, Pau G, Termine F (2022) An overview of augmented reality. Computers 11:28. https://doi.org/10.3390/computers11020028
2. Nee AYC, Ong SK (2023) Augmented reality. Cham
3. Lanier J (2001) Virtually there. Sci Am 284: 66–75
4. Gigante MA (1993) Virtual reality: definitions, history and applications. Academic Press Limited
5. Bown J, White E, Boopalan A (2017) Looking for the ultimate display: a brief history of virtual reality. Elsevier Inc.
6. Steuer J (1992) Defining virtual reality: dimensions determining telepresence. J Commun 42: 73–93. https://doi.org/10.1111/j.1460-2466.1992.tb00812.x
7. Thiele G, Mirica KA, Lau KC, Habig S (2023) AR, VR, and the metaverse in teaching: an advocacy for precise differentiation. J Chem Educ 100:4177–4180. https://doi.org/10.1021/acs.jchemed.3c00511
8. Craig AB (2013) Understanding augmented reality. Elsevier Inc., Waltham
9. Thiele G, Mirica KA, Habig S (2025) VeRidium: a fully immersive and interactive virtual reality educational environment. J Chem Educ 102:2364–2371. https://doi.org/10.1021/acs.jchemed.4c01256
10. Gonzalez-Franco M, Lanier J (2017) Model of illusions and virtual reality. Front Psychol 8: 1125. https://doi.org/10.3389/fpsyg.2017.01125
11. Blanchard C, Burgess S, Harvill Y et al (1990) Reality built for two: a virtual reality tool. Proc 1990 Symp Interact 3D Graph I3D 1990 35–36. https://doi.org/10.1145/91385.91409
12. Probst D, Reymond J-L (2018) Exploring DrugBank in virtual reality chemical space. J Chem Inf Model 58:1731–1735. https://doi.org/10.1021/acs.jcim.8b00402
13. Kingsley LJ, Brunet V, Lelais G et al (2019) Development of a virtual reality platform for effective communication of structural data in drug discovery. J Mol Graph Model 89:234–241. https://doi.org/10.1016/j.jmgm.2019.03.010
14. Berkmen MB, Balla M, Cavanaugh MT et al (2025) An idea to explore: use of the virtual reality app Nanome for teaching three-dimensional biomolecular structures. Biochem Mol Biol Educ 53:321–329. https://doi.org/10.1002/bmb.21892
15. Ferrell JB, Campbell JP, McCarthy DR et al (2019) Chemical exploration with virtual reality in organic teaching laboratories. J Chem Educ 96:1961–1966. https://doi.org/10.1021/acs.jchemed.9b00036
16. Sung R-J, Wilson AT, Lo SM et al (2020) BiochemAR: an augmented reality educational tool for teaching macromolecular structure and function. J Chem Educ 97:147–153. https://doi.org/10.1021/acs.jchemed.8b00691
17. Doak DG, Denyer GS, Gerrard JA et al (2020) Peppy: a virtual reality environment for exploring the principles of polypeptide structure. Protein Sci 29:157–168. https://doi.org/10.1002/pro.3752

18. Müller U (2006) Inorganic structural chemistry, 2nd edn. John Wiley & Sons Ltd, Chichester
19. Schmid JR, Ernst MJ, Thiele G (2020) Structural chemistry 2.0: combining augmented reality and 3D online models. J Chem Educ 97:4515–4519. https://doi.org/10.1021/acs.jchemed.0c00823
20. Schmid JR, von Randow CA, Thiele G (2021) Keine Kosten, viel Spaß. Nachr Chem 69:10–12. https://doi.org/10.1002/nadc.20214110213
21. Amirbekova E, Shertayeva N, Mironova E (2023) Teaching chemistry in the metaverse: the effectiveness of using virtual and augmented reality for visualization. Front Educ 8:1184768. https://doi.org/10.3389/feduc.2023.1184768
22. Abdinejad M, Talaie B, Qorbani HS, Dalili S (2021) Student perceptions using augmented reality and 3D visualization technologies in chemistry education. J Sci Educ Technol 30:87–96
23. Fatemah A, Rasool S, Habib U (2020) Interactive 3D visualization of chemical structure diagrams embedded in text to aid spatial learning process of students. J Chem Educ 97:992–1000. https://doi.org/10.1021/acs.jchemed.9b00690
24. Chen SS-C, Duh H (2019) Interface of mixed reality: from the past to the future. CCF Trans Pervasive Comp Interact 1:69–87. https://doi.org/10.1007/s42486-018-0002-8

Chapter 17

Molecular Architecture and Energy Visualization: Advancing Insights into Protein Structure, Interactions, and Activity

Luis A. Landrau Correa, Owen Borden, Oscar Claudio Ares, and Arthur D. Tinoco

Abstract

Protein biochemistry continues to progress rapidly through ongoing discoveries, from newly identified proteins to fascinating structural insights into mechanisms. Since proteins are the main machinery of cellular life, the desire to visualize their structure, whether through experimental or computationally obtained methods, has greatly increased. The ability to see proteins in three dimensions has become an essential tool in structural biology, protein design, drug development, and related fields. Among the available tools, PyMOL has become one of the most versatile and easy-to-learn platforms for structure visualization, allowing researchers to closely examine proteins and analyze their structural features. This chapter briefly explains how amino acids influence protein structure, then describes the hierarchical levels of protein organization. It illustrates how tertiary and potential quaternary structures form from energy landscapes that direct proper folding. Key regulatory interactions, such as allostery and metal binding, are highlighted. The chapter concludes by discussing PyMOL features for analyzing pH-dependent charge states, solvent accessibility, and structural or sequence alignments to better understand conformational changes and gain deeper structural insights.

Key words Protein structure, PyMOL, Protein electrostatics, Protein charge states, Allostery, Protein energy landscape, Protein sequence and structural alignment, Structure–activity relationship

1 Introduction

Proteins are a major cornerstone of life, partaking in a variety of processes crucial for survival, including respiration, DNA replication, biomolecular transport, catalysis, and more. The functions they carry out are directly tied to their structure and surface charges, and the dynamics of these properties are tuned by biochemical environment changes and different interactions. This chapter seeks to highlight the structural details of proteins by first reviewing the distinctive physicochemical features of amino acids, the building blocks of proteins. It next surveys the four levels of protein structure and then specifically focuses on tertiary and

Robert E. Dempski (ed.), *Transforming Molecular Biology with Emerging Technologies*, Methods in Molecular Biology, vol. 3043, https://doi.org/10.1007/978-1-0716-5320-3_17,

quaternary structure, in particular, the energy landscape profiling that drives protein folding for optimal stability and activity. It highlights key interactions, such as allostery and metal binding, that regulate protein function through structural changes and spotlights some proteins as classical case studies for emphasizing protein structure–activity relationships. The chapter then describes tools available through the PyMOL protein visualization program that allow one to calculate pH-dependent charge states, measure solvent accessibility, and perform sequence- and structure-based alignment to facilitate insight into protein conformation changes and associated crucial interactions.

2 Protein Structure

2.1 Fundamental Properties of Amino Acids

To truly understand the physicochemical properties and forces that govern protein structure, a discussion of amino acids is essential. Amino acids are the fundamental building blocks of peptides and proteins, each consisting of a carboxy group, an amino group, a hydrogen atom, and a variable side-chain residue ("R") [1]. All amino acids incorporated into proteins are L-amino acids, except for D-amino acids found in bacterial cell walls [1]. The 20 naturally occurring amino acids are commonly classified based on the different properties of the side chain into hydrophobic, polar, and charged groups [2]. Hydrophobic (apolar) amino acids include aliphatic amino acids like Glycine, Alanine, Valine, Leucine, Isoleucine, and Methionine, and aromatic amino acids like Phenylalanine, Tryptophan, and Tyrosine. Polar amino acids are subdivided into uncharged (Serine, Threonine, Asparagine, Glutamine, Proline, and Cysteine) and charged residues. Charged amino acids are the ones whose side chains may be fully ionized at neutral pH, with positively charged residues including Histidine, Lysine, and Arginine, and negatively charged being Aspartate and Glutamate [2].

The reason why there are different amino acids is that each serves a role in protein folding, stability, solubility, and function. For instance, the charged amino acids are important for proton transfers during enzymatic activity [1]. Ionizable groups help create ionic bonds, which help stabilize the protein's tertiary structure. Some amino acids, such as Aspartate, Glutamate, and Histidine, can bind metals such as iron (Fe), zinc (Zn), and calcium (Ca). This metal-binding capability is important for metalloproteins that need metal cofactors/prosthetic groups for their functionality. For instance, iron-based hemoglobin is a tetrameric protein that is crucial for breathing. It contains one heme group at each of its monomer subunits [1]. The Fe(II) ion at the heme groups has an accessible binding site that is useful for binding O_2 and transporting it through the circulatory system. This binding interaction is vital for respiration because O_2 is a nonpolar molecule and would otherwise not be soluble in blood, which is largely water-based.

Some amino acids provide crucial properties to proteins that can be exploited in the protein biochemistry laboratory setting. For example, aromatic residues, particularly Tryptophan, absorb ultraviolet light at 280 nm, allowing researchers to measure the protein's concentration [1]. Another widely used laboratory application involves Histidine. A short sequence of Histidine residues (His-tag) is bioengineered onto the N or C terminus of a protein via a plasmid design. Plasmids, circular DNA, serve as vectors to insert the gene encoding the desired protein into a host organism such as *Escherichia coli*. The His-tag facilitates protein purification due to its strong affinity for nickel(II) ions [3]. Using immobilized metal affinity chromatography (IMAC), the His-tagged protein binds to the column, hence allowing untagged proteins, peptides, and cellular debris to be washed away after the cell lysis.

2.2 Levels of Protein Structural Organization

Proteins are composed of peptide architectures (polypeptides) made up of amino acids. When multiple amino acids join through peptide bonds, a type of linkage that releases a water molecule, a polypeptide forms (Fig. 1) [4]. However, a polypeptide becomes a functional protein only when it folds into a specific three-dimensional structure.

Proteins possess four levels of structural organization. The primary structure is the linear sequence of amino acids in the polypeptide chain, linked together by peptide bonds [4]. The secondary structure forms when hydrogen bonds between backbone atoms lead to the creation of distinct shapes, such as alpha helices and beta-pleated sheets [5]. The tertiary structure is the three-dimensional conformation that arises when the polypeptide chain interacts with its environment, typically aqueous, to form a functional structure [5]. Lastly, the quaternary structure involves the

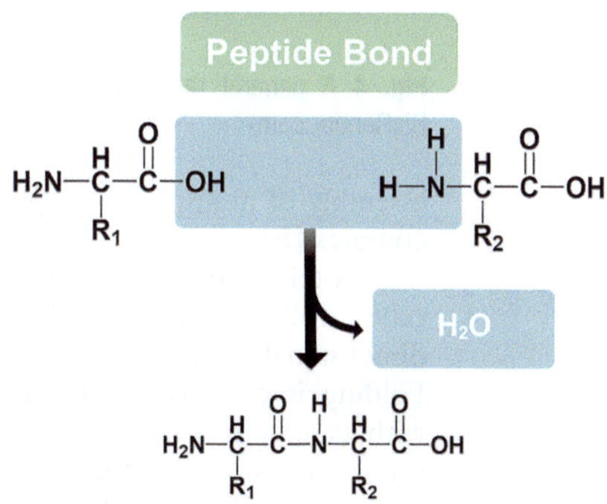

Fig. 1 Peptide bond formation scheme. (Created with BioRender.com)

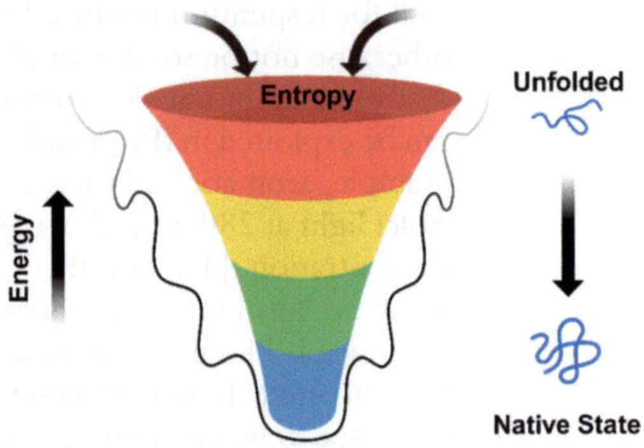

Fig. 2 Rugged energy landscape representation. (Created with BioRender.com)

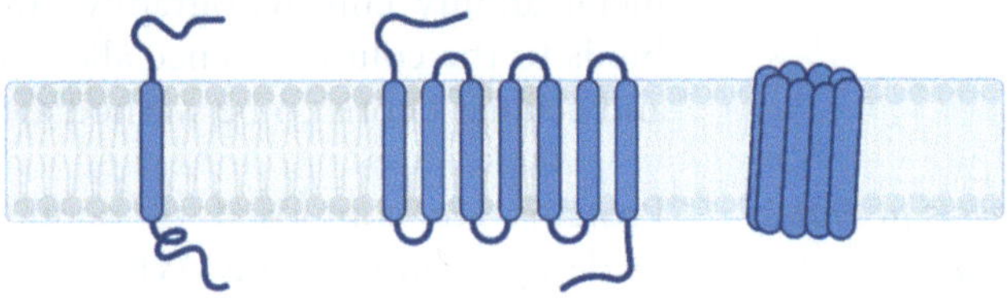

Fig. 3 Cartoon representation of transmembrane proteins associated with the lipid bilayer. The two structures on the left reveal transmembrane proteins that contain a domain outside of the membrane. (Created with BioRender.com)

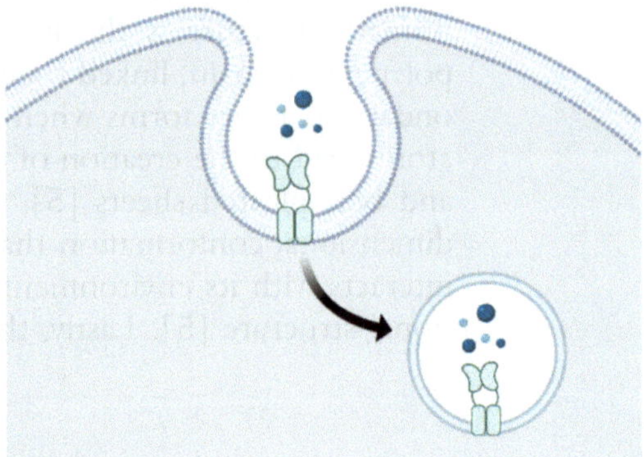

Fig. 4 A general representation of an endocytosis process. (Created with BioRender.com)

assembly of multiple polypeptide chains into a functional protein complex [5].

Proteins obtain their characteristic form through a process called folding, where a linear polypeptide adopts its three-dimensional structure, transforming into a functional protein. Folding is essential for the protein's function, which correlates with its roles in organisms. Proteins exist in two thermodynamically stable states: the folded (or native) state and the unfolded (or denatured) state [4]. In the native state, the protein has a

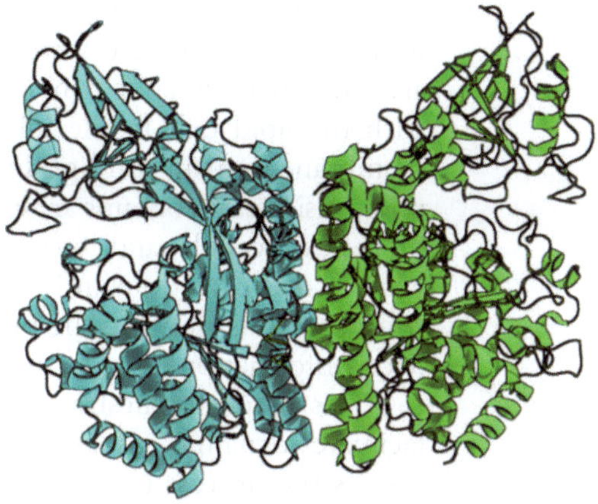

Fig. 5 Extracellular domain of human transferrin receptor protein-1 (TfR1) (PDB 1CX8)

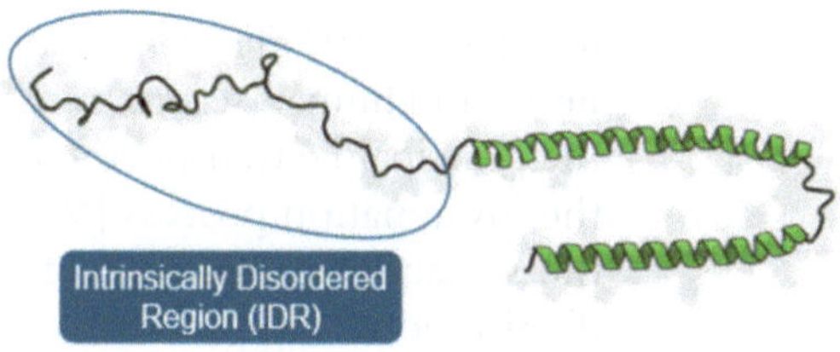

Fig. 6 Cartoon structure of Alpha-synuclein (PDB 1XQ8)

compact conformation, stabilized by hydrophobic interactions, hydrogen bonds between backbone atoms, salt bridges, and Van der Waals forces. These interactions lead to the bends and twists that create functional binding sites. In contrast, the unfolded state is less compact and organized, hence nonfunctional. The folding process begins with the primary structure (linear/unfolded and high-energy) progressing through secondary and tertiary structures to form the quaternary structure, which represents the low-energy, folded state. This folding process is often described by the energy landscape theory (Anfinsen's Dogma) [4]. This theory describes how proteins may find their final shape. Instead of following a defined path, proteins explore many possible conformations before settling into their most stable form. In other words, this theory portrays proteins as flexible molecules that navigate a rugged funnel-shaped energy landscape that guides them to their native state efficiently, which, through the process, decreases their Gibbs free energy and entropy (Fig. 2) [6].

Membrane proteins also navigate the energetic landscape of folding, seeking conformations that are thermodynamically favorable. These proteins are embedded in or associated with the membranes of cells or organelles [7]. Many of these proteins span the lipid bilayer entirely, forming transmembrane domains that anchor the protein within the membrane (Fig. 3) [7]. Due to the nature of folding and the drive toward reducing free energy, these proteins are amphipathic: They contain hydrophobic residues that interact with the lipid tails in the bilayer's interior, stabilizing the transmembrane domain through hydrophobic interactions. The hydrophilic residues it contains are exposed to the aqueous environments on the extracellular and cytosolic sides [7]. This spatial arrangement ensures that each domain is compatible with its surrounding environment, thus allowing the protein to adopt an energetically satisfied conformation.

This nature-orchestrated arrangement of membrane proteins facilitates dynamic cellular processes such as endocytosis. Endocytosis is the process by which a portion of the cell's plasma membrane invaginates, incorporating membrane proteins and lipids into an internal vesicle that enables nutrient uptake (Fig. 4) [8]. An example that integrates these concepts is the human transferrin receptor protein-1 (TfR1), a homodimeric transmembrane receptor (Fig. 5) [9]. TfR1 mediates the uptake of Fe into the cell from Fe-bound human serum transferrin (sTf). sTf is a glycoprotein, and the principal Fe transporter and its binding to the receptor triggers the invagination process [9]. Later in this chapter, we will explore the surface dynamics of sTf and its receptor, focusing on charge distribution and solvent accessibility before and after their interaction.

But, as in nature, there are exceptions. Close to 60% of the human proteome departs from the idea that folded states are functional, while unfolded states are nonfunctional. This percentage is composed of intrinsically disordered proteins (IDPs) and regions (IDRs) (Fig. 6). IDPs and IDRs are proteins that do not possess a defined structure or region [2]. These proteins, despite their lack of a defined structure, carry out biological roles such as cell signaling, stress response, and have even been correlated with the severity of a disease [10]. An example would be the protein Alpha-synuclein (A-syn), a small protein present in neurons, blood cells, and other tissues [11]. Its N-terminus region is amphipathic, which folds into an alpha-helical structure that interacts with membranes. In contrast, the C-terminus is the IDR portion, negatively charged, and responsible for Ca^{2+} ion binding and performing chaperone-like activity [11].

3 Allosteric Regulation of Protein Structure and Function

Molecular interactions are crucial to regulating protein function. For enzymes, their active sites are the primary functional site. Active sites are specific protein regions that interact with substrate molecules to catalyze their conversion into products or interact with other molecules to set forth the activation of key amino acids that

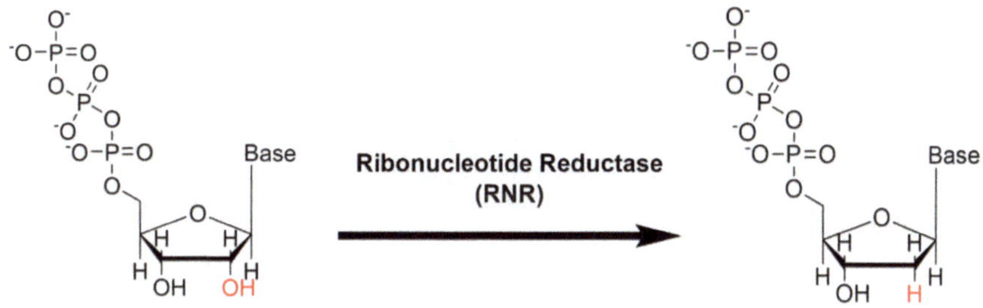

Fig. 7 Conversion of ribonucleotides to deoxyribonucleotides catalyzed by RNR

partake in the catalytic process. Some proteins also have allosteric sites, distinct, and distant from the active site [12, 13]. Allosteric sites bind specific molecules called allosteric effectors or modulator molecules that are not substrates but can influence the protein's behavior. When an allosteric effector binds to an allosteric site, it induces a domino effect transmitted through a conformational change, resulting in parts of the protein to move, twist, or rearrange. This conformational change alters the spatial arrangement, chemical environment, and accessibility of the active site [13, 14]. This has two effects: a positive allosteric effect (activation), which makes the active site more receptive to substrate binding and increases protein activity; and a negative allosteric effect (inhibition), which distorts the active site, making substrate binding more difficult and decreasing activity. This regulation allows cells to optimize metabolic processes without direct competition for the active site, providing a mechanism for feedback control in metabolic pathways [13, 14].

A highly studied example of allosteric effects is the protein ribonucleotide reductase (RNR). Ribonucleotide reductase is responsible for the biosynthesis of all DNA precursors by catalyzing the conversion of ribonucleotides to deoxyribonucleotides (Fig. 7) [15]. There are several types of this protein. Eukaryotic organisms and many aerobic bacteria use class Ia RNRs, which we will briefly focus on here. They are composed of two main subunits, α and β, which combine to form an active tetramer (α2β2). The α subunit contains three nucleotide-binding sites, while the β subunit contains a diiron(III) tyrosyl radical cofactor (Y176• in human RNR). The Y• cofactor in β oxidizes the conserved cysteine in the α active site, generating a sulfhydryl radical through a long-distance radical transfer, which triggers nucleotide reduction [16]. Class Ia RNRs can reduce any of the four ribonucleoside 5′-diphosphates (ADP, CDP, UDP, and GDP) bound to the active α site (site A). However, substrate specificity is controlled by the binding of deoxyribonucleoside 5′-triphosphates (dATP, TTP, and dGTP) and adenosine triphosphate (ATP) to an allosteric site (site S) on the α subunit; the so-called specificity site (Fig. 8). For RNR activity to occur, the β2 and α2 subunits must be in precise positions. This is crucial for

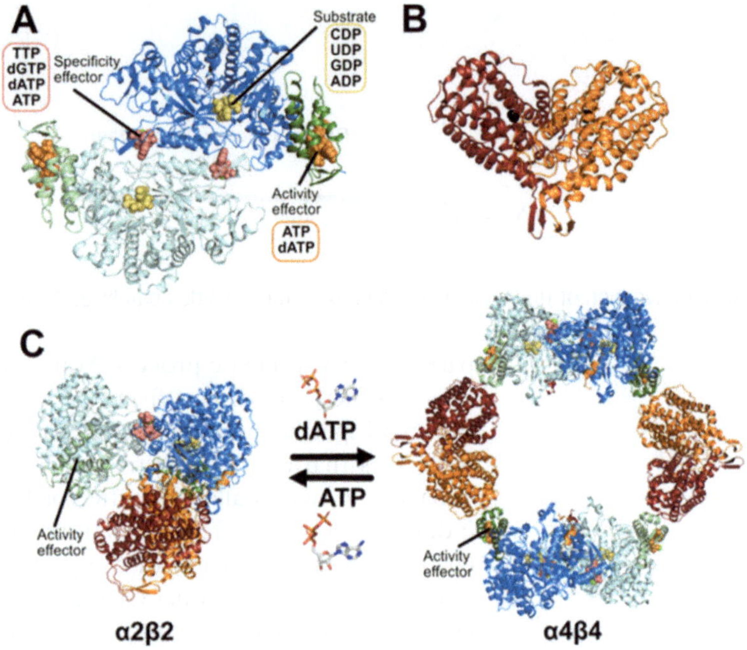

Fig. 8 The structure of *Escherichia coli* class Ia RNR. (**a**) The $\alpha 2$ subunit (cyan and blue, PDB: 8VHP) binds the NDP substrate in the active site and ATP and dNTP effectors in the allosteric site. The N-terminal domain containing the allosteric active site is shown in green. (**b**) Structure of the $\beta 2$ subunit (PDB: 6W4X). (**c**) The $\alpha 2\beta 2$ structure (PDB: 6W4X). The inactive $\alpha 4\beta 4$ complex (PDB: 5CNS) [17]. (Permission to use the figure is according to the terms of the Rightslink®by Copyright Clearance Center)

radical transfer to occur and for catalysis to proceed. Excess dATP inhibits RNR activity by forming an α-β interface. This alters its oligomeric structure and prevents β2 from occupying positions permissive for radical transfer. This effect can only be reversed by ATP coordination to the protein, which in turn alters the α-β interface, releasing β2 and restoring enzyme activity [17]. These allosteric interactions ensure the balanced production of deoxyribonucleotides (dNTPs) required for DNA replication and repair [17, 18]. Allosteric effects can be exploited to develop new and potentially more effective anticancer drugs. By binding to allosteric sites in the inactive site of a protein, protein function can be modulated, thereby inhibiting cancer cell growth and survival.

4 Serum Albumin: A Multifunctional Protein Due to Structural Diversity

Human serum albumin (HSA) is the most abundant serum protein. With a physiological concentration of 35–50 mg/mL in human plasma, HSA accounts for nearly 60% of the total intravascular

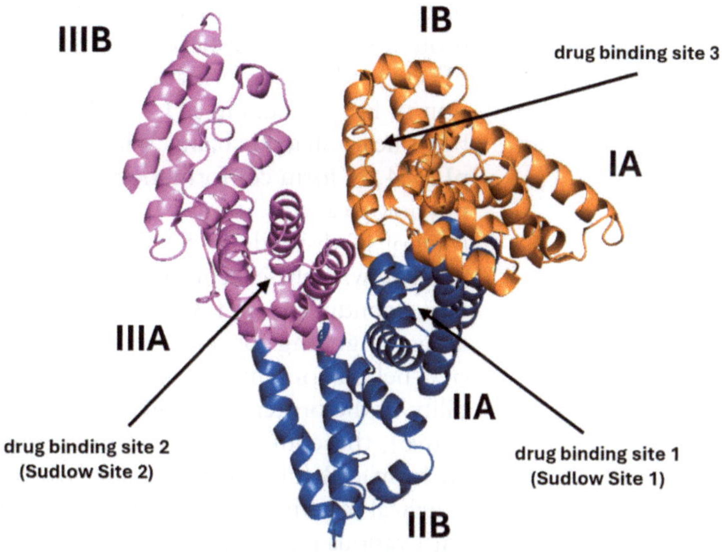

Fig. 9 Structure of a single chain of human serum albumin (HSA) in N-form conformation (PDB 1A06). Domain I (1–195 aa) is colored orange, domain II (196–383 aa) is colored in dark blue, and domain III (384–585 aa) is colored in pink. Subdomains A and B for each domain are indicated with a label. Three prevalent drug binding sites of the protein are denoted with the arrows

protein pool [19, 20]. HSA is produced in hepatocytes, the main functional cells of the liver, where only a small portion (<2 g) is retained and the rest is distributed into human plasma (~13 to 14 g per day) [20, 21]. While circulating within the bloodstream, HSA can extravasate across the vascular endothelium, entering surrounding tissue such as muscle and skin. It is believed that roughly 30–40% of the total HSA pool is within the plasma compartment of blood vessels at a given time, the remaining percent within surrounding cells or the interstitial space [19]. Extravascular HSA is recycled to the plasma via the lymphatic system, allowing for its continued circulation throughout the body. With a substantial half-life estimated at 19 days, HSA is able to recirculate in this manner around 28 times during its lifetime [22].

Structurally, HSA belongs to the globular protein family and consists of a single chain of 585 amino acids, totaling a molecular weight of 66,348 Da [23, 24]. The primary structure of HSA is comprised of 67% α-helices, 10% turns, 23% random coils, and no β-sheets [25, 26]. The protein is built of an N-terminus, three unique domains (labeled as I, II, and III, respectively), and an end by a C-terminus (Fig. 9). Domain I of the protein includes amino acid residues 1–195, domain II is composed of 196–383, and finally, domain III spans from 384 to 585. Each domain is further broken into two subdomains, labeled as A and B, which contain four and six α-helices, respectively [27]. Such classification of the subdomains is based on their unique contribution to the protein's overall function. HSA has three distinct known pH-dependent conformations. Neutral form (N-form) conformation of HSA occurs in pH 4.3–8.0 conditions and is characterized by its heart-shaped chains. With such a broad pH range inclusive of pH 7.4, N-form conformation is the predominant form of HSA in physiological conditions [25]. This conformation is facilitated by a total of 17 disulfide (S–S) bridges between cysteine residues of the protein, with the exception of the free Cys34 residue [28]. In more acidic conditions (pH 3.0–4.3), HSA takes a conformation known as fast migrating form (F-form), characterized by a significant loss of α-helical content, causing increased viscosity and reduced solubility of the protein [29, 30]. In more basic conditions (pH 8 and above), the conformation of the protein is known as basic form (B-form). B-form HSA is similarly distinguished by a loss of α-helical content but has been shown to have improved interactions with various ligands [30].

HSA is a remarkably multifunctional protein, fulfilling a variety of purposes within the body. Primarily, it helps to regulate colloid osmotic pressure and pH of the blood [23]. This is due to the protein greatly contributing to solute concentration within blood vessels, increasing water retention and preventing fluid from leaking into surrounding tissue. Additionally, HSA has a net negative surface charge in the N-form, causing it to draw in positive ions such as sodium, contributing to osmotic pressure via the Gibbs–Donnan effect [19]. Although constantly being synthesized, HSA production is regulated by variation from physiological colloid osmotic pressure, further evidencing its importance in the modulation of blood homeostasis [31]. The protein has also been shown to possess antioxidant effects against harmful reactive oxygen and nitrogen species (ROS/RNS). HSA possess a single free cysteine residue at position 34 (Cys34), which provides a major source of extracellular thiols [19]. With sulfur being more nucleophilic than oxygen and nitrogen, Cys34 donates electrons to neutralize ROS/RNS, reducing their threat of oxidative damage. Another prominent function of HSA is its role as a transport protein for various compounds within the bloodstream. Notably, HSA carries the large majority of fatty acids (99% of nonesterified fatty acids) within the blood, facilitating their solubility [23]. Similarly, it is used in the transport of biomolecules, including hormones, peptides, amino acids, and bilirubin, as well as metal ions [25].

Just as HSA contributes greatly to the transport of various endogenous compounds, it is a known effective carrier for many exogenous molecules, namely, pharmaceuticals. HSA's structure makes it extremely capable of interacting with small-molecule drugs in a manner that affects their pharmacokinetics/pharmacodynamics, including their solubility and uptake within various tissue types. One main contributing factor to this feature of HSA is its abundance of ionized residues. In total, HSA has 83 positively charged residues (Arg + Lys) and 98 negatively charged residues (Asp + Glu), plus 16 histidine residues, which can be protonated in physiological pH conditions to provide additional positive charges [32]. This array of positive and negative surface charges allows for HSA to bind and carry a wide spectrum of drug molecules. Additionally, N-form HSA folding creates numerous inner hydrophobic pockets within the protein that serve as common binding sites to improve the solubility of otherwise poorly water-soluble ligands or biomolecules.

HSA has seven uniquely identified high-affinity fatty-acid (FA) binding sites and more than 20 additional low-affinity FA binding sites [23]. Detailed structural analysis of the protein and its interaction with various ligands have revealed HSA to additionally possess three prominent drug binding sites [23, 33]. These sites are located within the subdomains IIA, IIIA, and IB. The drug binding sites in subdomain IIA and IIIA were first fully characterized by G. Sudlow et al. and are accordingly termed Sudlow site 1 and Sudlow site 2, respectively [33]. These different binding sites often favor specific types of drug interaction and can include both covalent and non-covalent attachments. Sudlow site 1 is characterized by numerous polar residues that are grouped at both the bottom of the site and at the opening of the binding pocket [23]. These residues include Tyr150, Lys195, Lys199, His242, Arg257, and His288 [34]. Sudlow site 1 has been shown to most favorably bind dicarboxylic acids as well as large heterocyclic, negatively-charged molecules [21]. Examples of drugs that bind with HSA at this site include warfarin, azidothymidine, and indomethacin. Sudlow site 2 is composed of residues of varying polarity and is mostly marked by its hydrophobic nature. Polar residues involved with drug binding at the site include Arg410, Tyr411, Lys414, and nonpolar residues include Leu407, Leu430, Leu453, and Phe488. Sudlow site 2 often binds aromatic carboxylic acids that have hydrophobic centers. Notable drugs that bind at this site include ibuprofen, diazepam, and benzodiazepine [21, 27]. It is worth noting that the binding of drugs is not exclusive to these sites; it often plays out that interactions between ligands and proteins in these regions are the most thermodynamically favorable.

As previously mentioned, HSA is capable of circulating outside of blood vessels into surrounding tissue and interstitial space, before being recycled by the lymphatic system. Through binding with drug molecules and serving as a carrier in this way, HSA can largely influence their absorption, distribution, metabolism, and excretion (ADME) profile. Mainly, HSA increases the specificity of bound drugs and thus lowers the occurrence and severity of adverse effects [35]. HSA's movement throughout the body is facilitated by its affinity for numerous cellular receptors, including glycoprotein Gp60 and secreted protein acidic and rich in cysteine

(SPARC). Also referred to as albondin, Gp60 is the primary receptor responsible for the transcytosis of HSA across vascular endothelial cells [27]. Binding of HSA to Gp60 activates a third protein, caveolin-1 (CAV1), which is an integral membrane protein that induces the formation of a caveolae, a specific type of endocytic vesicle. As a result of this interaction, HSA and its bound ligands are shuttled into the interstitial space between various kinds of cells [27]. Cellular uptake of HSA is then mediated by surface receptors on such cells. One such highly studied receptor of this nature for HSA is SPARC. Many clinical studies have confirmed that SPARC enhances the uptake of albumin-based drug systems, primarily in tumor cells as a consequence of its overexpression in cancer [36, 37]. Collectively, these receptors and associated pathways make it possible for HSA to readily reach tissues all throughout the body, even traditionally hard-to-access areas such as the brain [38]. The protein's natural biocompatibility and non-immunogenicity make it a powerful candidate for enhanced drug efficacy within the body.

5 Evaluating Protein States

Proteins constantly explore their energy landscape, transitioning between conformational states, mainly driven by dihedral flexibility of the backbone dihedral angles phi (N–C_{alpha} bond) and psi (C_{alpha}–C bond) (Fig. 10) [2]. Structural biologists use techniques like cryo-electron microscopy (cryo-EM) and X-ray crystallography to capture snapshots of proteins in distinct potential energy wells.

Let's consider protein X, of which two structures in different conformational states have been published: the apo state and the active state. One of the most straightforward ways to address

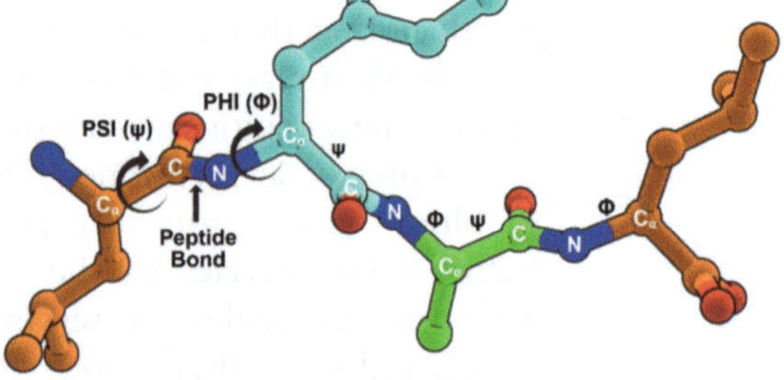

Fig. 10 The backbone dihedral angles, which describe the twist and rotation around specific bonds in the protein's polypeptide chain. The main angles are phi (Φ) around the N–C_{alpha} bond and psi (Ψ) around the C_{alpha}–C bond. These angles are important because they restrict protein conformation due to steric hindrance and the planarity of the peptide bond. (Created with BioRender.com)

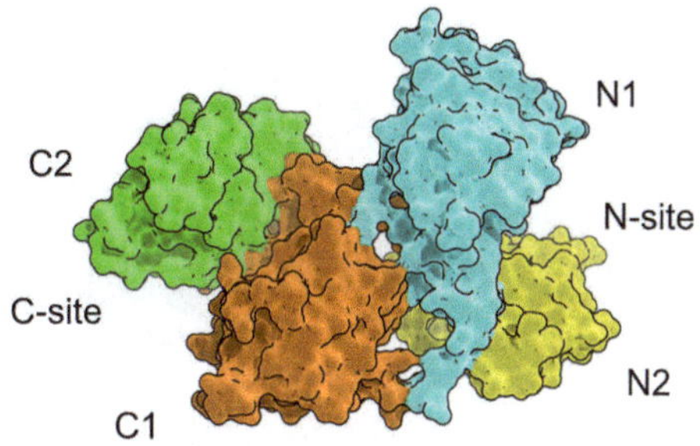

Fig. 11 Surface representation of bilobal serum transferrin (sTf) (PDB 2HAU) and its C-terminus and N-terminus lobes. Both protein lobes contain an Fe(III) binding site

whether there is a significant conformational change between these states is by performing structural alignment, meaning to overlay one structure over the other and observe whether there is a difference between the two. The metric commonly used to quantify these potential structural changes is the root mean square deviation (RMSD), which is a statistical measure that determines the difference between two entities, often employed to address similarity [39]. From a structural perspective, it is calculated based on the aligned pairs of the alpha carbon atoms between the two structures, and the result is a measurement reported in angstroms (Å) [40]. A low RMSD value tells us that the aligned structures are very similar to each other, thus suggesting minimal conformational change. On the other hand, High RMSD values indicate a difference between structures.

A good example of a protein with two distinctive conformational states is serum transferrin (sTf). This protein is responsible for transporting Fe ions to the cells [41]. The apo form of the protein is its metal-free form, and the holo "active" form of the protein is the Fe(III)-bound form. sTf is divided into the N-terminus lobe and the C-terminus lobe, which are further divided into the subdomains N1, N2, C1, and C2 (Fig. 11) [41]. When aligning the apo-sTf structure to its holo structure, the RMSD increases due to conformational changes in the N2 and C2 subdomains [42]. This lack of alignment is seen because these domains are responsible for lobe closure upon metal binding of two Fe(III) ions (Fig. 12), in contrast to their open conformation in the apo state of the structure (Fig. 13) [41]. The conformational change correlates with an increased stability of the holo form of the protein relative to the apo form as measured by thermal and chemical denaturation assays. The melting temperature (T_m) of the holo protein is 18–30 °C higher than the T_m of the individual lobes of the apo protein [43], which is a significant increase in stability. The holo protein also travels further vertically down a UREA

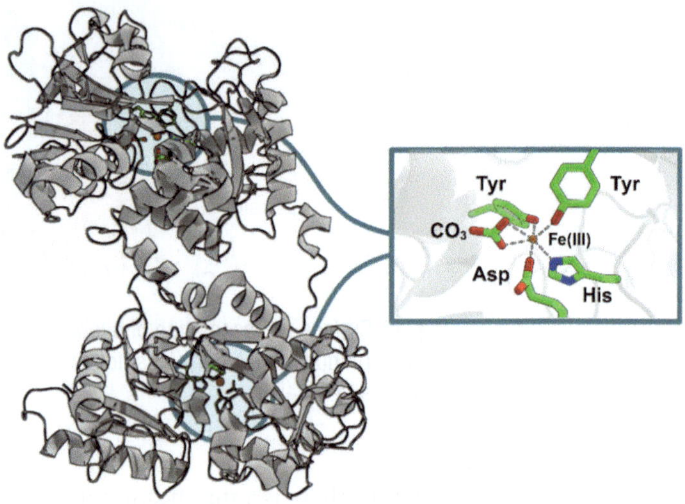

Fig. 12 Cartoon representation of human serum transferrin highlighting its Fe(III) binding sites (PDB 3 V83). (Created with BioRender.com)

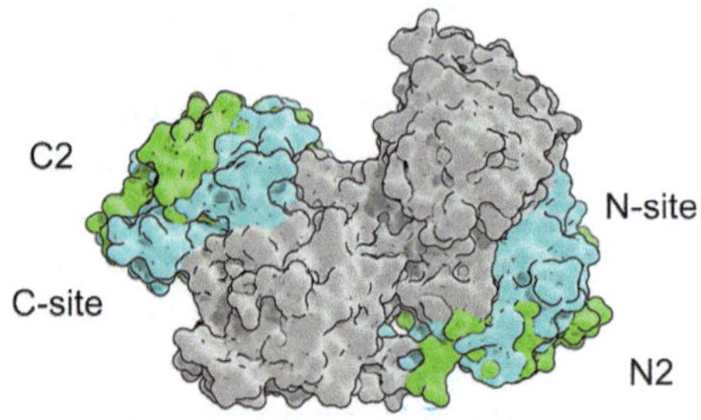

Fig. 13 Surface representation of apo-sTf (green) (PDB 2HAU) aligned with Holo-sTf (cyan) (PDB 3 V83)

PAGE gel than the apo protein [44]. These structural dynamics also influence the solvent exposure of some amino acids, thus having an influence on the protein's surface charge distribution.

6 Surface Charge Distribution

By principle of their physical nature, all solid materials possess a surface, and proteins are not the exception. As discussed earlier, amino acids can be neutral, positively charged, or negatively charged. When a protein folds into its three-dimensional structure, these residues become distributed across its surface, making a unique pattern of charges. This surface charge distribution is sensitive to the pH of the surrounding solvent, which determines the protonation and deprotonation of the ionizable side chains, as can be seen for the bovine version of serum albumin (>80% sequence

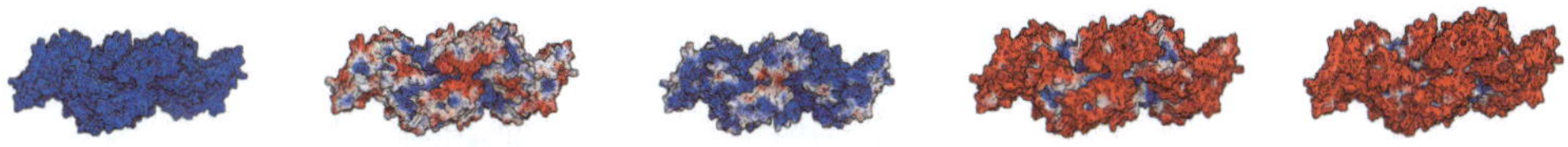

Fig. 14 Surface charge distribution at different pH values of bovine serum albumin (PDB 3 V03) protonation states calculated with PDB2PQR and APBS. The blue color represents positively charged regions, and the red represents negatively charged regions

homologous to the human version) in Fig. 14. Acidic residues such as Glutamate and Aspartate tend to lose protons at higher pH values, while basic amino acids like Histidine gain protons at lower pH values, hence becoming positively charged.

As the pH changes, so does the net charge of the protein, which leads to another important concept: the isoelectric point (pI). The isoelectric point is defined as the specific pH value at which the number of positively charged groups equals the number of negatively charged groups. This results in a net charge of zero. The absence of charge can significantly decrease protein solubility and result in precipitation from solution. The pI can be experimentally determined by titrations and 2D-gels, but it may also be determined computationally based on its sequence.

Due to computational analysis advances we are able to explore the surface charge distribution by using platforms such as PDB2PQR and APBS, which operate together. PDB2PQR serves as the preparatory framework, converting protein structure files, typically in PDB format, which we will describe further in this chapter, into a form suitable for the solvation calculations carried out by APBS [45, 46].

Utilizing these platforms, we may see, for instance, when the surface charge distribution is sensitive to any conformational change that may alter the solvent exposure to an amino acid. For example, as previously shown, apo-sTf, when aligned to Fe(III)-bound sTf, conveyed a high RMSD due to the change in conformation of subdomains C2 and N2. If we were to compare the charge surface distribution between these two proteins, we may appreciate that the surface charge in Fe_2-sTf is more negative than apo-sTf, revealing that the electrostatics of the protein are altered due to Fe(III) binding [47] (Fig. 15). Another valuable example for exploring surface electrostatics as a mechanism of cellular recognition is the human transferrin receptor protein-1 (TfR1). Upon binding two holo-sTf molecules, one per chain in the homodimer, the ectodomain of TfR1 becomes more negatively charged [47]. This electrostatic shift has been proposed as a critical prereq uisite for holo-sTf recognition and the subsequent initiation of endocytosis (Fig. 16). Holo-sTf exhibits high affinity for TfR1 (K_d ~ nM) at pH 7.4, whereas the apo form has much weaker affinity. However, at pH 5.5, the holo-sTf releases the Fe(III) ions

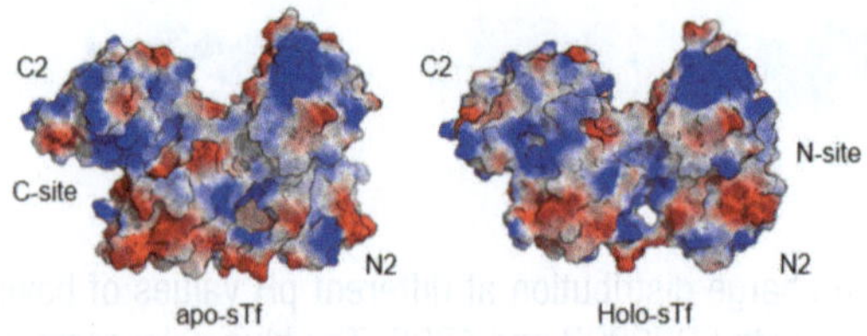

Fig. 15 Distribution of the surface charge for apo-sTf (PDB 2HAU) and for holo-sTf (Fe2-sTf) (PDB 3 V83) at pH 7.4, calculated using PDB2PQR and APBS

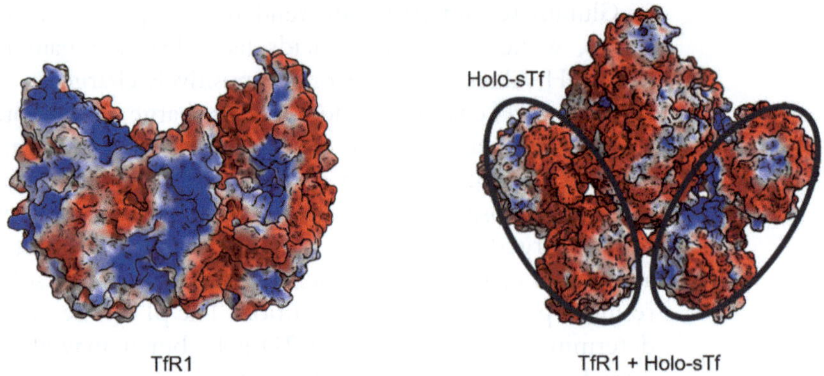

Fig. 16 Distribution of the surface charge for the ecto domain of human transferrin receptor protein-1(TfR1) (PDB 1CX8) and for TfR1 + holo-sTf (PDB 1SUV) at pH 7.4, calculated using pdb2pqr and APBS

in the endosome, and the affinity of the apo form of the protein is significantly higher for the TfR1 (K_d ~ nM), enabling sTf to return to the membrane to be recycled for more rounds of Fe cellular uptake [48].

7 Methods

7.1 Visualization (What Is PyMOL?)

You might be wondering how the previously shown protein structures were generated, and how modern biochemists and protein scientists visualize and perform preliminary analyses of structures. The answer is often PyMOL. PyMOL is a molecular visualization program that allows users to explore not only protein structures, but also nucleic acids and small molecules. In this section, we will introduce some fundamental PyMOL usage with a focus on protein structures beyond simple structure visualization, such as exploring structure alignment tools, solvent-accessible surface area (SASA), and surface charge distribution mapping.

In PyMOL, each object displayed in the viewer is manipulated by a command panel containing a series of lettered buttons, each corresponding to a specific category of functions (Fig. 17). The letter A stands for *Action*, and it includes commands related to modifying the object, such as renaming, duplicating, or deleting.

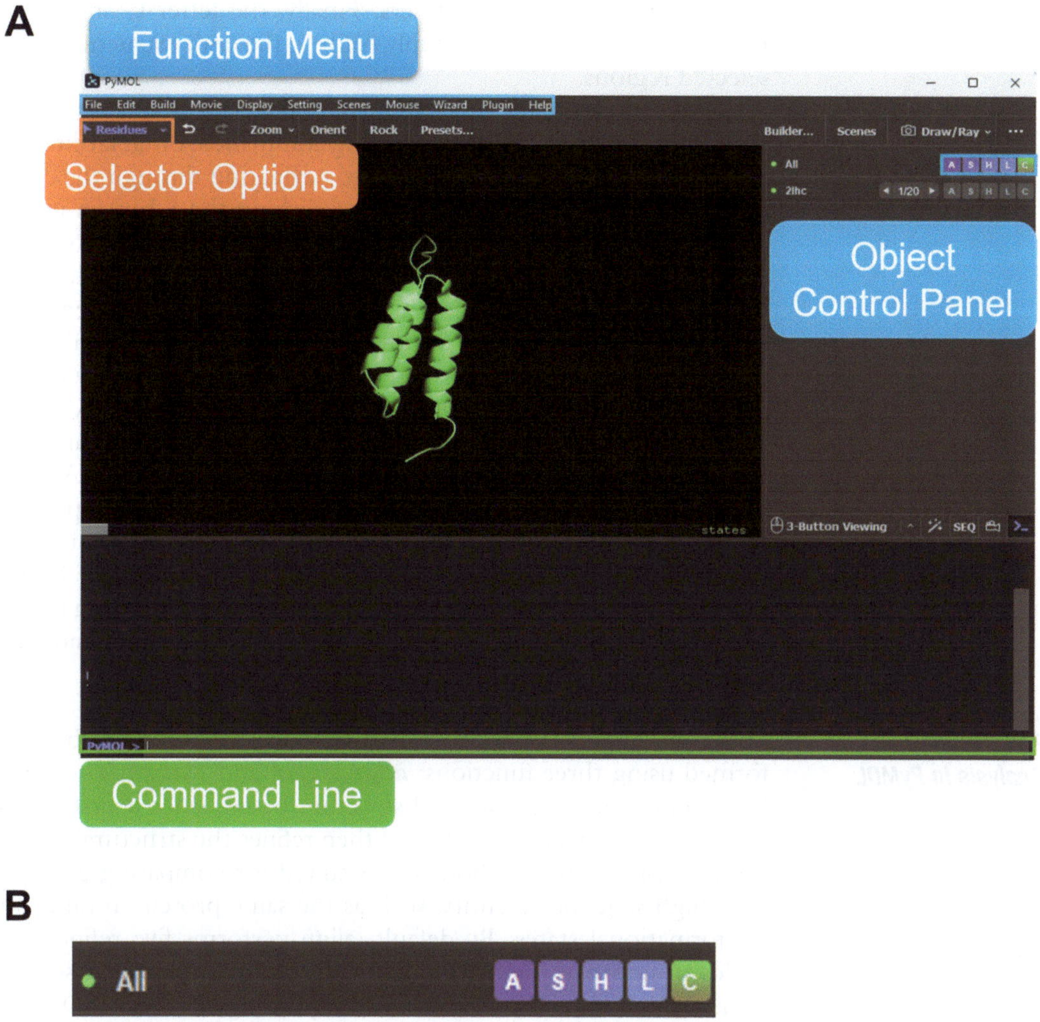

Fig. 17 PyMOL command interface. (**a**) The object control panel. (**b**) A zoom-in of the all command

It also gives access to view orientation tools like *zoom*, *orient*, and *center*, which help position the structure within the viewer. Also, under this menu, there is a *compute toggle* that allows users to perform preliminary analyses, such as calculating and visualizing at a per-residue level the solvent accessible surface area (SASA). This integrated feature will be explored later in this chapter. The letter S stands for *Show*, which controls how the structure is visually represented. By default, PyMOL displays proteins in cartoon form, but this menu enables users to switch to other representations, such as sticks to highlight side chains, surface to reveal the molecular envelope, among other representations. The letter H stands for *Hide*, and it is used to remove any visual elements added to the structure. The letter L stands for *Label*, and it enables the display of annotations like residue names (one-letter or three-letter format), atom names, or other identifiers. Finally, the letter C stands for *Color*, which allows users to apply color to the entire structure or to selected regions.

The command line in PyMOL is an interface that allows users to execute functions directly by typing commands, often facilitating the visualization process compared to navigating through the command panel. For example, instead of clicking the S button in the object command panel and selecting the surface toggle, you may simply type *show surface* in the command line to display the protein's surface. Many such commands exist, and learning them may significantly improve your workflow. One particularly useful command is *fetch*, which allows users to download protein structures directly from the Protein Data Bank (PDB), which is enabled when connected to the Internet. The PDB archives the 3D structure data for large biological molecules (proteins, DNA, and RNA). To implement it, you only need to know the PDB code of the protein structure of interest. For example, the human apo serum transferrin structure has a PDB code of 2HAU. Instead of visiting the RCSB PDB website, downloading the file, and manually importing it into PyMOL, you may simply type *fetch 2HAU* into the command line, and the structure will automatically load into the viewer.

7.2 Structural Analysis in PyMOL

Structural alignment between two protein structures can be performed using three functions: *align, super,* and *cealign*. The *align* function is both sequence and structure-based, since it first aligns residues by sequence identity and then refines the structural superposition [49]. This method is best suited for comparing proteins with high sequence identity, such as the same protein in different conformational states. By default, align performs five refinement cycles to improve RMSD by removing outlier atoms, but setting *cycles = 0* sets the cycles to zero. Thus, yielding a true RMSD value between the two structures. For example, aligning apo-sTf (PDB: 2HAU) and holo-sTf (PDB: 3 V83) using *align 2HAU, 3 V83, cycles = 0* results in an RMSD of approximately 8.29 Å, showing substantial conformational change upon Fe(III) binding as mentioned previously in the chapter. In contrast, the *super* function performs a sequence independent alignment, meaning that it solely relies on structure geometry, making it a suitable function when comparing two structures with moderate to low sequence identity [50]. Similarly, *cealign* is also sequence independent but relies on a combinatorial extension (CE) algorithm to detect structural similarities without considering sequence [51]. For instance, when comparing human holo-serum transferrin (PDB: 3 V83) to lactoferrin (PDB: 1BLF), which share structural features but lack high sequence identity, both *super* and *cealign* are appropriate (Fig. 18). By using *super 3 V83, 1BLF, cycles = 0* and/or *cealign 3 V83, IBFL*, you will obtain that for super there is an RMSD of 4.12 Å, while for

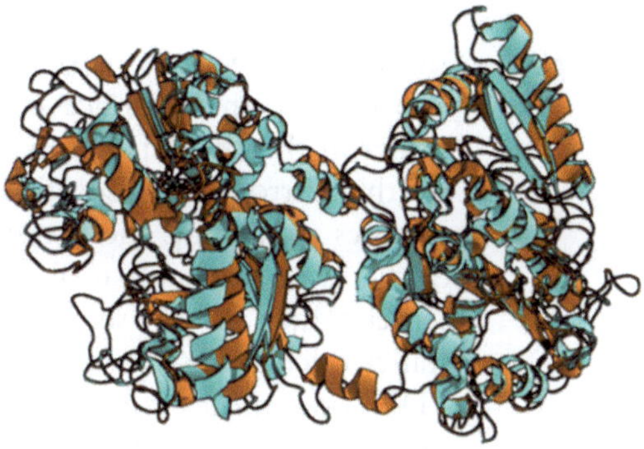

Fig. 18 Alignment of Holo-sTf (cyan) (PDB 3 V83) and Lactoferrin (orange) (PDB 1BLF)

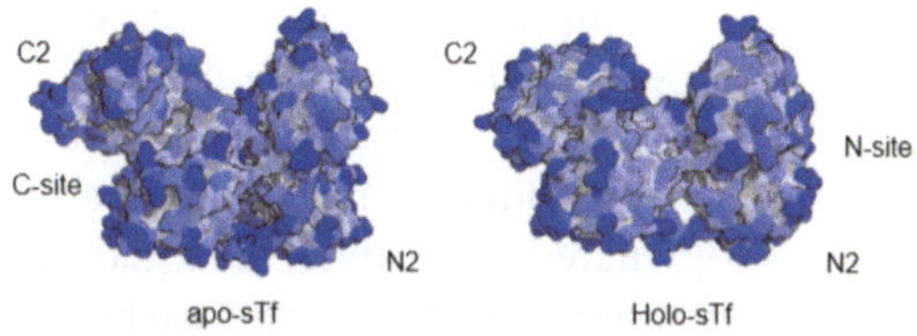

Fig. 19 Solvent-accessible surface area (SASA) representation for apo-sTf (PDB 2HAU) and for holo-sTf (Fe2-sTf) (PDB 3 V83)

cealign is 4.26 Å, indicating that there is a moderate structural similarity between the two.

We can also visualize the solvent-accessible surface area (SASA) at a per-residue level, as previously discussed. To access this in PyMOL, click the "A" menu next to the protein structure of interest, navigate to Compute, then select Surface Area → Per Residue (Relative Solvent Accessibility). Upon activation, each residue will be annotated with a numerical value representing its degree of solvent exposure. These values range from 0 to 1, where 1 indicates full exposure to solvent and 0 indicates complete burial. PyMOL calculates these values by simulating a water molecule probe across the protein surface. Wherever the probe can fit and make contact, the system assigns a corresponding accessibility score. The structure is also color-coded on a gradient from dark blue (high exposure) to white (low or no exposure), providing an intuitive visual cue. Additionally, the command line will display the solvent exposure percentage for each residue.

For optimal visualization, it's recommended to render the protein in surface mode. For example, when comparing apo-sTf and holo-sTf, the N2 and C2 subdomains (Fig. 19), which mediate closure upon Fe(III) binding, become visibly whiter in the holo form. This shift indicates that the residues in these regions become more buried, and the water probe can no longer access them—a clear structural consequence of metal coordination.

7.3 Surface Charge Distribution

In PyMOL, we can visualize protein structures in detail, but we can also explore their surface electrostatic properties, as introduced earlier in the chapter. Because surface charge depends on the pH of the surrounding solution, visualizing electrostatic changes requires simulating the protein's environmental conditions. One of the most accessible tools for this is the PDB2PQR/APBS web server, which allows users to prepare protein structures for electrostatics calculations. To use it, simply upload a PDB file or enter a PDB code, specify the desired pH, and leave the remaining parameters at their default settings [45]. After following the pipeline on the website, the final step offers a visualization of the processed structure. However, to integrate the results into PyMOL, go to the Export tab, select the PyMOL option, and download the three output files. These files should be opened together in the same PyMOL session to properly visualize the surface charge distribution. The resulting structure will be color-coded: red indicates protonated (positively charged) regions, blue indicates deprotonated (negatively charged) regions, and white represents neutral residues. When the surface appears mostly white, the protein is likely near its isoelectric point, where the net surface charge is minimal.

Acknowledgments

We are grateful for the support of the American Cancer Society Discovery Boost DBG-23-1156894-01-ET (A.D.T. and O.C.A.), MSEIP: TEES CHEM grant number PA120A240033 (L.A.L.C.), NSF REU: PR CLIMB grant 2349168(O.B.), and NIH RISE 5R25GM061151-22 grant (L.A.L.C.).

Conflicts of Interest The authors have no conflicts of interest to declare that are relevant to the content of this chapter.

References

1. Buxbaum E (2015) Fundamentals of protein structure and function. Springer International Publishing
2. Smetana JHC, Misra G (2017) Principles of protein structure and function. In: Introduction to biomolecular structure and biophysics. Springer Singapore, pp 1–32
3. Banks CA, Kong SE, Washburn MP (2012) Affinity purification of protein complexes for analysis by multidimensional protein identification technology. Protein Expr Purif 86(2): 105–119. https://doi.org/10.1016/j.pep.2012.09.007
4. Almeida P (2016) Proteins: concepts in biochemistry. Garland Science
5. Sanvictores T, Farci F (2025) Biochemistry: primary protein structure. StatPearls Publishing
6. Onuchic JN, Luthey-Schulten Z, Wolynes PG (1997) Theory of protein folding: the energy landscape perspective. Annu Rev Phys Chem 48:545–600. https://doi.org/10.1146/annurev.physchem.48.1.545

7. Alberts B, Johnson A and Lewis J (2002) Membrane Proteins. Garland Science, New York
8. Kumari S, Mg S, Mayor S (2010) Endocytosis unplugged: multiple ways to enter the cell. Cell Res 20(3):256–275. https://doi.org/10.1038/cr.2010.19
9. Cheng Y, Zak O, Alsen P, Harrison SC, Walz T (2004) Structure of the human transferrin receptor-transferrin complex. Cell 116(4): 565–576. https://doi.org/10.1016/s0092-8674(04)00130-8
10. Ball P (2025) Scientific American. https://www.scientificamerican.com/article/mysterious-blobs-found-in-cells-are-rewriting-how-life-works/
11. Calabresi P, Mechelli A, Natale G, Volpicelli-Daley L, Di Lazzaro G, Ghiglieri V (2023) Alpha-synuclein in Parkinson's disease and other synucleinopathies: from overt neurodegeneration back to early synaptic dysfunction. Cell Death Dis 14(3):176. https://doi.org/10.1038/s41419-023-05672-9
12. Horovitz A, Willison KR (2005) Allosteric regulation of chaperonins. Curr Opin Struct Biol 15(6):646–651. https://doi.org/10.1016/j.sbi.2005.10.001
13. Abhishek S, Deeksha W, Nethravathi KR, Davari MD, Rajakumara E (2023) Allosteric crosstalk in modular proteins: function fine-tuning and drug design. Comput Struct Biotechnol J 21:5003–5015. https://doi.org/10.1016/j.csbj.2023.10.013
14. Kobe B, Kemp BE (2010) Chapter 74 – Principles of kinase regulation. In: Bradshaw RA, Dennis EA (eds) Handbook of cell signaling, 2nd edn. Academic, San Diego, pp 559–563
15. Ahmad MF, Dealwis CG (2013) Chapter fourteen – The structural basis for the allosteric regulation of ribonucleotide reductase. In: Giraldo J, Ciruela F (eds) Progress in molecular biology and translational science, vol 117. Academic, pp 389–410
16. Licht S, Gerfen GJ, Stubbe J (1996) Thiyl radicals in ribonucleotide reductases. Science 271(5248):477–481. https://doi.org/10.1126/science.271.5248.477
17. Funk MA, Zimanyi CM, Andree GA, Hamilos AE, Drennan CL (2024) How ATP and dATP act as molecular switches to regulate enzymatic activity in the prototypical bacterial class Ia ribonucleotide reductase. Biochemistry 63(19):2517–2531. https://doi.org/10.1021/acs.biochem.4c00329
18. Ando N, Li H, Brignole EJ, Thompson S, McLaughlin MI, Page JE, Asturias FJ, Stubbe J, Drennan CL (2016) Allosteric inhibition of human ribonucleotide reductase by dATP entails the stabilization of a hexamer. Biochemistry 55(2):373–381. https://doi.org/10.1021/acs.biochem.5b01207
19. Garcovich M, Zocco MA, Gasbarrini A (2009) Clinical use of albumin in hepatology. Blood Transfus 7(4):268–277. https://doi.org/10.2450/2008.0080-08
20. Peters JT (1995) All about albumin: biochemistry, genetics, and medical applications. Academic, San Diego
21. Hoogenboezem EN, Duvall CL (2018) Harnessing albumin as a carrier for cancer therapies. Adv Drug Deliv Rev 130:73–89. https://doi.org/10.1016/j.addr.2018.07.011
22. Wang S, Liu S, Zhang Y, He J, Coy DH, Sun L (2020) Human serum albumin (HSA) and its applications as a drug delivery vehicle. Health Sci J 14
23. Mishra V, Heath RJ (2021) Structural and biochemical features of human serum albumin essential for eukaryotic cell culture. Int J Mol Sci 22(16). https://doi.org/10.3390/ijms22168411
24. Merlot AM, Kalinowski DS, Richardson DR (2014) Unraveling the mysteries of serum albumin-more than just a serum protein. Front Physiol 5:299. https://doi.org/10.3389/fphys.2014.00299
25. Asrorov AM, Mukhamedov N, Kayumov M, Sh. Yashinov A, Wali A, Yili A, Mirzaakhmedov SY and Huang Y (2024) Albumin is a reliable drug-delivering molecule: highlighting points in cancer therapy. Med Drug Discov 22: 100186 doi:https://doi.org/10.1016/j.medidd.2024.100186
26. Murayama K, Tomida M (2004) Heat-induced secondary structure and conformation change of bovine serum albumin investigated by Fourier transform infrared spectroscopy. Biochemistry 43(36):11526–11532. https://doi.org/10.1021/bi0489154
27. Larsen MT, Kuhlmann M, Hvam ML, Howard KA (2016) Albumin-based drug delivery: harnessing nature to cure disease. Mol Cell Ther 4: 3. https://doi.org/10.1186/s40591-016-0048-8
28. Paris G, Kraszewski S, Ramseyer C, Enescu M (2012) About the structural role of disulfide bridges in serum albumins: evidence from protein simulated unfolding. Biopolymers 97(11): 889–898. https://doi.org/10.1002/bip.22096
29. Muzammil S, Kumar Y, Tayyab S (1999) Molten globule-like state of human serum albumin at low pH. Eur J Biochem 266(1):26–32.

https://doi.org/10.1046/j.1432-1327.1999.00810.x

30. Ashraf S, Qaiser H, Tariq S, Khalid A, Makeen HA, Alhazmi HA, Ul-Haq Z (2023) Unraveling the versatility of human serum albumin – a comprehensive review of its biological significance and therapeutic potential. Curr Res Struct Biol 6:100114. https://doi.org/10.1016/j.crstbi.2023.100114
31. Nicholson JP, Wolmarans MR, Park GR (2000) The role of albumin in critical illness. Br J Anaesth 85(4):599–610. https://doi.org/10.1093/bja/85.4.599
32. Belinskaia DA, Voronina PA, Batalova AA, Goncharov NV (2021) Serum Albumin Encyclopedia. 1(1):65–75
33. Sudlow G, Birkett DJ, Wade DN (1975) The characterization of two specific drug binding sites on human serum albumin. Mol Pharmacol 11(6):824–832
34. Zhivkova ZD (2015) Studies on drug-human serum albumin binding: the current state of the matter. Curr Pharm Des 21(14):1817–1830. https://doi.org/10.2174/1381612821666150302113710
35. Yang F, Zhang Y, Liang H (2014) Interactive association of drugs binding to human serum albumin. Int J Mol Sci 15(3):3580–3595. https://doi.org/10.3390/ijms15033580
36. Desai N, Trieu V, Damascelli B, Soon-Shiong P (2009) SPARC expression correlates with tumor response to albumin-bound paclitaxel in head and neck cancer patients. Transl Oncol 2(2):59–64. https://doi.org/10.1593/tlo.09109
37. Zheng B, Chen Y, Niu L, Zhang X, Yang Y, Wang S, Chen W, Cai Z, Huang W, Huang W (2024) Modulating the tumoral SPARC content to enhance albumin-based drug delivery for cancer therapy. J Control Release 366:596–610. https://doi.org/10.1016/j.jconrel.2023.12.057
38. Liang J, Gao C, Zhu Y, Ling C, Wang Q, Huang Y, Qin J, Wang J, Lu W, Wang J (2018) Natural brain penetration enhancer-modified albumin nanoparticles for glioma targeting delivery. ACS Appl Mater Interfaces 10(36):30201–30213. https://doi.org/10.1021/acsami.8b11782
39. Kabsch W (1976) A solution for the best rotation to relate two sets of vectors. Acta Crystallogr Sect A 32(5):922–923. https://doi.org/10.1107/S0567739476001873
40. RCSB Protein Data Bank. https://www.rcsb.org/docs/tools/pairwise-structure-alignment. Accessed 17 Aug 2025
41. Benjamín-Rivera JA, Cardona-Rivera AE, Vázquez-Maldonado ÁL, Dones-Lassalle CY, Pabón-Colon HL, Rodríguez-Rivera HM, Rodríguez I, González-Espiet JC, Pazol J, Pérez-Ríos JD, Catala-Torres JF, Carrasquillo Rivera M, De Jesus-Soto MG, Cordero-Virella NA, Cruz-Maldonado PM, González-Pagan P, Hernández-Ríos R, Gaur K, Loza-Rosas SA, Tinoco AD (2020) Exploring serum transferrin regulation of nonferric metal therapeutic function and toxicity. Inorganics 8(9):48. https://doi.org/10.3390/inorganics8090048
42. Wally J, Halbrooks PJ, Vonrhein C, Rould MA, Everse SJ, Mason AB, Buchanan SK (2006) The crystal structure of iron-free human serum transferrin provides insight into inter-lobe communication and receptor binding. J Biol Chem 281(34):24934–24944. https://doi.org/10.1074/jbc.M604592200
43. Lin LN, Mason AB, Woodworth RC, Brandts JF (1994) Calorimetric studies of serum transferrin and ovotransferrin-estimates of domain interactions, and study of the kinetic complexities of ferric ion-binding. Biochemistry 33(7):1881–1888. https://doi.org/10.1021/bi00173a035
44. Levina A, Wang B, Lay PA (2022) Urea gel electrophoresis in studies of conformational changes of transferrin on binding and transport of non-ferric metal ions. Gels 8(1):19
45. Jurrus E, Engel D, Star K, Monson K, Brandi J, Felberg LE, Brookes DH, Wilson L, Chen J, Liles K, Chun M, Li P, Gohara DW, Dolinsky T, Konecny R, Koes DR, Nielsen JE, Head-Gordon T, Geng W, Krasny R, Wei GW, Holst MJ, McCammon JA, Baker NA (2018) Improvements to the APBS biomolecular solvation software suite. Protein Sci 27(1):112–128. https://doi.org/10.1002/pro.3280
46. Dolinsky TJ, Czodrowski P, Li H, Nielsen JE, Jensen JH, Klebe G, Baker NA (2007) PDB2PQR: expanding and upgrading automated preparation of biomolecular structures for molecular simulations. Nucleic Acids Res 35:W522–W525. https://doi.org/10.1093/nar/gkm276
47. Yajima H, Sakajiri T, Kikuchi T, Morita M, Ishii T (2000) Molecular modeling of human serum transferrin for rationalizing the changes in its physicochemical properties induced by iron binding. Implication of the mechanism of binding to its receptor. J Protein Chem 19(3):215–223. https://doi.org/10.1023/a:1007059820834
48. Lawrence CM, Ray S, Babyonyshev M, Galluser R, Borhani DW, Harrison SC (1999) Crystal structure of the ectodomain of human

transferrin receptor. Science 286(5440): 779–782. https://doi.org/10.1126/science.286.5440.779

49. Pymolwiki.org. https://pymolwiki.org/index.php/Align. Accessed 18 Aug 2025

50. Pymolwiki.org. https://pymolwiki.org/index.php/Super. Accessed 18 Aug 2025

51. Pymolwiki.org. https://pymolwiki.org/index.php/Cealign. Accessed 18 Aug 2025

Chapter 18

Integrating ChatGPT into Biochemistry Education: A Practical Guide to Developing Interactive Learning Applications

Yannick Belo Nuapia and Hlanganani Tutu

Abstract

To improve the accessibility and ease of use of rigorously derived enzyme-kinetics routines for learners with limited programming experience, we developed a graphical, web-based Michaelis–Menten and inhibition simulator in Python using Streamlit. The interface was assembled through a human-in-the-loop workflow in which concise, goal-directed prompts to ChatGPT yielded small, behavior-preserving code patches via an error-message feedback loop, while all quantitative results were computed by audited, deterministic functions rather than the language model. This approach enabled students and instructors without extensive coding backgrounds to explore substrate saturation, inhibition classes, and linearized diagnostics within a stable, classroom-ready environment. In addition to interface scaffolding, ChatGPT supported practical tasks such as environment configuration, Streamlit setup, and brief, equation-aware explanatory text aligned with course materials. The work serves as a template for undergraduate and postgraduate courses in biochemistry and chemical education, demonstrating how conversational assistance can lower the barrier to creating domain-specific teaching tools without requiring advanced software training. Sample lesson scaffolds illustrate undergraduate activities on varying substrate and inhibitor concentrations and graduate exercises involving Lineweaver–Burk or Eadie–Hofstee analysis and parameter interpretation.

Key words Biochemistry education, Enzyme kinetics, Michaelis–Menten, Enzyme inhibition, Streamlit, ChatGPT, Large language models, Deterministic modeling, Prompt engineering, Educational software

1 Introduction

ChatGPT is a large language model that engages in conversational exchanges to produce text. Following its public release in 2022, the tool has been associated with increased efficiency in research workflows and with support for individualized study [1]. At the same

Supplementary Information The online version contains supplementary material available at https://doi.org/10.1007/978-1-0716-5320-3_18.

Robert E. Dempski (ed.), *Transforming Molecular Biology with Emerging Technologies*, Methods in Molecular Biology, vol. 3043, https://doi.org/10.1007/978-1-0716-5320-3_18,

time, it has prompted justified concern regarding factual inaccuracy, fabricated citations, hallucination, and fragile quantitative reasoning when used without oversight [2]. Within biochemistry education, the question has therefore shifted from whether such systems should be used to how they may be integrated in ways that are pedagogically productive and epistemically sound [3].

Biochemistry provides a particularly instructive setting for this inquiry. The discipline requires mastery of dense terminology, interconnected metabolic pathways, and three-dimensional molecular structure [4]. Appropriately constrained, interactive tutoring systems can deliver timely explanations, analogies, and formative practice that align with course objectives, and early instructional pilots report gains in learner confidence and engagement [5]. Nevertheless, model limitations are well documented. Output may be plausible but incorrect, context may be omitted, and performance can degrade on authentic, open resource assessments. Student perceptions are likewise mixed, with some learners preferring artificial intelligence (AI) generated study materials and others favoring traditional resources [6–8]. These observations support a cautious integration strategy; language models should augment rather than replace conventional teaching, and their operation should be governed by clear academic oversight [9]. To safeguard integrity, institutions and instructors require transparent policies that define acceptable use, ensure equitable access, and position AI as a study aid rather than a source of ready-made answers [7]. Assessment should be designed to emphasize process, reasoning, and application, thereby discouraging reliance on verbatim output [10].

In parallel, practical experience in courses has demonstrated that the same tools can be useful for code authoring and debugging, particularly in Python, where brief, carefully framed prompts can yield working scaffolds that would otherwise demand specialized programming skills [11–13]. This is important in biochemistry, where many learners have limited formal training in programming but increasingly encounter computational tasks in data analysis, visualization, and model deployment [14]. Prior reports have leveraged conversational assistance to assemble beginner-friendly interfaces and have emphasized a human-in-the-loop error message feedback cycle to refine code iteratively [15]. However, a gap remains between rigorously derived scientific models that demand numerical correctness and the student-facing tools needed to make those models accessible to nonexperts [16].

The present study sought to assist in addressing that gap for biochemistry. It developed a reproducible pathway for a Michaelis–Menten and inhibition simulator that separates concerns: all numerical results are produced by audited, deterministic Python functions, while ChatGPT is confined to assist in deployment logistics of the developed application. The chapter describes materials and the setup of the environment, construction of a compact knowledge pack from textbooks and peer-reviewed literature, implementation of deterministic kernels, and a minimal Streamlit interface suitable for headless or local use. It then presents verification procedures, operational guidance for teaching contexts, and governance for responsible model use. The intent is to offer an actionable template that can be adapted to topics such as cooperativity, basic thermodynamics, or pathway energetics, while preserving a clear boundary between conversational support and code-based numerical computing.

2 Methodology

This methodology adapts the workflow reported for ChatGPT-assisted GUI construction in chemical education [1] to a biochemistry learning application while preserving a separation between numerical computation and conversational support. In brief, deterministic Python code implements the kinetics and transforms, whereas ChatGPT is used only to scaffold the Streamlit interface, resolve environment issues, and generate concise, course-aligned explanations. The structure, emphasis on novice accessibility, and use of an error-message feedback loop mirror prior work that paired human prompts with ChatGPT to produce beginner-friendly interfaces in the Jupyter and Anaconda ecosystems.

2.1 Biochemistry Content Sources for Equations

In this project, every equation the app presents, whether a rate law, a linear transform, or a thermodynamic identity, originates from these sources: biochemistry textbooks and a small, curated set of peer-reviewed articles. We transcribe each relation verbatim, normalize symbols to a course-wide convention and record units, admissible domains, and assumptions, including steady state, initial rate, and temperature. Each equation is stored in dual form, LaTeX for student-facing display and a callable SymPy/Python expression for deterministic evaluation. To prevent drift between print and code, we attach reference oracles and correct low or high substrate limits that run as unit tests.

2.2 Software and Computing Environment

Development is performed in Python 3.13 within an isolated virtual environment (e.g., Conda or venv). The core stack comprises numpy and sympy for deterministic mathematics, pandas for tabular data, and matplotlib for plotting. The interface uses Streamlit, a viable alternative where route-level control is required. Optional components include Graphviz for pathway schematics and rdkit for structure thumbnails. Environments are captured with conda env export or pip freeze, and version pins are retained in `requirements.txt` to ensure portability across machines and academic terms. A laptop or desktop with at least 8 GB RAM suffices; no graphical processing unit (GPU) is required.

2.3 Deterministic Kinetics Module

The scientific core consists of pure Python functions implementing initial-rate Michaelis–Menten kinetics:

$$\upsilon([S] : V_{\max}, K_m) = \frac{V_{\max}[S]}{K_m+[S]}$$

Where υ is the reaction rate; S is substrate concentration; $V_{\max}$ is the maximum reaction rate when the substrate concentration is maximum, and K_m is the Michaelis constant. With domain guards $V_{\max} \geq 0$, $K_m > 0$, $[S] \geq 0$. Inhibition is applied via apparent parameter transformations:

Competitive: $\upsilon = \frac{V_{\max}[S]}{\alpha K_m+[S]}$, where $\alpha = 1+\frac{I}{K_i}$

Pure noncompetive: $V_{\max}^{\mathrm{eff}} = \frac{V_{mx}}{\alpha}$, $\upsilon = \frac{V_{\max}^{\mathrm{eff}}[S]}{K_m+[S]}$

Uncompetition: $V_{\max}^{\mathrm{eff}} = \frac{V_{mx}}{\alpha}$, $K_m^{\mathrm{eff}} = \frac{K_m}{\alpha}$, $\upsilon = \frac{V_{\max}^{\mathrm{eff}}[S]}{K_m+[S]}$

Where I is the inhibitor concentration; K_i is the inhibition constant; $V_{\max}^{\mathrm{eff}}$ is the effective maximum velocity in the presence of inhibitor; K_m^{eff} is the effective Michaelis constant in the presence of the inhibitor; α is the inhibition factor.

Vectorization allows S to be a scalar or NumPy array, enabling dense sampling for plots and linearization. Diagnostic transforms (Lineweaver–Burk and Eadie–Hofstee) are computed only when their domain constraints hold (e.g., $[S] > 0$, $\upsilon > 0$). The module consists of pytest cases that assert analytic limits and invariants, including $\upsilon([S] = K_m) = \frac{1}{2}V_{\max}$, $\lim_{[S]\to 0} \upsilon = 0$, and $\lim_{[S]\to\infty} \upsilon = V_{\max}$. These tests run in continuous integration and after any prompt or interface change.

3 Prompt-by-Prompt Development Protocol

To ensure that ChatGPT's assistance remains precise, auditable, and aligned with pedagogical constraints, the application was developed through a staged, prompt-by-prompt workflow. Each prompt targeted a single, well-scoped objective; responses were required to be minimal and behavior-preserving, with numerical logic reserved exclusively for deterministic functions authored and tested by the instructor. This approach reduced the risk of uncontrolled refactoring, made changes easy to review, and supported reproducibility across sessions.

Prompt 1: Ground Rules

The interaction began with a framing message that defined the assistant's role: produce only small, localized patches; never invent secrets; do not compute final numeric answers; and keep all numerics in deterministic code. Stating these constraints upfront materially improved response discipline.

Prompt 1: "Create *requirements.txt and a minimal app.py skeleton... Only code blocks*" was used to set the ground rules.

ChatGPT output

```
requirements.txt
streamlit>=1.34
numpy>=1.24
matplotlib>=3.7
pandas>=2.0
openai>=1.30
app.py

python
Copy code
import streamlit as st

st.set_page_config(page_title="Biochemistry Teaching App",
layout="wide")
st.title("Biochemistry Teaching App")
st.write("Hello, Streamlit")
```

Prompt 2: Deterministic Kernels

The next prompt added pure functions for the Michaelis–Menten rate, inhibition transforms, substrate grids, and linearizations. The assistant was instructed to return just the new functions, leaving existing code untouched. Unit-ready signatures and explicit domain guards (for example, nonnegative concentrations and positive constants) were required.

Prompt 2: "*Add pure functions: mm_rate, inhibition_rate, make_grid, lineweaver_burk, eadie_hofstee with guards. Return only added code*"

ChatGPT output (append to `app.py`)

```
from __future__ import annotations
import math
from dataclasses import dataclass
from typing import Literal, Tuple
import numpy as np

@dataclass
class KineticsParams:
    Vmax: float
    Km: float
    inhibitor: Literal["None", "Competitive", "Pure noncompe-
titive", "Uncompetitive"]
    I: float = 0.0
    Ki: float = 1.0
```

```
def _validate_params(Vmax: float, Km: float, S: np.
ndarray, Ki: float | None = None) -> None:
    if Vmax < 0:
        raise ValueError("Vmax must be ≥ 0.")
    if Km <= 0:
        raise ValueError("Km must be > 0.")
    if np.any(S < 0):
        raise ValueError("[S] must be ≥ 0.")
    if Ki is not None and Ki <= 0:
        raise ValueError("Ki must be > 0.")

def mm_rate(Vmax: float, Km: float, S: np.ndarray | float) ->
np.ndarray | float:
    S_arr = np.asarray(S, dtype=float)
    _validate_params(Vmax, Km, S_arr)
    v = (Vmax * S_arr) / (Km + S_arr)
    if np.isscalar(S):
        return float(v)
    return v

def inhibition_rate(params: KineticsParams, S: np.ndarray |
float) -> np.ndarray | float:
    S_arr = np.asarray(S, dtype=float)
    _validate_params(params.Vmax, params.Km, S_arr, params.Ki)
    if params.inhibitor == "None":
        Veff, Keff = params.Vmax, params.Km
    else:
        alpha = 1.0 + params.I / params.Ki  # α ≥ 1
        alpha = max(alpha, 1.0)
        if params.inhibitor == "Competitive":
            Veff, Keff = params.Vmax, params.Km * alpha
        elif params.inhibitor == "Pure noncompetitive":
            Veff, Keff = params.Vmax / alpha, params.Km
        elif params.inhibitor == "Uncompetitive":
           Veff, Keff = params.Vmax / alpha, params.Km / alpha
        else:
            Veff, Keff = params.Vmax, params.Km
    return mm_rate(Veff, Keff, S_arr)

def make_grid(Smin: float, Smax: float, n: int, logscale:
bool) -> np.ndarray:
    Smin = max(Smin, 0.0)
    Smax = max(Smax, Smin + 1e-9)
    if logscale:
        lo = max(Smin, 1e-6)
       return np.logspace(math.log10(lo), math.log10(Smax), n)
    return np.linspace(Smin, Smax, n)
```

```
def lineweaver_burk(S: np.ndarray, v: np.ndarray) -> Tuple[np.
ndarray, np.ndarray]:
    mask = (S > 0) & (v > 0)
    return 1.0 / S[mask], 1.0 / v[mask]
def eadie_hofstee(S: np.ndarray, v: np.ndarray) -> Tuple[np.
ndarray, np.ndarray]:
    mask = (S > 0)
    return v[mask] / S[mask], v[mask]
```

Prompt 3: Minimal Streamlit Layout with Sidebar

A subsequent prompt layered a two-pane Streamlit layout: parameter controls in the sidebar and a primary v –[S] plot in the main view. The request emphasized clear labels, LaTeX for equations, and no stylistic embellishments beyond what is necessary for readability.

Prompt 3: "*Add sidebar inputs and a primary v–[S] plot. Code only.*"

ChatGPT output (append to app.py)

```
import matplotlib.pyplot as plt

# Sidebar controls
with st.sidebar:
    st.header("Parameters")
    Vmax = st.number_input("Vmax (rate units)", min_value=0.0,
value=100.0, step=1.0, format="%.6f")
    Km = st.number_input("Km (same units as [S])",
min_value=1e-9, value=10.0, step=0.5, format="%.6f")
    inhibitor = st.selectbox("Inhibition model", ["None",
"Competitive", "Pure noncompetitive", "Uncompetitive"])
    c1, c2 = st.columns(2)
    with c1:
        I = st.number_input("[I] (inhibitor conc.)",
min_value=0.0, value=5.0, step=0.5, format="%.6f")
    with c2:
        Ki = st.number_input("Ki (inhibition constant)",
min_value=1e-9, value=10.0, step=0.5, format="%.6f")
    st.markdown("---")
    st.subheader("Grid / Plot")
    Smin = st.number_input("[S] min", min_value=0.0, va-
lue=0.0, step=0.1, format="%.6f")
    Smax = st.number_input("[S] max", min_value=1e-6,
value=100.0, step=1.0, format="%.6f")
    npts = st.slider("Points", min_value=50, max_value=2000,
value=400, step=50)
    logx = st.checkbox("Log-scale [S]", value=False)
    overlay = st.checkbox("Overlay uninhibited baseline",
value=True)
```

```
params = KineticsParams(Vmax=Vmax, Km=Km, inhibitor=inhibi-
tor, I=I, Ki=Ki)
Sgrid = make_grid(Smin, Smax, npts, logx)
v_cur = inhibition_rate(params, Sgrid)

fig, ax = plt.subplots()
ax.plot(Sgrid, v_cur, label="Current")
if overlay and inhibitor != "None":
    v_base = mm_rate(Vmax, Km, Sgrid)
    ax.plot(Sgrid, v_base, "--", label="Baseline")
    ax.legend(frameon=False)
ax.set_xlabel("[S]")
ax.set_ylabel("v")
ax.set_title("Rate vs Substrate")
if logx:
    ax.set_xscale("log")
st.pyplot(fig)
```

Prompt 4: Instantaneous Metric and Transforms

Focused prompts then added (i) an instantaneous-rate metric at a user-selected [*S*] and (ii) optional Lineweaver–Burk and Eadie–Hofstee plots, each gated by domain checks. The prompts specified that invalid domains must suppress plots with an explanatory message rather than fail silently.

Prompt: "*Add optional Lineweaver–Burk and Eadie–Hofstee plots with domain checks. Minimal changes*"

ChatGPT Output (Append to app.py After the Main Plot)

```
cA, cB = st.columns(2)
with cA:
      S_inst = st.number_input("Instantaneous [S]", min_va-
lue=0.0, value=float(Km), step=0.5, format="%.6f")
with cB:
    try:
        v_inst = float(inhibition_rate(params, S_inst))
        st.metric("Instantaneous v", f"{v_inst:.6g}")
        st.caption("At [S] = Km, expected v = 1/2 · Vmax.")
    except Exception as e:
        st.error(f"Instantaneous v unavailable: {e}")

show_transforms = st.checkbox("Show transforms (Lineweaver-
Burk, Eadie–Hofstee)", value=True)
if show_transforms:
    # Lineweaver–Burk
    LBx, LBy = lineweaver_burk(Sgrid, v_cur)
    if LBx.size == 0:
        st.warning("Lineweaver–Burk suppressed (requires [S]>0
and v>0).")
```

```
    else:
        fig2, ax2 = plt.subplots()
        ax2.plot(LBx, LBy)
        ax2.set_xlabel("1/[S]")
        ax2.set_ylabel("1/v")
        ax2.set_title("Lineweaver–Burk")
        st.pyplot(fig2)

    # Eadie–Hofstee
    EHx, EHy = eadie_hofstee(Sgrid, v_cur)
    if EHx.size == 0:
           st.warning("Eadie–Hofstee suppressed (requires [S]
>0).")
    else:
        fig3, ax3 = plt.subplots()
        ax3.plot(EHx, EHy)
        ax3.set_xlabel("v/[S]")
        ax3.set_ylabel("v")
        ax3.set_title("Eadie–Hofstee")
        st.pyplot(fig3)
```

Prompt 5: Glycolysis Explorer

A single prompt created a deterministic, didactic table of glycolytic steps with energy/redox annotations and a context toggle (aerobic versus anaerobic). The request explicitly prohibited any language-model calls in this view.

Prompt: "*Create a deterministic table + context toggle. No model calls.*"

ChatGPT Output (Append to app.py After the Main Plot)

```
def view_glycolysis():
    st.header("Glycolysis explorer")
     context = st.selectbox("Physiological context", ["Aero-
bic", "Anaerobic (lactate)"], index=0)

    rows = [
           {"Step": "1. Hexokinase", "Substrate → Product":
"Glucose → G6P", "ATP": "-1", "NADH": "0", "Control": "↓ by
G6P (tissue-specific)"},
         {"Step": "3. PFK-1", "Substrate → Product": "F6P →
F1,6BP", "ATP": "-1", "NADH": "0", "Control": "↑ AMP/F-2,6-BP;
↓ ATP/citrate"},
         {"Step": "6. GAPDH", "Substrate → Product": "GAP →
1,3-BPG", "ATP": "0", "NADH": "+2", "Control": "Sensitive to
NAD+/NADH"},
         {"Step": "7. PGK", "Substrate → Product": "1,3-BPG →
3-PG", "ATP": "+2", "NADH": "0", "Control": "Substrate-level
```

```
phosphorylation"},
        {"Step": "10. Pyruvate kinase", "Substrate → Product":
"PEP → Pyruvate", "ATP": "+2", "NADH": "0", "Control": "↑
F-1,6-BP; ↓ ATP/alanine"},
    ]
    df = pd.DataFrame(rows)
    if context.startswith("Anaerobic"):
            df.loc[df["Step"] == "6. GAPDH", "NADH"] = "0
(reoxidized via LDH)"
        note = "Anaerobic: cytosolic NADH is reoxidized by LDH;
ATP yield per glucose remains 2."
    else:
            note = "Aerobic: NADH can support oxidative
phosphorylation; ledger reflects cytosolic generation."

    st.dataframe(df, use_container_width=True)
    st.caption(note)
    st.info("Deterministic view: no language-model calls.")
```

Prompt 5: Navigation and Status

Another prompt added a sidebar module selector and an "About" page explaining the separation of concerns. A small status routine reported whether the hosted model path was active or whether the app had fallen back to offline hints.

Prompt: "*Add sidebar module selector and About page; route views*"

ChatGPT Output (Append to `app.py`)

```
def view_kinetics():
    st.header("Enzyme kinetics")
    # (Reuse the existing controls/plots already defined above;
this is a named wrapper.)

def view_hints():
    st.header("Practice & Hints")
    st.write("See tabs above for Practice & Hints, Knowledge
Pack, and Governance.")

def view_about():
    st.header("About")
    st.markdown(
        "Deterministic Python functions compute all numerical
values. "
         "When available, the language model provides brief,
equation-aware hints only; "
        "it does not perform grading or arithmetic."
    )
st.sidebar.markdown("---")
```

```
module = st.sidebar.selectbox("Module", ["Enzyme kinetics",
"Glycolysis explorer", "Practice & Hints", "About"], index=0)

if module == "Enzyme kinetics":
    # assumes the earlier plotting code is in the main script;
optionally move it inside view_kinetics()
    pass
elif module == "Glycolysis explorer":
    view_glycolysis()
elif module == "Practice & Hints":
    # the tabbed hints section already exists above; nothing
additional required
    pass
else:
    view_about()
```

Prompt 6: Deployment Notes

A final configuration prompt appended concise run-mode guidance (local classroom versus headless server), environment variables, and logging locations, emphasizing that deployment changes do not alter scientific behavior because all computation is deterministic.

Prompt: "*Append concise run-mode guidance; comments only*"

ChatGPT Output (Append to End of app.py as Comments)

```
#
--
--
--
--
--
--
--
--
--
--
--
--
------------------------------------------------------------
------------------------------------------------------------
# Run modes (deployment notes)
# Local/classroom:
#   streamlit run app.py
#
# Headless/server (fixed port):
#   streamlit run app.py --server.headless true
--server.port 8501
#
```

```
                # Optional tunnel/proxy for external access;
never hard-code secrets.
               # Environment:
                #    OPENAI_API_KEY (optional: enables hosted
hints)
               #    OPENAI_MODEL   (optional: default 'gpt-4o-
mini')
             # All numerics are deterministic; hosting choice
does not change results.
                                                           #
- -
- -
- -
- -
- -
- -
- -
- -
- -
- -
- -
- -
--------------------------------------------------------------
```

ChatGPT was also instrumental in diagnosing and resolving debugging errors encountered during script development. The example below illustrates how the assistant identified the cause of a runtime failure and proposed a minimal, behavior-preserving fix.

Box 1: Example of Debug and Resolution with ChatGPT's Assistance **"Prompt to ChatGPT (verbatim, with Colab error copy-paste)**

Error copied from Colab:

```
- -
- -
- -
- -
- -
- -
- -
- -
- -
- -
- -
- -
```

```
-------------------------------------------------------------
NameError                              Traceback (most recent
call last)
/tmp/ipython-input-3770823485.py in <cell line: 0>()
     50
     51 # 1) Ensure cloudflared is present
---> 52 ensure_cloudflared()
     53
     54 # 2) Start Streamlit

/tmp/ipython-input-3770823485.py in ensure_cloudflared()
      6 def ensure_cloudflared():
     7     """Download the cloudflared Linux binary if it's not
already available."""
----> 8     if shutil.which("cloudflared"):
      9         return
     10     print("Installing cloudflared binary...")

NameError: name 'shutil' is not defined
Minimal failing code region:
def ensure_cloudflared():
      """Download the cloudflared Linux binary if it's not
already available."""
    if shutil.which("cloudflared"):  # NameError here
        return
    print("Installing cloudflared binary...")
    # download + chmod steps elided
# 1) Ensure cloudflared is present
ensure_cloudflared()
# 2) Start Streamlit (omitted here)
Request to the assistant:
Please provide a minimal patch that:
adds any missing imports,
shows the corrected ensure_cloudflared() function (preserving
its behavior and return value), and
includes a one-sentence rationale.
```

Box 2: Assistant Response from ChatGPT ChatGPT diagnosed the issue and suggested the response below:

Cause The standard-library module shutil is not imported, the name is undefined at runtime.

- Patch Add import shutil at the top of the file. If you also reference Path, urllib.request, stat, or subprocess in the omitted install steps, import those explicitly.
- Rationale shutil, which is a function in the shutil module; without importing the module, the global name shutil does not exist.

Minimal, targeted patch

```
# imports grouped at top of file
import os
import stat
import shutil          # ← required for shutil.which
import subprocess
from pathlib import Path
from urllib.request import urlretrieve

def ensure_cloudflared() -> Path:
    """
    Ensure the 'cloudflared' binary is available.
    Returns the resolved Path to the executable.
    """
    found = shutil.which("cloudflared")
    if found:
        return Path(found)

    print("Installing cloudflared binary...")
     target = Path("/usr/local/bin/cloudflared") if os.access
("/usr/local/bin", os.W_OK) \
           else Path.home() / ".local" / "bin" / "cloudflared"
    target.parent.mkdir(parents=True, exist_ok=True)

     # Example download URL for Linux x86_64; adjust for your
platform if needed.
    url = "https://github.com/cloudflare/cloudflared/releases/
latest/download/cloudflared-linux-amd64"
    urlretrieve(url, target)
    target.chmod(target.stat().st_mode | stat.S_IEXEC)

    return target
Post-patch verification
Unit smoke test
p = ensure_cloudflared()
assert isinstance(p, Path)
assert os.access(p, os.X_OK)
```

From the above example of the debug error, ChatGPT handled the failure by reading the traceback, mapping the unbound symbol to its standard-library source, and proposing the smallest possible fix. Specifically, it diagnosed that the "library module shutil" was being called without importing the library; the python script recommended adding the missing (library) and returned a behavior-preserving patch without refactoring unrelated code. The response also anticipated adjacent dependencies used in the elided installer and explained the cause in one concise sentence, making the correction easy to audit and apply. This exchange exemplifies an effective error-message feedback loop: A tightly scoped prompt yields a precise, minimal remedy that restores execution while maintaining the original program structure.

4 Interface Construction

The application exposes three modules via a sidebar selector:

4.1 Enzyme Kinetics

Inputs for $[S]$, $V_{\max}$, K_m and, when relevant, $[I]$ and K_i. The main pane renders the primary v –$[S]$ curve, optional overlays comparing inhibited to baseline behavior, and linear transforms when valid (Fig. 1).

4.2 Glycolysis Explorer

The Glycolysis explorer (Fig. 2) presents a compact, instructor-curated table listing each step of glycolysis (reaction name, enzyme, and substrates/products) alongside concise regulatory notes (allosteric activators/inhibitors and energy charge sensitivity) and an energy/redox ledger that tallies ATP investment/production and NADH generation by stage. A small selector allows toggling physiological context (aerobic vs anaerobic), which updates the ledger and commentary for pyruvate fate (e.g., lactate vs acetyl-CoA) and highlights implications for redox balance.

4.3 Practice and Hints

A text area for learner questions and a hint engine selector. Controls echo units; governing equations are rendered in LaTeX beneath

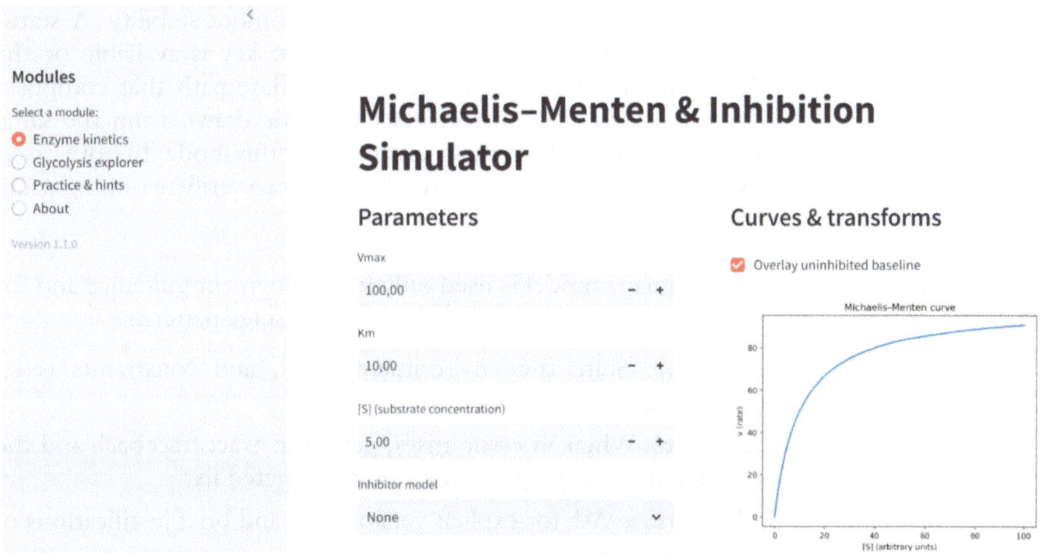

Fig. 1 Michaelis–Menten and inhibition simulator showing enzyme kinetics interface

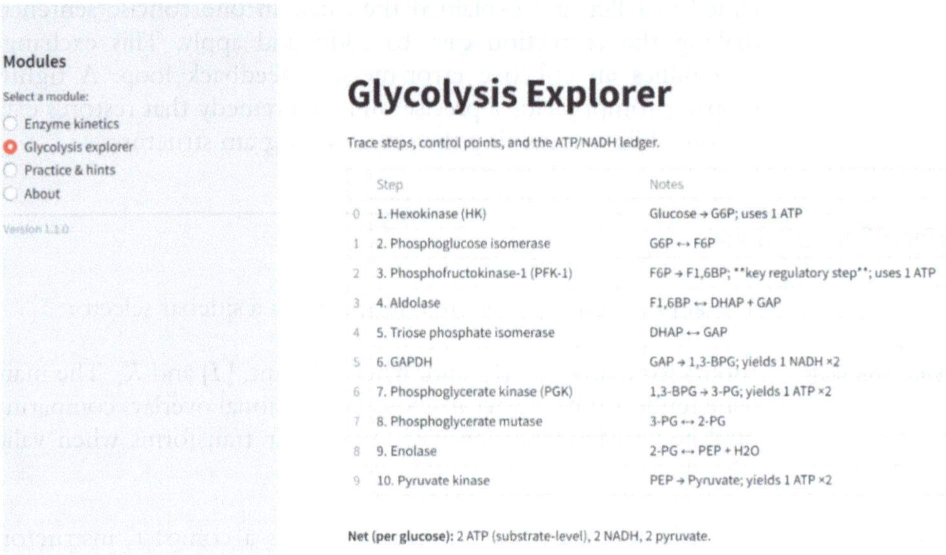

Fig. 2 Michaelis–Menten and inhibition simulator showing Glycolysis explorer interface

inputs. Plot functions return Matplotlib figures that Streamlit renders via st.pyplot. Where appropriate, overlays and captions summarize model signatures (e.g., competitive inhibition increases the apparent K_m).

The practice and hints module accepts free-form learner questions and offers two hint engines (Fig. 3). When OPENAI_API_KEY is present, the OpenAI path uses a system prompt that prohibits final numeric answers and requests brief, equation-aware guidance. The user message concatenates the learner's question with a small set of vetted facts from the knowledge pack; the sampling temperature is kept low to promote stability. A status banner indicates the model in use. If no key is available or the API call fails, the app switches to an offline path that composes short hints via a keyword-to-fact heuristic drawn from the same knowledge pack; the status banner reflects this mode. In both cases, the hint concludes by directing the learner to verify reasoning using the deterministic calculators.

4.4 Prompting Protocol and Error-Message Feedback Loop

The language model is used only for deployment guidance and for explanatory text. Prompts follow a three-stage pattern:

1. *Setup*: State the environment, goal, and constraints (e.g., Streamlit layout, headless execution).
2. *Patch*: When an error arises, paste the exact traceback and the minimal code region to request a targeted fix.
3. *Harden*: Ask for explicit version pins and brief justifications of dependencies.

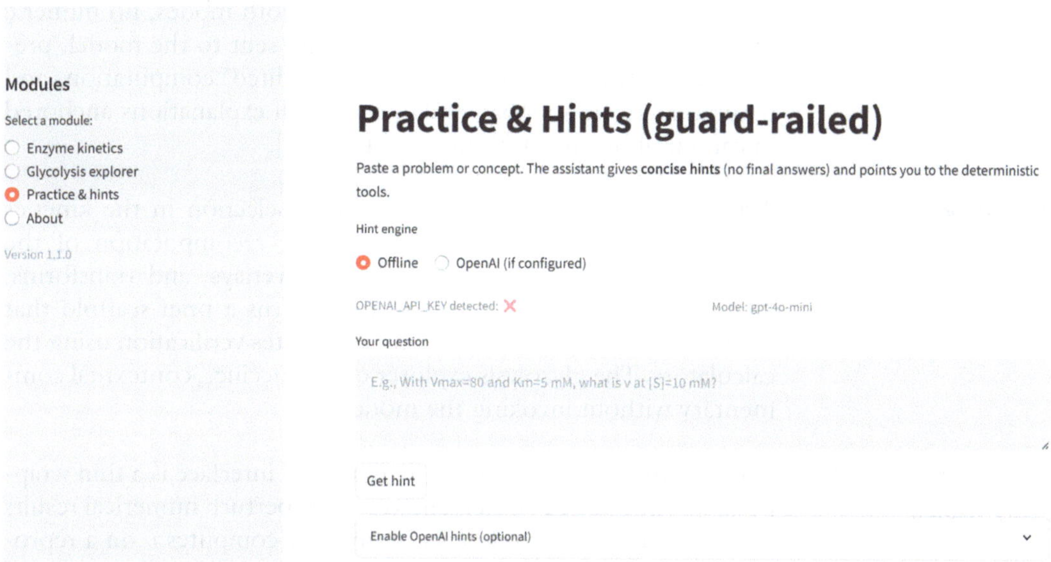

Fig. 3 Michaelis–Menten and inhibition simulator showing practice and hints interface

Each patch is tested immediately; if a proposed change touches numeric code paths, the unit suite is rerun before acceptance. Conversations are kept task-focused to limit drift.

4.5 Knowledge Pack and Hint Governance

The knowledge pack is a compact, instructor-curated repository that anchors every explanation the app provides. Each entry pairs a plain-language definition with the corresponding equation (LaTeX), symbol and unit notes, typical assumptions, common misconceptions, and a one-sentence rationale. Equations and terminology are drawn only from the course textbook and standard enzymology sources to ensure fidelity to instruction, for example, *Lehninger Principles of Biochemistry* for core definitions and rate laws [17], and peer-reviewed research articles underlying Michaelis–Menten kinetics and the common linear transforms for context and notation [18–21]. When pathway or macromolecular context is helpful, brief identifiers and summaries are taken from instructor-approved databases such as KEGG, UniProt, and the Protein Data Bank, used strictly for orientation rather than for equations [20]. The pack is versioned alongside code, so changes remain auditable and aligned with syllabus updates.

Hint governance constrains how the language model may respond [10]. A fixed system prompt limits outputs to short, equation-aware guidance and explicitly forbids final numeric answers or any grading action (approximately 150–200 words), instructing the model to defer all calculations to the on-page deterministic calculators and to reference only the vetted snippets from the knowledge pack [8, 9]. If the API is unavailable, an offline heuristic selects and formats the same snippets, so tone and content remain consistent with course policy. In both modes, no numeric outputs from the computation engine are sent to the model, preserving a strict separation between audited computation and natural-language support while keeping all explanations anchored in canonical biochemical sources [4, 7, 22].

4.6 Student Workflow

A typical session begins with parameter selection in the kinetics module. Any change triggers immediate recomputation of the instantaneous rate, the primary curve, overlays, and transforms. Learners can request a hint; the app returns a brief scaffold that references the displayed equations and invites verification using the calculators. The glycolysis explorer offers succinct, contextual commentary without invoking the model.

4.7 Verification and Parity Testing

Parity testing establishes that the graphical interface is a thin wrapper over the audited kernels and does not perturb numerical results [9]. We begin with a reference script that computes v on a reproducible $[S]$ grid (linear and, where appropriate, logarithmic) across a Cartesian product of $V_{\max}$, K_m and inhibition settings, then serialize both inputs and outputs. The GUI is exercised with the same parameter tuples and grid, and it emits raw arrays prior to plotting; element-wise comparisons must agree to machine precision (within 10^{-12} absolute or 10^{-9} relative tolerance), with a looser bound only where platform float libraries differ [8]. In addition to pointwise parity, we assert analytic invariants v-$([S] = K_m) = ½V_{\max}$, monotonicity in $[S]$, and the expected asymptotes and verify that linearizations (Lineweaver–Burk and Eadie–Hofstee) match the reference after domain masking. Discrepancies trigger a short diagnostic sequence; confirm unit consistency and parameter echo in the UI, dump the exact $[S]$ grid used by each path, rerun the pytest suite, and bisect recent changes to isolate regressions; if the root cause is nontrivial, changes are rolled back and retested. The full parity suite runs in continuous integration on at least two operating systems to detect environment-specific drift, and a fixed random seed is used wherever sampling is involved to guarantee reproducibility [10].

4.8 Deployment Pathways

A simplified web deployment of the Michaelis–Menten and inhibition simulator was produced to reinforce a no-code, click-to-learn experience, consistent with current directions in educational technology that leverage advances in AI and large language models. The intent was to spare learners the cognitive overhead of Jupyter and environment management while preserving the full deterministic behavior of the underlying kinetics engine. The web version can be accessed here: https://enzylab.streamlit.app (last visited date: 06 September 2025).

5 Application of Michaelis–Menten and Inhibition Simulator

5.1 Workflow Demonstration

A typical classroom session begins with the enzyme kinetics module. The instructor sets baseline parameters (for example, $V_{\max} = 100$, $K_m = 10$) and sweeps $[S]$ to visualize saturation. Students then activate competitive inhibition by specifying $[I]$ and Ki, overlaying inhibited and baseline curves to observe the characteristic rightward shift in the v –$[S]$ relationship. The interface reports the instantaneous rate at the current $[S]$ and regenerates diagnostic linearizations when domain assumptions are satisfied. These steps foreground the connection between parameter choices and qualitative changes in curve shape, establishing a concrete basis for subsequent analytic exercises.

In a second phase, learners engage the practice and hints panel. They pose domain questions in free text and receive brief, equation-aware guidance drawn from a vetted knowledge pack. The system prompt forbids the provision of final numeric answers and instructs learners to verify reasoning using the on-page calculators. Instructors may conclude with a short formative check, such as computing v at $[S] = K_m$ or diagnosing the inhibition class from overlaid curves, with correctness determined solely by the deterministic engine.

Example Dialogs and Student Interactions

Student: How does competitive inhibition change K_m and V_{max}

App: Competitive inhibition increases the apparent K_m because additional substrate is required to reach the same rate, while $V_{\max}$ at saturating $[S]$ remains unchanged. Overlay the uninhibited curve to confirm the differing substrate requirement for equivalent v.

Student: I calculated v at $[S] = 10$ with $V_{\max} = 80$ and $K_m = 10$ as 32. Is that correct?

Hint: At $[S] = K_m$, $v = ½\,V\_max$. Re-evaluate with $V_{\max} = 80$

5.2 Visualization Features

The app renders a primary v –[S] curve and optional overlays for inhibited cases. Lineweaver–Burk and Eadie–Hofstee transforms are displayed in separate axes to avoid domain conflation and to make model signatures visually salient. For pathway context, the Glycolysis explorer provides a concise table of steps, regulatory notes, and energy or redox ledgers; optional pathway schematics or structure thumbnails can be added without affecting computational logic.

6 Guidelines

6.1 Maintaining Scientific Accuracy

All quantitative outputs originate from audited, deterministic code. The language model is used exclusively for explanations, not for arithmetic or grading purposes. Prompts are constrained by a system message that defines role and scope and are supplied with short, instructor-vetted facts that mirror course materials. Responses are intentionally brief to reduce verbosity and drift. Assertions that extend beyond introductory content should refer learners back to official sources used in the course.

6.2 Managing Hallucinations and Prompt Tuning

Reliability is improved by restricting the model to hinting and rephrasing [4]. The system message encodes an explicit path for uncertainty, instructing the model to acknowledge limits and direct learners to the app calculators. A small regression set of representative prompts is maintained; any behavioral drift after model or prompt changes triggers targeted edits to the system message or to the knowledge pack, followed by retesting.

6.3 Ethical Considerations in AI-Assisted Teaching

To protect assessment integrity, the hints module is disabled during graded work. Prompts contain no student identifiers, and confidential information is stored in environment variables, not in code or logs. Course documentation should disclose the model's role and emphasize that all numeric values shown in the interface are generated by transparent equations under instructor control. Human editorial oversight is retained for any model-generated text that becomes part of official materials [6].

7 Application Programming Interface and Connectivity

When authentication or network issues arise, begin by confirming that the environment contains a valid OPENAI_API_KEY and restart the session so the client can read it; if calls return unauthorized errors, the key is missing or malformed, and if requests time out or fail TLS verification, test basic connectivity from the same shell, configure proxy variables where required, and retry. Rate limiting manifests as intermittent failures; mitigate by caching stable hints, debouncing rapid requests from widget changes, and using exponential backoff with jitter. If the service reports that context is too long, trim prompts and include only the most relevant facts. The interface should degrade gradually to the offline hint engine and surface a clear status message whenever remote calls are unavailable.

8 Deterministic Engine

Nonphysical outputs usually reflect domain violations or unit mismatches, so enforce guards on parameters such as nonnegative concentrations, positive K_m and K_i, and treat zero substrate as yielding zero rate [13]. Show linearized views only when their domains hold, for example, positive substrate and positive rate, and echo current parameter values and units near plots to prevent misinterpretation [14]. If the graphical interface disagrees with a reference script, run the unit tests that assert analytic limits and compare arrays to a tight numerical tolerance to localize the fault.

9 Troubleshooting

Blank or sluggish plots typically arise from missing render calls, excessive sampling, or figure reuse; create a fresh figure for each view, render explicitly with st.pyplot, and reduce the substrate grid density or cache stable curves to improve responsiveness. Separate the transform views into distinct axes to avoid domain confusion, keep labels concise and legible, and render equations with LaTeX to maintain alignment between visuals and formal statements. Further details on troubleshooting are provided in the supplementary file.

10 Mechanism to Assess the Developed Approach

Students complete a short concept quiz immediately before and after using the app. The quiz focuses on three pillars of enzyme kinetics: K_m, V_{max}, and the Hill coefficient η. Comparing the two scores provides a direct estimate of learning gains. To assess application rather than recall, students also complete two compact performance tasks. First, they inspect baseline and inhibitor overlaid curves and identify the inhibition class with a one-sentence, parameter-based justification. Second, they estimate K_m and V_{max} from plotted data and briefly explain which representation they trusted and why. A concise rubric with criteria for correctness, parameter-linked reasoning, and clarity ensures consistent scoring across cohorts.

During the activity, the app records lightweight interaction signals such as time spent, slider adjustments, and plot toggles. No personal content is stored. These indicators help instructors see whether certain controls cause friction or whether students meaningfully explore the parameter space. At the end, a 1-minute micro survey with five to seven questions asks about clarity of controls, readability of plots, and confidence in interpreting curves, plus an open prompt: "What changed in how you reason about the curve?" Responses guide small, high-impact refinements to the interface and teaching materials.

Before classroom use and after any change to code or libraries, the app's outputs are compared against a small reference script at predefined test points for parity testing. In addition, we check analytic invariants that should always hold in enzyme kinetics. When $[S] = K_m$, the rate v must equal $\frac{1}{2} V_{max}$. The function v-([S]) should increase smoothly with [S] and remain bounded by Vmax. Linearized plots, such as Lineweaver–Burk and Eadie–Hofstee must appear only when inputs are in valid domains. These checks ensure the interface never conceals a numerical error.

To preserve assessment integrity, the hint panel is disabled in summative contexts. Explanations shown within the app are drawn from a curated, citation-anchored knowledge pack, and all numerical outputs are produced by audited, deterministic Python functions rather than the language model. This separation keeps the model in a supportive role while reserving quantitative authority for verified code.

Results can be summarized in a short paragraph: average pre- and post-gain on the concept quiz with an effect size, median rubric scores for the two performance tasks, basic usage statistics such as median time and interaction counts, usability ratings with one or two representative comments, and a technical note confirming that parity tests and invariant checks passed after the final build. Presented together, these elements offer a complete picture of learning, usability, and reliability that readers can replicate or adapt to related topics such as cooperativity modules, thermodynamics calculators, and pathway energetics.

Conflicts of Interest

The authors have no conflicts of interest to declare that are relevant to the content of this chapter.

References

1. Hasrod T, Nuapia YB, Tutu H (2024) ChatGPT helped me build a chemistry app, and here's how you can make one also. J Chem Educ 101(2):653–660. https://doi.org/10.1021/acs.jchemed.3c01170
2. Rhee KY, Jansen RS, Grundner C (2022) Activity-based annotation: the emergence of systems biochemistry. Trends Biochem Sci:785–794. https://doi.org/10.1016/j.tibs.2022.03.017
3. Doğru MS, Faulconer EK (2025) ChatGPT as a Virtual Laboratory Teaching Assistant in Undergraduate Biology. Res Sci Educ. https://doi.org/10.1007/s11165-025-10271-z
4. Tong Y, Zhang L (2023) Discovering the next decade's synthetic biology research trends with ChatGPT. In: Synthetic and systems biotechnology. KeAi Communications Co, pp 220–223. https://doi.org/10.1016/j.synbio.2023.02.004
5. Kaspar RL (2002) Integrating Internet assignments into a biochemistry/molecular biology laboratory course. In: Biochemistry and molecular biology education, pp 36–39. https://doi.org/10.1002/bmb.2002.494030010007
6. Guo Y, Lee D (2023) Leveraging ChatGPT for enhancing critical thinking skills. J Chem Educ 100(12):4876–4883. https://doi.org/10.1021/acs.jchemed.3c00505

7. Kasneci E, Sessler K, Küchemann S, Bannert M, Dementieva D, Fischer F, Gasser U, Groh G, Günnemann S, Hüllermeier E, Krusche S, Kutyniok G, Michaeli T, Nerdel C, Pfeffer J, Poquet O, Sailer M, Schmidt A, Seidel T, Stadler M, Weller J, Kuhn J, Kasneci G (2023) ChatGPT for good? On opportunities and challenges of large language models for education. Learn Individ Differ 103:102–274. https://doi.org/10.1016/j.lindif.2023.102274
8. Sirisathitkul C, Jaroonchokanan N (2025) Implementing ChatGPT as tutor, tutee, and tool in physics and chemistry. Substantia 9(1): 89–101. https://doi.org/10.36253/Substantia-2808
9. Roman A, Simaitis M, Sheely K, Roth YM, Tansey JT, Cogan J (2025) Learning tools using ChatGPT in the biochemistry class: creating notes and performance on exams. Biochem Mol Biol Educ 53(4):381–388. https://doi.org/10.1002/bmb.21904
10. Bolgova O, Shypilova I, Mavrych V (2025) Large language models in biochemistry education: comparative evaluation of performance. JMIR Med Educ 11. https://mededu.jmir.org/2025/1/e67244
11. Al Shloul T, Mazhar T., Abbas Q, Iqbal M, Ghadi YY, Shahzad T, Mallek F, Hamam H (2024) Role of activity-based learning and ChatGPT on students' performance in education. In: Computers and education: artificial intelligence. Elsevier B.V. https://doi.org/10.1016/j.caeai.2024.100219
12. Lo CK, Hew KF, Jong, Yung MS (2024) The influence of ChatGPT on student engagement: a systematic review and future research agenda. Comput Educ 219. https://doi.org/10.1016/j.compedu.2024.105100
13. Mai DTT, Da CV, Hanh NV (2024) The use of ChatGPT in teaching and learning: a systematic review through SWOT analysis approach. Front Educ (Frontiers Media SA) 9. https://doi.org/10.3389/feduc.2024.1328769
14. Durgungoz A, Kharrufa A (2025) ChatGPT is like a study buddy, a teacher and sometimes just a friend': a longitudinal exploration of students' interactions, perception and acceptance. Interact Learn Environ. https://doi.org/10.1080/10494820.2025.2509276
15. Gasaymeh AM, Qbeita AA, AlMohtadi R, Beirat M (2025) Exploring education students' use of ChatGPT for academic and personal purposes: insights from a developing country context. Front Educ 10. https://doi.org/10.3389/feduc.2025.1580310
16. Vieriu AM, Petrea G (2025) The Impact of Artificial Intelligence (AI) on Students' Academic Development. Educ Sci *15*(3):343. https://doi.org/10.3390/educsci15030343
17. Nelson DL, Cox MM, Hoskins AA (2021) Lehninger principles of biochemistry, 8th edn. Macmillan Learning, Austin/Boston/New York/Plymouth
18. Valencia PL, Sepúlveda B, Gajardo D, Astudillo-Castro C (2020) Estimating the product inhibition constant from enzyme kinetic equations using the direct linear plot method in one-stage treatment. Catalysts 10(8):853. https://doi.org/10.3390/catal10080853
19. Saganuwan SA (2021) Application of modified Michaelis—Menten equations for determination of enzyme inducing and inhibiting drugs. BMC Pharmacol Toxicol 22(1):57. https://doi.org/10.1186/s40360-021-00521-x
20. Murugan R (2024) Theory on the rate equations of Michaelis-Menten type enzyme kinetics with competitive inhibition. PLoS One 19(7):1–66. https://doi.org/10.1371/journal.pone.0302679
21. Mohanasundaraganesan M, Guirao JLG, Rathinam S (2025) Theoretical analysis of amperometric biosensor with substrate and product inhibition involving non - Michaelis - Menten Kinetics. Match Math Comput Chem 93:319–347. https://doi.org/10.46793/match.93-2.319M
22. Mahat RK, Jantikar AM, Rathore V, Panda S (2023) Assessing the performance of ChatGPT to solve biochemistry question papers of university examination. Adv Physiol Educ 47:528–529. https://doi.org/10.1152/advan.00076.2023

Chapter 19

Automated Genetic Engineering in the Laboratory

Zekun Li and Eric M. Young

Abstract

In today's rapidly evolving research landscape, improving experimental throughput and reproducibility are essential. Furthermore, given that biological systems are dependent on many variables, stepwise optimization is not an efficient or effective method for exploring the multidimensional biological design space. Laboratory automation enables increased throughput, reproducibility, and combinatorial optimization of biological designs by integrating hardware and software platforms with synthetic biology tools. Here, we describe an automated DNA assembly protocol that may be used for the construction of large, diverse libraries of genetic designs. This combination of laboratory infrastructure, software, and expertise in molecular biology and genetic engineering is foundational for the "biofoundry" concept and can support a wide range of research thrusts in biological engineering.

Key words Automation, Automated DNA Assembly, GoldenGate, Automated transformation and plating, Echo liquid handler

1 Introduction

Biofoundries are integrated automation platforms designed to accelerate the engineering of biological systems for both research and biotechnology applications [1–5]. An enabling technology is automated high-throughput DNA assembly based on modular cloning. While several modular cloning methods have been developed, this chapter describes the automation of Type IIS restriction enzyme cloning, often referred to as "GoldenGate" [6]. Unlike traditional restriction enzymes that cleave DNA within the recognition sequence, Type IIS enzymes cut DNA outside of the recognition sequence, which means that the sticky end can be designed to be any sequence. Type IIS assembly can produce scarless junctions or introduce scars depending on how the sticky end sequences are designed.

The seminal work for using standardized sticky ends for controlling the assembly of multi-part DNA fragments was by Marillonnet and colleagues, termed *Modular Cloning (MoClo)*

Robert E. Dempski (ed.), *Transforming Molecular Biology with Emerging Technologies*, Methods in Molecular Biology, vol. 3043,
https://doi.org/10.1007/978-1-0716-5320-3_19,

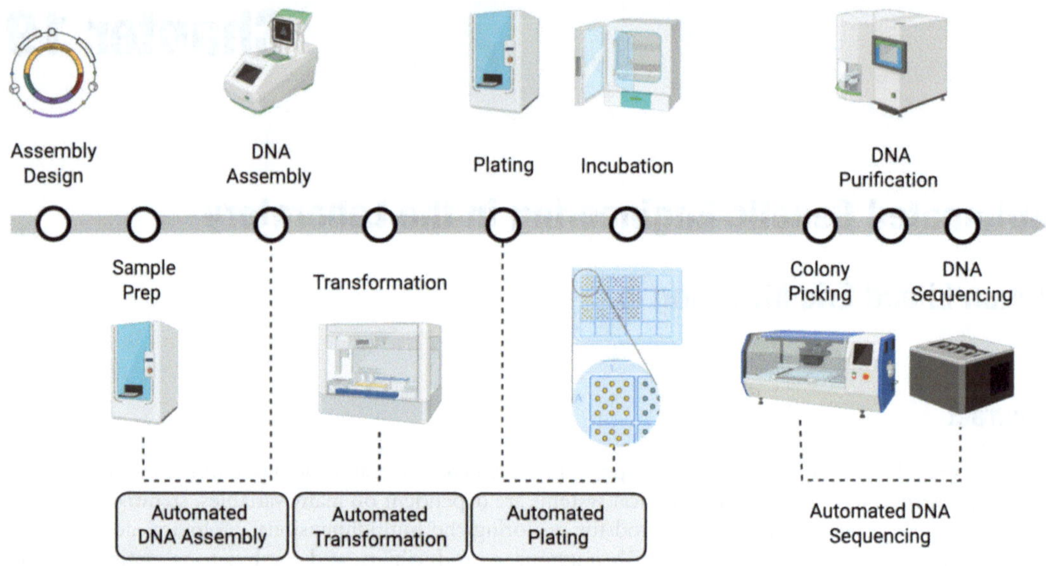

Fig. 1 Automated molecular biology engineering workflow. An integrated robotic workflow for molecular biology automates the design, assembly, transformation, and plating of synthetic DNA constructs. (1) *Automated DNA Assembly:* Labcyt Echo liquid handlers combine standardized DNA parts and enzymes in precise volumes for Golden Gate reactions, followed by thermocycling to generate plasmids. (2) *Automated Transformation:* Eppendorf epMotion systems introduce assembled plasmids into chemically competent cells via heat shock, with automated recovery in growth media. (3) *Automated Plating:* Echo liquid handlers spot transformed cells onto agar plates in an 18-spot format, enabling high-throughput colony screening

[7]. The MoClo system employs a hierarchical and standardized assembly scheme to facilitate scalable construction of transcriptional units and multigene constructs. Individual DNA parts—such as promoters, RBS, CDS, and terminators—are first cloned into Level 0 (L0) plasmids with a single Type IIS enzyme (e.g., BbsI) and standardized flanking overhangs that dictate assembly order. These L0 parts can then be efficiently combined into Level 1 (L1) transcription units using a single Type IIS enzyme (e.g., BsaI). The system's scalability allows for flexible recombination of L0 components into higher-level constructs, enabling the rapid generation of combinatorial libraries of genetic circuits and metabolic pathways [8, 9].

Since, many synthetic biology "parts kits" have been created for many different organisms, all using standardized scars and hierarchical assembly [6, 10–16]. These support the design, execution, and analysis of complex DNA assemblies in many genetic engineering and synthetic biology research groups [17–19]. Modular, hierarchical assembly has also enabled the development of automated procedures using liquid handling robots (Fig. 1). Here, we describe automated protocols for the three steps of DNA assembly: mixing of modular DNA parts and assembly (Fig. 2), heat shock transformation (Fig. 3), and plating (Fig. 4). These protocols are designed

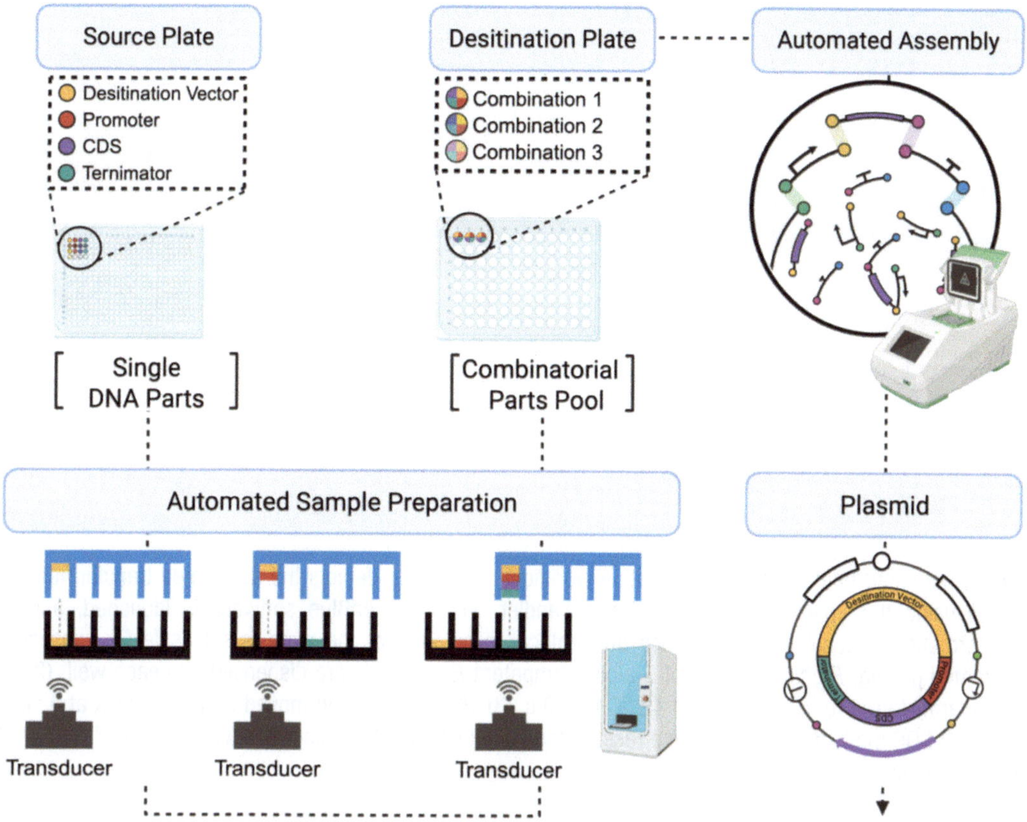

Fig. 2 Automated DNA assembly workflow. The modular DNA assembly process using the Echo 525. *Automated Sample Preparation:* An Echo Plate Reformat transfer protocol is loaded to define source/destination wells and transfer volumes, and the Echo 525 transfers reagents from the LDV plate to a 96-well PCR destination plate following the defined protocol. *Automated Assembly:* The PCR plate is placed into a thermocycler to perform Golden Gate Assembly, and the final assembled plasmids are stored at 4 °C for downstream applications

to be used with the Beckman Coulter Echo 525 acoustic liquid handler and Eppendorf epMotion deck-based liquid handler equipped with a heat block, cooling block, and shaker. They may be adapted for use with other liquid handlers.

In summary, the integration of modular cloning with automated liquid handling platforms enables automated high-throughput, precise, and reproducible DNA assembly in the laboratory.

2 Materials

Use ultrapure water and analytical-grade reagents for the preparation of all solutions. Prepare and store all reagents at 4 °C unless

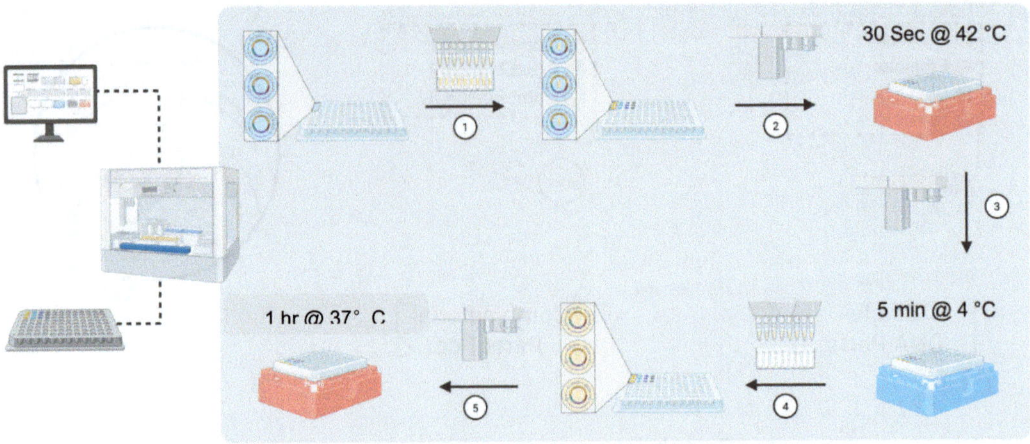

Fig. 3 Automated heat shock transformation workflow. This figure depicts the automated transformation of *E. coli* using the Eppendorf epMotion liquid handler. *Left:* The epBlue software is launched, and the transformation protocol is loaded. Once the protocol is completed, the plate is set aside for subsequent automated plating. *Right:* (1) 10 μL of 5-alpha competent *E. coli* cells are dispensed into each well. (2) The plate is transferred to a heat block at 42 °C for 30 s. (3) The plate is then moved to a cold block at 4 °C for 5 min. (4) SOC medium is added from a reservoir to each well. (5) The cells are incubated at 37 °C for 1 h to enable recovery

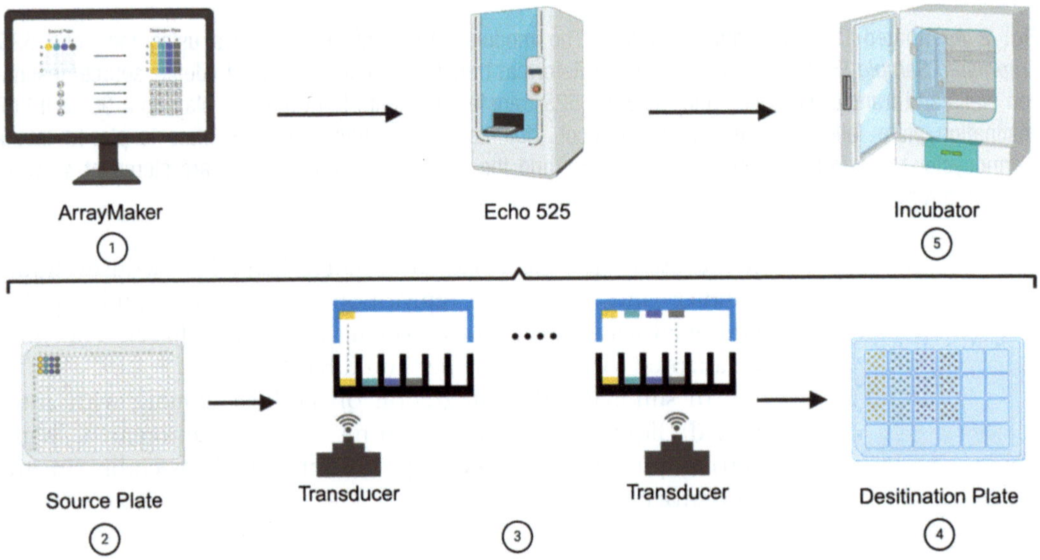

Fig. 4 Automated plating workflow. Automated plating of transformed cells using the Echo 525 liquid handler. (1) A cell culture transfer file is generated in Echo Array Maker. (2) The source and destination plates are loaded onto the Echo 525. (3) The Echo 525 is run to dispense cell culture in an 18-spot format. (4) Both plates are removed after plating. (5) The destination plate is transferred to the incubator for further growth

otherwise specified. Ensure strict compliance with all applicable waste disposal regulations when discarding waste materials.

2.1 Automated Type IIs DNA Assembly

1. Nuclease-free distilled water (DNase/RNase-Free).
2. Type IIs Restriction Enzyme: Store at −20 °C (*see* **Note 1**).
3. T4 DNA Ligase and Ligase Buffer: Store at −20 °C.
4. 20 fmol/μL DNA Working Solutions: Using a nanodrop spectrophotometer test the concentration of the DNA stock solution (in units of ng/μL). Use NEBioCalculator (*see* **Note 2**) to convert the units to fmol/μL. Dilute the stock solutions to 20 fmol/μL with nuclease-free water. Store Parts DNA and destination DNA working solution at 4 °C.
5. Master Mix Preparation.

Component	Per-well volume (μL)	Total volume for 96 wells (μL)
Restriction enzyme[a]	0.2	19.2
T4 DNA ligase	0.08	7.68
Ligase buffer	0.2	19.2
Nuclease-free water	0.52–0.1 × N[b]	49.92—9.6 × N

[a]Restriction enzyme (BbsI for Level 0 or Level 2, BsaI for Level 1)
[b]Number of DNA parts, including the destination vector

6. Echo 384 LDV Source plate.
7. 96-Well PCR plate.
8. 96-Well PCR plate parafilm.
9. Thermocycler used for PCR.

2.2 Automated Transformation

1. 5-Alpha competent *E. coli* cells: Store at −80 °C. Thaw the competent cells on ice for 10 min before transformation.
2. Super Optimal Broth with Catabolite Repression (SOC) Recovery Media: 2% vegetable peptone, 0.5% yeast extract, 10 mM NaCl, 2.5 mM KCl, 10 mM $MgCl_2$, 10 mM $MgSO_4$, 20 mM glucose.
3. Benchtop centrifuge with plate fittings.
4. Echo 384 LDV Source plate.

2.3 Automated Plating

1. Echo 384 LDV source plate.
2. Autoclave for making sterile agar media.
3. Nunc OmniTray Single-Well Plate with appropriate Luria-Bertani (LB) agar media supplemented with the appropriate antibiotic selection(s): A total of 500 mL of deionized water

was added to a clean 1 L glass bottle, followed by 10 g of agar, 5 of peptone, 2.5 g of yeast extract, and 5 g of NaCl. The bottle was capped loosely (½ turn) and sealed with autoclave tape before autoclaving at 121 °C and 15 psi for 30 min. After autoclaving, the cap was tightened, and the bottle was placed in a 60 °C water bath for 15 min to cool. Subsequently, 500 μL of the appropriate antibiotic stock solution was added to the medium (*see* **Note 3**). The molten agar was poured into Nunc OmniTray Single-Well Plates until the bottom was fully covered and allowed to solidify overnight at room temperature. Plates were then labeled and stored at 4 °C until use.

3 Methods

Carry out all procedures at room temperature unless otherwise specified.

3.1 Automated DNA Assembly

Create Reaction Procedure Using Echo Plate Reformat

1. Create a transfer map in Excel and save in CSV. format.
   Example Format:

Source plate	Source well	Destination plate	Destination well	Transfer volume (nL)
1	A1	1	B1	200
1	B1	2	C1	200

2. Open the Echo Plate Reformat software package.
3. Identify your source and destination plate types (source: Echo 384 LDV, destination: 96-well PCR plate).
4. Load the transfer file into Echo Plate Reformat.

LDV Source Plate Preparation with epMotion Liquid Handling

5. Determine the total quantity of Master Mix and DNA to be added to each well. (*see* **Note 4**).
6. Prepare the required volume of Master Mix by multiplying the total volume required by the component volume per μL.
7. Open the Eppendorf epblue procedure.
8. Load the master mix solution and DNA parts to microcentrifuge tube holder.
9. Simulate the procedure and record the volumes transferred.

PCR Plate Preparation with Echo Liquid Handling

10. Open the protocol using Echo plate reformat.
11. Run the protocol and follow the guide to load the source plate (384-well LDV plate) and destination plate (96-well PCR plate) into the source and destination plate holder when prompted.
12. Cover both plates with adhesive film on ice and move to run the DNA assembly when completed.

DNA Assembly on a Thermocycler

13. Spin down the PCR plate briefly in the centrifuge.
14. Place in the thermocycler and run the TypeIIS Assembly Protocol (*see* **Note 5**).

    Step 1: 37 °C for 2 min.

    Step 2: 16 °C for 2 min.

    Step 3: 50 °C for 5 min.

    Step 4: repeat steps 1–3 for 30 cycles.

    Step 5: 80 °C for 10 min.

    Step 6: Hold at 4 °C.
15. Cover the PCR plates and store in the 4 °C fridge.

3.2 Automated Transformation

Transformation with epMotion

1. Turn on the epMotion machine and open epBlue on the computer.
2. Open the application editor and start to build the "96-wells *E. coli* Transformation" procedure as described below:
   - Add 10 μL 5-alpha Competent *E. coli* cells to each well.
   - Shake at 500 RPM for 15 s.
   - Incubate for 15 min in a block pre-frozen at −20 °C.
   - Heat shock at 42 °C for 30 s.
   - Moved to cold block for 5 min.
   - Add 100 μL SOC.
   - Incubate at 37 °C for 1 h.
3. Load the following labware onto the epMotion deck as shown in the worktable layout:
   - 50 μL filter tips.
   - 300 μL filter tips.
   - Reservoir rack with 10 mL reservoir.
   - Waste bag.
   - LDV plate (remove adhesive film).

4. Remove lid and transfer SOC media to 10 mL reservoir using serological pipette. Work within epMotion enclosure with HEPA filter activated.
5. Remove PCR ThermoBlocks from freezer and place on temperature modules.
6. Place the PCR plate on the ThermoBlock and remove the adhesive film.
7. Remove the PCR plate from the thermocycler or fridge.
8. Spin it down briefly in a centrifuge equipped with plate fittings.
9. The 96-well plate was then transferred to an Eppendorf epMotion 5075 t liquid handler.
10. Run the "96-well *E. coli* transformation" procedure (*see* **Note 6**). Select the following settings:

Detect volume	ON
Detect labware	ON
Check tips	ON
Detect tube lid	ON
Detect tube lid	OFF

11. When prompted, spin down competent cells and place in position on PCR block.

3.3 Automated Plating

Pouring Luria-Bertani Antibiotic Agar OmniTray Plates

1. Weigh 10 g of agar and 12.5 g of LB powder per 500 mL of Milli-Q water.
2. Add the dry components to a 1 L glass media bottle suitable for autoclaving.
3. Pour 500 mL of sterile Milli-Q water into the bottle and swirl to suspend the solids evenly.
4. Loosely cap the bottle and secure it with autoclave tape.
5. Autoclave the bottle at 121 °C and 20 psi for at least 30 min.
6. While autoclaving, thaw a 1000× stock solution of the desired antibiotic.
7. Set up a 60 °C water bath deep enough to submerge ~75% of the agar bottle.
8. After autoclaving, carefully transfer the bottle to the water bath to cool the molten agar to ~60 °C.
9. Using sterile technique near a flame, add the appropriate volume of antibiotic to the cooled agar and swirl gently to mix.

10. Pour warm agar into each OmniTray plate, ensuring even coverage of the surface.
11. Let plates solidify overnight at room temperature.
12. Store at 4 °C until use.

Making an Array Maker Document to Identify Plate Arrangement

13. Create a transfer map using Excel and save as CSV format (see below).

    Example Format:

Source plate	Source well	Destination plate	Destination well	Transfer volume (nL)
1	A1	1	B1	25
1	B1	2	C1	25

14. Import the CSV file into Labcyte Array Maker.
15. Confirm source plate type: 384 LDV plate, destination plate: OminiTray, Spot format: custom (*see* **Note 7**).

Plating Transformed Cells Using the Echo 525 Acoustic Liquid Handler

16. Place antibiotic selection plates in a 37 °C incubator to pre-warm.
17. Transfer 10 μL of the transformation mixture into unused wells of the Echo-compatible 384-well source plate.
18. Run the method on the Echo 525 array Maker.
19. Load both the source plate (with transformed cells) and the destination OmniTray plate onto the Echo 525 deck as a guide.
20. Initiate the transfer. The Echo will use focused ultrasound to eject 25 nL droplets of cells into the destination plate in a pseudo-96-well pattern, with 18 droplets per pseudo-well location.
21. After spotting, incubate the destination plates at 37 °C overnight.

4 Notes

1. Type IIS restriction enzymes, cutting DNA at a defined distance away from their recognition site, are essential tools for Golden Gate Assembly. This allows for the creation of custom overhangs that do not include the recognition site itself. Each

DNA fragment is flanked by specific overhangs (scar) that determine the order in which fragments are joined. The reaction can be one-pot and one-step, combining digestion and ligation simultaneously using a Type IIS enzyme (e.g., BsaI, BbsI, and Esp3I) and T4 DNA ligase.

2. NEBioCalculator is a free online tool developed by New England Biolabs (NEB) that provides various calculations relevant to molecular biology experiments. It simplifies complex calculations, especially those involving DNA and RNA manipulations, offering modules for mass-to-moles conversions, ligation planning, and more.

3. Preparation of Antibiotic Stock Solution: The appropriate amount of antibiotic powder was weighed and dissolved in the corresponding volume of sterile water, according to the concentrations listed in the table below. The solution was sterilized by filtration through a 0.22 μm membrane filter, aliquoted into 1.5 mL microcentrifuge tubes (500 μL per tube), and stored at −20 °C.

Antibiotic	Antibiotic powder (g)	Sterile water (mL)	Stock concentration (mg/mL)
Ampicillin	4	40	100
Chloramphenicol	1	40[a]	25
Kanamycin	2	40	50

[a]For the Chloramphenicol stock solution, use 95% ethanol instead of sterile water

4. A practical guideline is to use the required transfer volume plus 4.5 μL for dead volume and plus 1 μL for overage, rounding up as necessary to simplify calculations. If the final volume exceeds 14 μL, the procedure needs to be modified to use two wells and recalculate the total volume.

5. To ensure successful Type IIS Golden Gate assembly, several key parameters should be verified: (1) confirm that only one pair of internal Type IIS restriction sites is present in both the insert fragments and the destination vector; (2) check that the number of random base pairs between the restriction sites and scars is correctly designed; (3) verify that scar sequences are complementary; (4) ensure that no two overhangs in the assembly are identical to avoid ligation ambiguity; (5) confirm that all DNA parts are present at equimolar concentrations.

6. Wipe the deck down with 70% ethanol before starting the "96well *E. coli* transformation pre-chill" procedure.

7. Creating a New Array Type in Array Marker: Start the Array Marker software. From the top menu, select Tools, then choose

Add New Array Type. Enter the desired number of rows and columns and save the array configuration.

Acknowledgments

This work was supported by the National Science Foundation under awards 1944046 (CAREER award to EMY) and 2400302 (NSF CREATE Biofoundry).

Conflict of Interest This study was supported by the National Science Foundation Award #2400302.

References

1. Dudley QM, Cai Y-M, Kallam K et al (2021) Biofoundry-assisted expression and characterization of plant proteins. Synth Biol (Oxf) 6: ysab029. https://doi.org/10.1093/synbio/ysab029
2. Holowko MB, Frow EK, Reid JC et al (2021) Building a biofoundry. Synth Biol 6:ysaa026. https://doi.org/10.1093/synbio/ysaa026
3. Carbonell P, Le Feuvre R, Takano E, Scrutton NS (2020) In silico design and automated learning to boost next-generation smart biomanufacturing. Synth Biol 5:ysaa020. https://doi.org/10.1093/synbio/ysaa020
4. Moffat AD, Elliston A, Patron NJ et al (2021) A biofoundry workflow for the identification of genetic determinants of microbial growth inhibition. Synth Biol 6:ysab004. https://doi.org/10.1093/synbio/ysab004
5. Farzaneh T, Freemont PS (2021) Biofoundries are a nucleating hub for industrial translation. Synth Biol 6:ysab013. https://doi.org/10.1093/synbio/ysab013
6. Engler C, Youles M, Gruetzner R et al (2014) A golden gate modular cloning toolbox for plants. ACS Synth Biol 3:839–843. https://doi.org/10.1021/sb4001504
7. Engler C, Kandzia R, Marillonnet S (2008) A one pot, one step, precision cloning method with high throughput capability. PLoS One 3: e3647. https://doi.org/10.1371/journal.pone.0003647
8. Weber E, Engler C, Gruetzner R et al (2011) A modular cloning system for standardized assembly of multigene constructs. PLoS One 6:e16765. https://doi.org/10.1371/journal.pone.0016765
9. Werner S, Engler C, Weber E et al (2012) Fast track assembly of multigene constructs using Golden Gate cloning and the MoClo system. Bioeng Bugs 3:38–43. https://doi.org/10.4161/bbug.3.1.18223
10. Iverson SV, Haddock TL, Beal J, Densmore DM (2016) CIDAR MoClo: improved MoClo assembly standard and new E. coli part library enable rapid combinatorial design for synthetic and traditional biology. ACS Synth Biol 5:99–103. https://doi.org/10.1021/acssynbio.5b00124
11. Keating KW, Young EM (2023) Systematic part transfer by extending a modular toolkit to diverse bacteria. ACS Synth Biol 12:2061–2072. https://doi.org/10.1021/acssynbio.3c00104
12. Lee ME, DeLoache WC, Cervantes B, Dueber JE (2015) A highly characterized yeast toolkit for modular, multipart assembly. ACS Synth Biol 4:975–986. https://doi.org/10.1021/sb500366v
13. Stuttmann J, Barthel K, Martin P et al (2021) Highly efficient multiplex editing: one-shot generation of 8× Nicotiana benthamiana and 12× Arabidopsis mutants. Plant J 106:8–22. https://doi.org/10.1111/tpj.15197
14. Grützner R, Martin P, Horn C et al (2021) High-efficiency genome editing in plants mediated by a Cas9 gene containing multiple introns. Plant Commun 2:100135. https://doi.org/10.1016/j.xplc.2020.100135
15. Obst U, Lu TK, Sieber V (2017) A modular toolkit for generating Pichia pastoris secretion libraries. ACS Synth Biol 6:1016–1025. https://doi.org/10.1021/acssynbio.6b00337

16. Moore SJ, Lai H-E, Kelwick RJR et al (2016) EcoFlex: a multifunctional MoClo kit for E. coli synthetic biology. ACS Synth Biol 5: 1059–1069. https://doi.org/10.1021/acssynbio.6b00031
17. Young EM, Zhao Z, Gielesen BEM et al (2018) Iterative algorithm-guided design of massive strain libraries, applied to itaconic acid production in yeast. Metab Eng 48:33–43. https://doi.org/10.1016/j.ymben.2018.05.002
18. Hillson NJ, Rosengarten RD, Keasling JD (2012) j5 DNA assembly design automation software. ACS Synth Biol 1:14–21. https://doi.org/10.1021/sb2000116
19. Nielsen AA, Voigt CA (2014) Multi-input CRISPR/Cas genetic circuits that interface host regulatory networks. Mol Syst Biol 10: 763. https://doi.org/10.15252/msb.20145735

Index

Robert E. Dempski (ed.), *Transforming Molecular Biology with Emerging Technologies*, Methods in Molecular Biology, vol. 3043, https://doi.org/10.1007/978-1-0716-5320-3,

G

I

J

K

L

M

O

P

S

T

U

V

W

MIX
Papier aus verantwortungsvollen Quellen
Paper from responsible sources
FSC® C105338

If you have any concerns about our products,
you can contact us on
ProductSafety@springernature.com

In case Publisher is established outside the EU,
the EU authorized representative is:
Springer Nature Customer Service Center GmbH
Europaplatz 3, 69115 Heidelberg, Germany

Printed by Libri Plureos GmbH
in Hamburg, Germany